William Whewell

Science and Culture in the Nineteenth Century

BERNARD LIGHTMAN, EDITOR

William Whewell

VICTORIAN POLYMATH

EDITED BY

Lukas M. Verburgt

University of Pittsburgh Press

Published by the University of Pittsburgh Press, Pittsburgh, Pa., 15260
Copyright © 2024, University of Pittsburgh Press
All rights reserved
Manufactured in the United States of America
Printed on acid-free paper
10 9 8 7 6 5 4 3 2 1

Cataloging-in-Publication data is available from the Library of Congress

ISBN 13: 978-0-8229-4829-2
ISBN 10: 0-8229-4829-X

Cover art: Portrait of William Whewell, by George Francis Joseph, 1836. *Source:* TC
Oils P 208, Trinity College Library, Cambridge.
Cover design: Melissa Dias-Mandoly

Contents

Acknowledgments

My research for and preparation of this volume was made possible by grants from the Dutch Research Council (NWO) and the Gerda Henkel Stiftung. The book would not have materialized without the support of the staff of Trinity College Library, Cambridge, upon whose holdings it draws heavily. I am especially thankful to Nicolas Bell, Adam Green, James Kirwan, Diana Smith, and Jonathan Smith. I am grateful to the following libraries and archives for permission to cite and quote from manuscript material in their collections: Cambridge University Library, Cambridge; Royal Society, London; St. Andrews University Library, St. Andrews; and Trinity College Library, Cambridge. Many thanks go to Abby Collier, for her encouragement and support of this project; to Amy Sherman, for her guidance, and—on behalf of all contributors—to Judy Loeven, for her amazing copyediting; to the two readers, for their valuable comments; to Christopher Stray, for making scholarly life so enjoyable; and to all the contributors, for their willingness to work with me on this book.

Brief Chronology of Whewell's Life

For a more extensive chronology see the online supplement "Chronology of Whewell's Life."

1794	Born in Lancaster, May 24, the eldest son of John Whewell and Elizabeth Bennison
1812	Entered Trinity College, Cambridge, as a subsizar
1815	Elected scholar, Trinity College, Cambridge
1816	Graduated second wrangler, second Smith's Prizeman. Death of his father
1817	Elected fellow of Trinity College, Cambridge. President of the Cambridge Union Society
1818	Appointed assistant tutor (mathematical lecturer), Trinity College, Cambridge. Founding member of the Cambridge Philosophical Society
1819	Publication of *Elementary Treatise on Mechanics*
1820	Elected fellow of the Royal Society of London. Moderator on the Mathematical Tripos, University of Cambridge
1823	Appointed (head) tutor, Trinity College, Cambridge. Publication of *A Treatise on Dynamics*
1825	Ordained priest
1826	Start of experiments at Dolcoath mine, Cornwall, with George Biddel Airy, to determine the density of the earth

1828   Elected professor of mineralogy, University of Cambridge. Publication of *An Essay on Mineralogical Classification and Nomenclature*. Moderator on the Mathematical Tripos, University of Cambridge

1830   Publication of *Architectural Notes on German Churches: With Remarks on the Origin of Gothic Architecture*. Publication of first of four papers on political economy, "Mathematical Exposition of Some Doctrines of Political Economy." Supported John Herschel as a candidate for the office of president of the Royal Society

1831   Coined the words *Miocene* and *Pliocene* in a letter to Charles Lyell, dated January 31

1832   Resigned as professor of mineralogy

1833   Publication of *Astronomy and General Physics Considered with Reference to Natural Theology*, the third Bridgewater Treatise. Publication of the first of fourteen papers on the tides, "Essay towards a First Approximation to a Map of Cotidal Lines." Cofounder of the statistics section (Section F) of the British Association for the Advancement of Science (BAAS)

1834   Review of Mary Somerville's *On the Connexion of the Physical Sciences*, where Whewell used the word *scientist*—coined by him—for the first time in print. Coined the words *anode, cathode*, and *ion* in letters to Michael Faraday, dated April 25 and May 5

1835   Vice-president of the BAAS

1836   Stood as (unsuccessful) candidate for the Lowndean Professorship of Astronomy and Geometry, University of Cambridge

1837   Publication of *The History of the Inductive Sciences*. Publication of *On the Principles of an English University Education*, placing classics and mathematics at the core of a liberal education. Publication of *On the Foundations of Morals*

1837–1839   President of the Geological Society

1838   Elected Knightbridge Professor of Moral Theology or Casuistical Divinity, University of Cambridge. Took degree of Bachelor of Divinity (BD)

1839   Resigned as tutor, and briefly considered taking a parish

1840   Publication of *The Philosophy of the Inductive Sciences*

1841   Married Cordelia Marshall, October 12. Appointed master of Trinity College, Cambridge. President of the BAAS

1842–1843   Vice-chancellor, University of Cambridge

1843   President of the Cambridge Philosophical Society. Advocated, alongside George Peacock, for the establishment of a board of mathematical studies at the University of Cambridge

1844   Took degree of Doctor of Divinity (DD)

1845   Publication of *Of a Liberal Education*. Publication of *The Elements of Morality, including Polity*

1846   Publication of *Lectures on Systematic Morality*

1847   Publication of *Sermons Preached in the Chapel of Trinity College, Cambridge*. Publication of *English Hexameter Translations*, with John Herschel, Julius Hare, Edward Hawtrey, and J. G. Lockhart. Supported election of Prince Albert as chancellor of the University of Cambridge, hoping that it would avert the appointment of a (royal) commission

1848   Publication of *Butler's Three Sermons on Human Nature*. Helped establish the Natural and Moral Sciences Triposes, University of Cambridge

1849   Publication *Of Induction, with Especial Reference to Mr. J. Stuart Mill's* System of Logic

1850–1852   Answered inquiries of the commissioners of the Royal Commission on the University of Cambridge, appointed in 1850, prefacing each with a protest against government interference

1852   Publication of *Lectures on the History of Moral Philosophy in England*

1853   Publication *Of the Plurality of Worlds: An Essay*

1855   Death of Cordelia Whewell (neé Marshall). Resigned as Knightbridge Professor, succeeded by John Grote

1855–1856   Vice-Chancellor, University of Cambridge

1858   Married Lady Everina Affleck (neé Ellis), widowed sister of former pupil Robert Leslie Ellis. Publication of first two parts of third edition of the *Philosophy*: *History of Scientific Ideas*, 2 vols., and *Novum Organon Renovatum*

1860   Publication of third part of third edition of the *Philosophy*: *On the Philosophy of Discovery*

1860–1861   Publication of *The Platonic Dialogues for English Readers*, 3 vols.

1862   Publication of *Six Lectures on Political Economy*, which resulted from a set of lectures delivered to the Prince of Wales, the future King Edward VII

1865   Death of Lady Everina Frances Whewell (née Ellis, and formerly Affleck)

1866   Died in Cambridge, March 6, at the age of seventy-one. Established by his will the Whewell Professorship and Scholarships in International Law, maintained from rents of Whewell's Court, Trinity College, Cambridge, built at his expense

William Whewell

William Whewell, Victorian Polymath

LUKAS M. VERBURGT

William Whewell was born on May 24, 1794, as the son of a Lancaster master carpenter.[1] He came up to Trinity College, Cambridge, in 1812 as a sub-sizar, the lowest rank of student, whose duties were to wait the tables and clean the rooms of his wealthier peers. Some fifty-four years later, on March 6, 1866, he died as master of Trinity, one of the leading figures in the scientific community of the early Victorian era, and a towering defender of the moral, social, and political status quo in the midst of Britain's "Age of Reform."

During the uninterrupted period he spent at Trinity College—the college, as he often reflected, of Isaac Newton and Francis Bacon—Whewell's rise was spectacular and his accomplishments many and impressive.[2] His family's chosen representative in an elite milieu, a young Whewell wrote that "*we* have reason to be proud" when telling his father of his first place in every subject in 1814.[3] Two years later he graduated second wrangler in 1816, beaten by a student of Gonville and Caius College, who was also first Smith's Prizeman, with Whewell taking second place. He was elected fellow of his college in October 1817, thus securing "a comfortable establishment for life,"[4] and afterward served as assistant tutor (1818–1823), tutor (1823–1839), and master (1841–1866) of his college. At the University of Cambridge, he was appointed professor of mineralogy (1828–1832), Knightbridge Professor of Moral Philosophy (1838–1855), and vice-chancellor (1842–1843, 1855–1856) and was a key figure in establishing the Natural and Moral Sciences Triposes. In these various roles Whewell was a dominating, inescapable presence in Cambridge's period of transition from a mathematical seminary to a leading center of scientific education and research. He was also a member of the Council of the Royal Society (1831, 1858), vice president (1835) and president (1841) of the British Association for the Advancement of Science (BAAS)—in which founding he was involved—president (1837–1839) of the Geological Society, and in 1834 he helped establish the Statistical Society of London, later renamed the Royal Statistical Society.[5] From his self-created position of inside-outsider, Whewell set out to shape Victorian science—from its disciplinary boundaries and its methodology to its organizational structure and the ethos of its practitioners—in his own towering image.

Both inside and outside the walls of the university, Whewell used his growing power and influence to "keep up, constantly alive" that "formative spirit which makes *reform* unnecessary." What guided his activities as a conservative Anglican at Cambridge was twofold. First, an unshakeable belief in "our National Constitution and our National Religion" and the role of Oxbridge in protecting this holy marriage.[6] Second, he held a deep-seated dislike of anything—whether Lockean sensationalism, Ricardian political economy, French analytical mathematics, Paleyean natural theology, or the *Westminster Review* utilitarianism—that could be seen as a potential ground for those seeking "to destroy the church and democratize the nation."[7] Whewell was not a simple reactionary who favored the old over the new. His favorite image was that of a hand transmitting a torch to another hand, with the motto "Holding torches they will pass them on one to another." He was in favor of change, whether in society, culture, morality, education, or science, but only if this change was of a particular kind. The new had to "include and rest upon that which has been true up to the present time." For such truths were not just necessary and permanent—in the sense that, once disclosed, nothing could disprove them. They were also partial fulfillments of that greater "Truth" that God "has given us the means of seeing," such that "we *must* . . . accept them." Whewell looked at the status quo, not as a "dead, stationary, immovable thing," but as the gradual "unfolding" of God's providence on earth. This is why it had to be defended: to make sure that "all that is really good and great" was neither destroyed nor prevented from happening.[8]

One of, if not the most prolific authors of his time, Whewell took novel and authoritative issue with almost every major topic on the early nineteenth-century agenda. Many of these topics were put on there by Whewell himself, both through book-length publications like his Bridgewater Treatise, *Astronomy and General Physics Considered with Reference to Natural Theology* (1834), *The History of the Inductive Sciences* (1837), and *The Philosophy of the Inductive Sciences* (1840), as well as through "suggestions, comments, experiments, measurements, linguistic improvements, and . . . critiques of the work of others" in reviews, papers, reports, addresses, and sermons.[9] His monumental oeuvre—totaling some ten thousand pages at a very modest estimate—included textbooks on physics and mathematics; original scientific research on mineralogy and the tides; periodical reviews of books by Charles Lyell, Richard Jones, Mary Somerville, and John F. W. Herschel; numerous sermons and public addresses; translations of Greek philosophy and German and Latin poetry; and publications on political economy, church architecture, natural theology, history and philosophy of science, language, university education, ethics, and law. Over the course of the nineteenth century, many of Whewell's books went through numerous editions and were translated into several languages, including French, German, Russian, and Chinese.

In his fifty-year-long career Whewell assumed many roles and identities. He was a mathematician, a textbook author, a mineralogist, a physicist, a reviewer, a natural theologian, a meta-scientist or looker-on of science, a historian and philosopher of science, a scientific organizer, a college head, a prominent university figure, an educationalist, a preacher, a poet of sorts, a translator, an editor, and a moral philosopher. He was also a social climber; a husband; a two-time widower; a colleague; a candid and sensitive friend; a polemicist; a member of some twenty-five scientific societies in Britain, Germany, Austria, France, and Belgium; a conservative; a frequent traveler; and an avid note taker and letter writer (some seven thousand letters from or to some thousand correspondents).[10] Some of these identities were transient; others were more stable. Sometimes they overlapped; at other times they existed side by side or stood in some tension. Above all, as most commentators agree, Whewell was a *polymath*. This label, however, must not be taken to be self-explanatory, as it took on different layers of meaning. For it to capture something important, let alone essential, about Whewell's life and work it should arguably stimulate more questions than it answers.

Whewell himself grappled with his scholarly identity almost his entire working life, which suggests that his polymathy was sui generis and changed over time. This already started in 1815, when as a nineteen-year-old undergraduate he suspected that "not much good would be likely to come of me if I were to remain in such an all-reading, all-learning mood for ever."[11] As the diary of his readings, which he kept from 1817 to 1830, shows, Whewell remained in this mood for most of the 1820s, devouring everything from Kant's *Kritik der reinen Vernunft* (first read in 1825) and De Ségur's *Histoire de Napoleon* to Boccaccio's *Life of Dante* and William Henry's *An Epitome of Chemistry*.[12] He, in fact, never stopped being his all-reading, all-learning self. What changed was that he eventually made the rigorous systematization of what he read and learned one of his main occupations. Tellingly, from 1838 on Whewell started referring to himself as a "system-maker."[13]

There are different ways to bring out the trajectory of Whewell's polymathy—his great "odyssey from mathematics to moral philosophy."[14] The most insightful is to develop a down-to-earth look at his oeuvre as a whole (see "Appendix A: List of Whewell's Published Works") and to divide his career (see "Appendix B: Chronology of Whewell's Life") into three periods: 1810s–1820s, 1830s, 1840s–1860s. This makes it possible to recognize both major transitions and long-term continuities in his intellectual and professional development. It also shows that what Whewell is mostly remembered for, and what has often been taken to be key to his polymathy—metascience and especially philosophy of science—represented a fraction, albeit a very important one, of his oeuvre and just one period in his long career. While Whewell remained active as a scientist throughout the 1830s–1840s, he effectively ceased to be a philosopher of science by 1840, and he himself found his *History* and *Philosophy* less important than his *Elements of Morality*. Perhaps most crucially, Whewell's metascience was itself part of, and stood in the service of, his "meta-religion"—the lifelong zeal to (re)conceptualize everything in the world as either weakening or strengthening, as either a danger to or support of, Christian faith.

Whewell's Journey

During the first period, which started with his appointment as an assistant mathematics tutor at Trinity in 1818, Whewell emerged as a prolific author of textbooks.[15] Among several other works, he published *An Elementary Treatise on Mechanics* (1819), which went through seven editions by 1847, and *A Treatise on Dynamics* (1823), which reached a third edition in 1834. Whewell met John F. W. Herschel, Charles Babbage, George Peacock, and other members of the Analytic Society, founded in 1811 to advocate Continental notation and algebraic analysis at Cambridge. He was at first impressed by their plans, and in the *Mechanics* and *Dynamics* treated his subjects from an analytic perspective. However, subsequent editions emphasized the value of traditional geometrical, intuitive methods characteristic of Euclid and Newton, as they represented what he called "permanent" rather than "progressive" knowledge. Whewell would use this view—oriented around the belief that mathematics was a means and not an end in itself—to condition the reform of the Cambridge mathematical curriculum, which he did in various ways: through textbooks, tutorial lectures, pedagogical tracts—such as *Thoughts on the Study of Mathematics as Part of a Liberal Education* (1835, reprinted in *On the Principles of English University Education* in 1837)—and involvement in numerous syndicates.

Whewell was also involved in various scientific activities, which supported his election to the Royal Society (1820) and admission to the Geological Society (1827). Many of these started as onetime events, aimed at self-education: a geological expedition with Adam Sedgwick in 1821; three months of instruction in mineralogy and crystallography in Berlin, Freiburg, and Vienna under Friedrich Mohs in 1825; experiments in a Cornwall mine in 1826 and 1828, with George Biddell Airy and Richard Sheepshanks, to discover the mean density of the earth; occasional architectural tours in Normandy (1823), Cumberland (1824), Germany (1829), Devonshire and Cornwall (1830), and Picardy and Normandy (1832); and the development of an anemometer to measure the velocity of the wind, first devised in 1837. Other pursuits were more sustained, each resulting from the application of mathematics to get to a subject's theoretical core. Between 1821 and 1827, the year in which he was elected professor of mineralogy, Whewell published several papers and an eighty-page books on crystallography and mineralogy. He provided a nomenclature, a taxonomy, and a simple mathematical foundation, thereby paving the way for mineralogy's development for the rest of the century. Having no ambition to pursue its details further, Whewell resigned the chair in 1832, never to return to the subject again, though he would forever remain associated with it through the mineral named after him (whewellite). Another field to which he applied mathematics was political economy, about which he learned in conversations with his close friend Richard Jones.[16] Whewell produced various publications, in which he attacked Ricardian (deductive) economics by reformulating its principles into mathematical equations, deducing consequences, finding them erroneous, and then concluding that political economy needed to be founded on induction from facts, a task he left for Jones to complete. These outputs appeared over a thirty-year period, in 1830, 1831, 1856, and 1862. Whewell also took an interest in church architecture, to which he brought a structural-analytic approach reinforced by careful geometrical descriptions.[17] In 1830 he published *Architectural Notes on German Churches, with Remarks on the Origin of Gothic Architecture*, enlarged and republished in 1835 and 1842—his last publication on the subject appearing as late as 1863 in the form of a lecture to the Royal Institute of British Architects. Whewell's goal was to point out and interpret the form and evolution of Gothic architecture and to propose a theory of why the Gothic's pointed arches came into existence. This had some bearing on his work on the history and philosophy of science, for instance, because it introduced the notion of a "Fundamental Idea"—in this case "verticality."

Having finished off mechanics, mineralogy, architecture, and political economy, at least temporarily, Whewell decided that it was time for something else. His most significant scientific research was his study of the tides, or tidology, which he sought to place on a new footing.[18] He took it upon himself to amass a body of data, leaving the theorizing largely to others. It earned him the Royal Society's Queen's Medal in 1837. Whewell published fourteen memoirs on the subject in the *Philosophical Transactions of the Royal Society* between 1833 and 1850 (about three hundred pages in total), wrote several occasional papers, outlined practical instructions for making tidal observations in articles in the Admiralty's *Manual of Scientific Inquiry* and the *Journal of the Asiatic Society of Bengal*, and presented reports to the Royal Society, the BAAS, and the Cambridge Philosophical Society. For his "Great Tide Experiment" of 1835—a pioneering historical example of citizen science—Whewell directed thousands of people to take tidal readings simultaneously at 650 stations in nine countries, extending to the farthest edges of the Empire. One of his ambitions was to establish a global map of cotidal lines passing through all the points where high water occurs at the same time, and to discover the law of these cotidal lines. Much to his own disappointment, this experiment failed, though Whewell did manage to work out cotidal lines for restricted areas and to establish laws for the diurnal inequality. His work was not entirely in vain. But the fact that he relied entirely on the equilibrium theory, rather than Pierre-Simon Laplace's dynamical theory, placed him at odds with the "bolder and stronger mathematicians"[19]—George Biddell Airy, George Gabriel Stokes, and William Thomson (Lord Kelvin)—who continued British tidal studies.

Despite his scientific achievements and the recognition he received, Whewell did not consider himself "an eminent man of science." One reason might have been that his contributions to mineralogy and tidology, though important, failed to meet his own criteria for major scientific breakthroughs. More likely is that he compared himself to the prominent discoverers around him, which made him feel very modest about his own work. "In the study of the tides," Whewell observed in 1840, "I have voluntarily given up all the profounder parts of the subject, and confined myself to collecting laws of phenomena in such a manner as it could be done with little of my own labor."[20] Another reason was that he simply had too many other interests and felt, with a mix of enthusiasm and regret, that his strengths lay elsewhere. As he wrote to Herschel in an almost apologetic letter in 1818: "There is another point of view [on recent research on the properties of light] which occurs to us lookers on, who, not making a single experiment to further the progress of science, employ ourselves with twisting the results of other people into all possible speculations mathematical, physical, and metaphysical."[21] It would take Whewell over a decade to accept that he himself was such a looker-on—a delay at least partly due to the low opinion of metaphysics (that "*poor word*"[22]) common among his circle of friends. And it would take several more years to work himself into a condition, with regard to "worldly affairs," to "be a philosopher and nothing else."[23] However cerebral, being a polymath was also a financial matter.

At the core of Whewell's metascience, developed in the 1830s, stood two distinct, yet mutually supportive projects: his "theology" and his "induction," that is, natural theology and the history and philosophy of science.[24] The first project ran from *Astronomy and General Physics* (1833) and *Indications of the Creator* (1845) to *Of the Plurality of Worlds* (1855).[25] It was aimed to demonstrate, and to flesh out the consequences of, the harmony of modern science with Christianity. An immediate "best seller"[26] and his first book for an audience outside Cambridge, in the *Astronomy* Whewell argued that the latest scientific knowledge of various phenomena—from the solar system to heat, electricity, sound, and light—offered proof of intelligent design. Although the book contradicted some of his statements in the *Astronomy*, in the *Plurality of Worlds* Whewell would also draw on scientific evidence, especially geology, to make a theological point: in this case, to counter the popular claim that all planets must be inhabited because, were that not the case, God's creation of them would have been wasted. What stood out particularly from the *Astronomy* was book 3 ("Religious Views") where Whewell introduced a theme running through his entire metascientific oeuvre: the contrast between inductive and deductive "habits of the mind," between reasoning by *ascending* from facts and by *descending* from principles. Whewell argued that the former was not just of greater value to science but also had a stronger tendency to religiosity. This was illustrated by the bold claim that "original discoverers" like Johannes Kepler and Newton had been men of Christian faith, whereas "mere mathematicians" such as Laplace and Joseph-Louis Lagrange were more inclined to atheism.[27]

The second project ran from the *History* and *Philosophy*—as well as various original papers included in later editions—to the spin-off *Of Induction, with Especial Reference to Mr. J. Stuart Mill's System of Logic* (1849). It took the form of a renovation, in light of the history of modern science, of Bacon's inductive philosophy. The two projects were explicitly brought together in *Indications of the Creator*, quickly compiled in response to Robert Chambers's controversial *Vestiges of the Natural History of Creation* (1845). Whewell's subtitle was *Extracts, Bearing upon Theology, from the History and the Philosophy of the Inductive Sciences*. From this book, as well as from several of Whewell's numerous sermons,[28] it becomes abundantly clear that the motivation behind all his metascientific work was deeply religious. Whewell did not think that science could bring people to faith. At a time when its meaning was not yet settled, his ambition was to define science such that it was understood to fall in fully and entirely with Christianity. It was no mere verbal coincidence that Whewell spoke of induction as *"the true faith."*[29]

The first step of Whewell's induction project was taken around 1830–1831, when he started carving out a new space for himself in the emerging British scientific world. Whewell privately began writing notebooks (1830–1834), scribbling hundreds of pages of "scraps and snatches"—some actually of book length—on the inductive nature and methodology of science.[30] He also offered public statements on the subject in an address to the BAAS (1833), in two reports to that body on the state of particular sciences (1833, 1836), in a lengthy appendix to the textbook *Mechanical Euclid* (1837), and in major reviews of Herschel's *Preliminary Discourse on the Study of Natural Philosophy* (1831), Lyell's *Principles of Geology* (1831, 1832), Jones's *Essay on the Distribution of Wealth, and on the Sources of Taxation* (1831), and Somerville's *On the Connexion of the Sciences* (1834). Whewell's decision to devote himself to "that higher philosophy . . . which legislates for the sciences"[31] did not come out of nowhere. He was acting on a wish he had held since his student days, the "ancient subject of [his] early liking"; the "Inductive Method of Philosophising."[32] This Baconian outlook helped shape his myriad activities from the 1820s to the 1830s, in which he slowly but gradually, and bit by bit, developed his own mature philosophical views. Already in 1825 Whewell had boasted that he would use the Lucasian Professorship in Mathematics—which had fallen vacant—to "make very grand lectures on the principles of induction," and had sworn to accept the position of professor of mineralogy only because the field was "one of the very best occasions to rectify and apply our general principles of [inductive] reasoning."[33] By that time, however, most ingredients of his mature position, such as the key notion of conceptions, were still lacking.

Whewell's decision to become a meta-scientist did come with a significant transformation of his polymathy. During the second period of his career it became "simultaneous" in a different—more "clustered," less "fiddle faddle"—way, and altogether more "centripetal."[34] Whewell no longer primarily sought to reform each and every science in which he himself was active on the basis of a roughly Baconian notion of induction. Instead, he now set out to reform Bacon's inductive philosophy itself, looking almost exclusively at the physical sciences as a "connected and systematic body of knowledge."[35] Sometimes these ambitions overlapped. About his work on the tides, Whewell remarked in 1833, "I wish I could explain to you how useful my philosophy is in shewing me how to set about a matter like this, and how good a subject this one . . . is to exemplify it."[36] The induction project itself, however, was closest to his heart. "My induction 'invites my steps' every half hour that I am left to my own thoughts. If I am ever to do any good, I must set about it soon (I shall be forty in half a year)."[37] What stood in the way were his duties (or "business") as head tutor at Trinity, which he found increasingly tiresome. Over the course of the 1830s Whewell frequently toyed with the idea of retiring from his post. By 1836 he admitted to Herschel that he had "very serious thoughts of [giving up] my share in the active business of the College, and of giving *the rest of my life* to the formation and exposition of a Philosophy . . . such as we ought to have." For this Whewell realized he needed more time and focus:

> I do not know how you [Herschel] manage to carry on so many speculations at once; but for my own part I begin to find that I have set myself a task, which is hardly consistent with my other employments here [at Cambridge]. . . . I have tried for several years, and I cannot combine these two employments to my own satisfaction; and I think it is more wise and right to transfer to other hands occupations, which I am conscious of being unfit for, and duties, which I discharge imperfectly, than to go on with an impossible struggle, and to endanger the attainment of a great object.[38]

Whewell did not reach his decision lightly, knowing it was a leap in the dark. Besides, his work at Trinity was not "ungrateful, either as a chance of doing good or of making money."[39]

Between 1830 and 1834 Whewell pursued his project without a clear writing plan, moving back and forth between philosophical theorizing and historical illustration or "exemplification." A breakthrough came in the summer of 1834:

> I am to consist [sic] of three Books. Book 1, *History* of Inductive Science . . . historiographized in a new and philosophical manner. Book 2, *Philosophy* of Inductive Science. . . . It will be dry and hard . . . as it must contain most of the metaphysical discussions . . . , but it must also contain all the analysis of the nature of Induction and the Rules of its exercise, including Bacon's suggestions. Book 3, *Prospects* of Inductive Science. The question of the possibility . . . of applying Inductive processes, as illustrated in . . . Book 2, to other than material sciences; as philology, art, politics, and morals.[40]

From their inception, the *History* and *Philosophy* were conceived as mutually supportive, yet independent parts of a single inquiry: to complete, that is, to "renovate and extend" the "Reform of the Methods and Philosophy of Science" initiated by Francis Bacon.[41] The two books even stood to each other as Bacon's *Advancement of Learning* and *Novum Organum* (1620). The *History* appeared in 1837 in three volumes, totaling 1,600 pages. On the basis of a study of primary sources, but borrowing heavily from other writers' histories of specific disciplines, it gave an overview from the ancients to the present of the physical sciences—including astronomy, mechanics, acoustics, optics, "thermotics," "atmology," chemistry, mineralogy, botany, zoology, physiology, anatomy, and geology. One of Whewell's innovations was to label and classify the different sciences.[42] Another was to introduce a three-stage pattern, a historiographical novelty informed by Whewell's own philosophy:[43] a crucial period of discovery—"inductive epoch"—was marked by a convergence of distinct facts and clear ideas in the mind of a great scientist; it was preceded by a "prelude" in which these facts and ideas were gradually clarified, and succeeded by a "sequel" in which the discovery was accepted and consolidated by the scientific community.

The *History*'s survey made it possible for Whewell to do what Bacon had been unable to do: to ground inductive philosophy on historical knowledge of the actual development of modern science. Whewell explicitly thought of his own theory of induction as adding to Bacon's nothing more and nothing less than "such new views as the advances of later times cannot fail to produce or suggest."[44] The *Philosophy* was published in 1840 in two volumes, totaling some 1,200 pages. At its heart stood the insight—on which Whewell had stumbled in 1833–1834—that all knowledge requires both facts and ideas or, that is, has both an ideal, subjective and an empirical, objective dimension. He called this the "Fundamental Antithesis" of knowledge, with which he carved out his own middle way between Immanuel Kant and the German idealists and John Locke and the sensationalists. What emerged was twofold. First, an antithetical epistemology, which said that observation is "idea-laden": the phenomena studied by scientists are first made possible by certain "Fundamental Ideas" supplied by the mind itself. For example, the Fundamental Idea of "Cause," together with the concept of "force" implied in it, is what allows physicists to obtain knowledge of mechanical phenomena. Second, a quasi-idealist philosophy of science in the form of a new theory of induction: "Discoverers' Induction." Whewell rejected the standard view of induction as the mere generalization from particulars. Instead, he argued, in induction "there is a New Element added to the combination [of particulars] by the very act of thought by which they were combined."[45] Whewell called this act of thought "colligation," that he defined as the mental operation of bringing together a number of facts by "superinducing" on them a concept that unites them and thus renders them capable of being expressed by a general law. More than Bacon, Whewell emphasized the creative role of the mind in science, and more than Kant, he insisted that the ideas that made the construction of scientific knowledge possible unfolded gradually over time.

Whewell believed that this Kant-inspired Baconianism—as opposed, for instance, to Herschel's empiricist alternative—made him Bacon's legitimate heir. Others were less convinced. Unlike the *History*, the *Philosophy* was met with almost universal rejection, including from Herschel, who reviewed the book in 1841. This did not come as a surprise to Whewell, who was aware of the widespread antipathy to metaphysics among his English countrymen. Neither did the negative reactions to his philosophy position despair him completely. Whewell himself wrote that the reform of inductive philosophy, "when its Epoch shall arrive, will not be the work of any single writer, but the result of the intellectual tendencies of the age."[46] He expressed a similar sentiment about his metascience—a mix of humility and self-confidence—in 1840 when he received an invitation to serve as president of the BAAS:

> My only pretensions to such a position are what I may have done as a cultivator of science, and my constant attendance upon the business of the Association. With regard to the former point, . . . I know perfectly well that there is nothing of such a stamp, in what I have attempted, as entitles me to be considered an eminent man of science. . . . My *History* and *Philosophy* of Science are disqualifications, not qualifications, for my being put at the head of the scientific world; for I cannot expect, I know it is impossible, that men of science should assent to my views *at present*: and those who have laboured hard in special fields will naturally feel indignant at having a person put at their head, recommended only by what they think vague and false general views.[47]

Whewell found himself in a peculiar situation: to become a meta-scientist, he had had to abandon his work as a man of science, but now that his project was finished, his position in the scientific world (that of "*lay* speculator") counteracted its potential influence. Already in 1836, around the time of his career switch, he was aware of this situation: "I shall do this with some regrets; . . . because I think I perceive that any improvement in our academical studies (and of course a reform of philosophy ought to improve *them*) may be introduced with greater advantage by a person actively engaged in them than by an insulated spectator."[48] The *History* and *Philosophy* were Whewell's crowning achievements as commentator on, and aspiring judge of, science, establishing his reputation for posterity. Lyell, Herschel, Adam Sedgwick, and James Clerk Maxwell, among other leading men of science, praised their value for understanding how science developed and how it might proceed. The books also capped the central part of Whewell's induction project, without sending the shock waves through the scientific community their author had hoped for. The *History* was "too crabbed for the general reader," too limited for the specialist, and not scholarly enough for the historian; the *Philosophy*, in turn, was too "dry and hard" for almost everyone.[49] Perhaps for this reason, the books (which saw a first print run of 1,500) reached their third and final editions only some twenty years after they first appeared. Neither of them was ever significantly revised.[50] This is most telling in the case of the *History*. Whewell did incorporate new discoveries into subsequent editions, which the young Charles Darwin found very useful. But already in the second edition of 1847, Whewell introduced no new branches of science and accepted that the book was no longer (as its title still promised) a history up to the present. It stood in marked contrast to the original ambition of creating "a platform on which we might stand and look into the future."[51] A decade later, this was simply no longer feasible, and not just because of the rapid developments in the physical sciences. The fact was that Whewell himself had moved on to other pursuits.

The next step of Whewell's induction project was taken around 1840. Whewell had been appointed Knightbridge Professor at Cambridge in 1838 and resigned as tutor of Trinity College the next year, thereby completing his career switch from man of science to philosopher. At the beginning of 1841 he briefly considered leaving Cambridge to take up a country parish. A few months later things looked decidedly different: he was married to Cordelia Marshall and installed as master of Trinity, a Crown appointment recommended by Robert Peel, the Tory prime minister. An extraordinary feat of social elevation—accompanied by a new, formal, and stiff demeanor—it enabled Whewell to put into practice the conservative Anglican vision that underpinned his metascientific project. For instance, in 1844 Whewell revised the college statutes to limit the system of private tuition—a utilitarian, commercial practice that he believed weakened the moral and theological dimension of a liberal education. In 1848 he introduced the Moral and Natural Sciences Triposes, which widened the traditional Cambridge curriculum for the first time in its history. The aim was not to break the dominance of mathematics and classics, which Whewell continued to defend in educational writings and textbooks. Instead, it was done from a wish for university reform to be internally directed rather than externally imposed, in the form of a Royal Commission (appointed, much to Whewell's chagrin, in 1850).[52]

Upon his election to the Knightbridge Chair, Whewell, ready to take off his "mathematical boxing gloves, and go on with arms of wider range,"[53] changed its title from "Moral Theology or Casuistry" to "Moral Philosophy." It marked the start of his pursuit of the final part of his system, the subject of "Book 3" from his original 1834 plan: the application of induction to "other than material sciences." For the next two decades or so, interrupted only by his work as vice-chancellor (1842–1843, 1855–1856), Whewell made moral philosophy the main subject of his reading, lectures, and sermons. He published several books on the topic—*On the Foundations of Morals* (1837), the four-hundred-page *Elements of Morality, including Polity* (1845), *Lectures on Systematic Morality* (1846), and *Lectures on the History of Moral Philosophy in England* (1852)—and also edited other ethical and legal writings, explaining their significance in prefaces, including James Mackintosh's *Dissertations on the Progress of Ethical Philosophy* (1836), Joseph Butler's *Three Sermons on Human Nature* (1848), and Hugo Grotius's *De Jure Belli et Pacis* (1853).

From 1840 Whewell, at least in his own mind, ceased to be a historian and philosopher of science. He shifted his "centripetal" polymathy to the moral sciences, where it became more "limited" and focused on the humanities (or "moral sciences"). As he wrote in a letter to Herschel on April 22,

1841, in the *Philosophy* he had put down his philosophical views on science "once and for all."[54] This does not mean that the *History* and *Philosophy* did not have a moral. Quite the contrary, the books were written precisely to "get a scientific *moral* out of" them. The John Stuart Mill–Whewell debate from the 1840s to the 1850s makes this feature of Whewell's metascience abundantly clear.[55] Both men thought of their different views on induction as a struggle literally between good and evil. Mill believed Whewell's idealistic philosophy, *"physical as well as moral,* . . . to serve as a support and justification to any opinions which happen to be established."[56] This is why, for Mill, to defeat Whewell was "no mere matter of abstract speculation," but rather to defeat conservatism and, as such, "full of practical consequences."[57] Whewell, for his part, believed that the empiricism of Mill's *System of Logic* was "entangled in the prejudices of a bad school"[58]—the utilitarian ethics of Jeremy Bentham and William Paley, whose writings Whewell successfully managed to remove from the curriculum at Cambridge. This link traced back to the 1820s, when Whewell had lumped together Ricardians and utilitarians as the "irreligious school," hoping that an inductive methodology would create an "ethical" political economy based on a view of humans as divine creatures rather than selfish beasts.[59] This argument was continued in the *Philosophy,* which Whewell presented as a continuation of the fight, initiated in the field of ethics by Sedgwick in 1833, against the "ultra-Lockian school" that falsely advocated the "exclusive authority of the senses."[60] The book's key message was that humans can obtain absolutely certain knowledge of the God-created world and that this knowledge is made possible by God-given ideas, existing independently from experience. This view was important as a philosophical insight on the nature of science, and Whewell had strategically chosen to illustrate it by drawing examples from the physical sciences that were generally accepted as certain and true. "But," Whewell scribbled in a notebook in 1833:

> It is a subject of a far higher and deeper interest when we include in our survey all branches of human knowledge, those which concern his moral and religious condition as well as those which refer us to the material world. And it is not only allowable but necessary, to consider all the branches of human knowledge as having before them the same prospect of . . . perfection, till we have discovered how and why the rules and processes under which the physical sciences flourish and advance are incapable of being applied, with some modifications, to other parts of our knowledge.[61]

Whewell had long wondered whether what held for knowledge of the physical world also applied to knowledge of the moral world. By 1840 he was ready to take up the challenge and create a new system of non-utilitarian ethics.[62] Whewell set out on two missions, one historical and the other philosophical, each building on his metascientific outlook. First, he traced the history of moral philosophy in England, seen as a centuries-long struggle between the secular and the religious, between the "low morality" of pleasure and consequences and the "high morality" of "necessary, universal, and eternal" principles derived from mankind's innate moral conscience.[63] Whewell ridiculed proponents of the former, especially Bentham, and took what he considered valid from advocates of the latter, such as Butler. Second, for his own system of "independent morality," Whewell put forward five "Ideas" ("Benevolence," "Justice," "Truth," "Purity," and "Order")—corresponding to the five elements of human nature: love, mental desires, speech, bodily appetites, and reason—as the ground of virtues, to be realized in social rules, duties, laws, and institutions. Whewell's reasoning was rather similar, as he himself recognized, to Plato's argument in the *Republic* for finding the cardinal virtues by examining the different elements of the soul. For Mill, whose criticism of Whewell's moral philosophy was harsh, the system amounted to an arrangement of dominant opinions put into an obscure framework.

Whewell's central aim was to show that morality must be understood in terms that also apply to science. He initially compared morality to geometry and, somewhat later, to mechanics, drawing an analogy between moral and scientific "Ideas" and "Axioms" and suggesting that both are necessary and unfold over time. It is possible to obtain moral and scientific knowledge, since God created the physical and moral world and gifted humans with the powers and Ideas that make them knowable. At the same time, because of the limited powers of the human mind, the self-evident truths (or axioms) that follow from these Ideas may neither be immediately self-evident nor seen to be necessarily true by everyone at all times. This is why the "intuition" through which humans come to realize such truths, and come to recognize that they are implied in the Ideas, develops "progressively."[64] The process starts by using an implicit apprehension of an Idea to organize certain observed facts; and these organized facts, in turn, help to arrive at a more explicit awareness of the Idea. Whewell admitted there were significant differences between scientific and moral knowledge—in morality, for instance, a crucial role is played by conscience, understood as reason applied to moral subjects. But he always maintained that progress in the latter occurs in the same way as in the former. It is clear that Whewell's thinking on what this progress consists of changed over time. By the 1850s it was less Kantian and more Platonist and Romantic, placing the emphasis much more heavily on intuiting, or "guessing," Ideas in the mind of God.

Whewell's moral philosophy—the grand finale of his religious-moral-philosophical project—offered an alternative to utilitarianism but was received largely negatively, and would soon be forgotten, buried by Mill and Henry Sidgwick under a heavy tombstone. Even his biographer, Isaac Todhunter, writing in the late 1870s, believed that it was best laid to rest. From Whewell's own, Cambridge-centered perspective, things looked decidedly different. By the 1850s he had introduced his *History, Philosophy,* and *Elements of Morality* as well as his editions of Mackintosh and Butler into various Trinity examinations and into the Moral Sciences Tripos (first offered in 1851). Because he felt he had accomplished his "great object," Whewell resigned the Knightbridge Chair in 1855 at age sixty-one.

During his second professorship, which he combined with his work as master and his vice-chancellorships, Whewell took up various other subjects, none of which was scientific or directly related to his metascience. He now gave free reign to a more leisurely polymathy, allowing it to become more "centrifugal,"[65] and shifted his focus to the general English reader. In addition to moral philosophy, natural theology, sermons, college management, and university politics, language became one of Whewell's main occupations. There are several senses in which language had stood high on his agenda ever since the 1820s: as a mineralogist, he had been concerned with nomenclature and classification and as a philosopher with scientific terms ("instruments of thought"), many of which he coined himself, such as *ion, anode,* and *Pliocene.*[66] It suggests that the languages of nature and culture— indeed, nature and culture themselves—were for Whewell deeply integrated: they were divine co-creations, both designed according to God's plan. At a later point in his career, it was language itself that became an object of interest and study. Over the course of the 1840s–1860s, Whewell, who was well versed in French, German, Latin, and Greek, published various poems and was for some time a member of the Cambridge Etymological Society and the London Philological Society. Some of his linguistic undertakings were important but sporadic, such as in etymology.[67] Others were more sustained and took years of dedicated labor. All of them were aimed, in one way or another, to promote the standing of the English language.

One of Whewell's projects was to introduce hexameters, popular in Germany, into England.[68] Against the tide, and unsuccessfully, Whewell hoped to "naturalize" the hexameter: to show that English hexameters were different from Latin or Greek hexameters, and that English was a suitable language for poetry in this meter. He published adaptations into hexameters of English-language works (e.g., Thomas Carlyle's *Chartism* [1840]) as well as numerous translations of German-language prose into English hexameters. His most important work was *English Hexameter Translations from Schiller, Goethe, Homer, Callinus, and Meleager* (1847). It contained, among other texts, his own translation of Johann Wolfgang von Goethe's *Hermann and Dorothea* (1798) and translations by others, including Herschel, who at one point also translated the entire *Iliad* into hexameters. Whewell's work on hexameters points to the influence of German Romanticism on his mature outlook—he came to greatly admire Friedrich Schiller's *Der Spaziergang*—in addition to the usual Kantianism. What he liked about hexameters was the way in which they allowed for the expression of "the simplicity and truth of reality."[69] Unlike his Romanticist peers, such as Julius Hare, Whewell brought his typical approach to the subject: he went for its foundation, in this case the grammatical technicalities of spondees and trochees, on which he wrote several articles and reviews.

Another significant project was the translation and editing of the *Platonic Dialogues*, published in three volumes in 1860–1861. Here, the rationale was to "naturalize" ancient philosophy. Whewell, by combining translation and comment, sought to make the dialogues "intelligible and even interesting to the ordinary readers of English literature." But he also had a scholarly purpose in mind: "It seems not unreasonable to require," Whewell wrote, "that if Plato is to supply a philosophy for us, it must be a philosophy which can be expressed in our own language."[70] Plato was Whewell's "new love" from the 1850s, by which time his "old love,"[71] Bacon, had been taken care of. He presented five papers on aspects of Plato's philosophy to the Cambridge Philosophical Society. These appeared together in a booklet in 1855, and some were reproduced in *On the Philosophy of Discovery: Chapters Historical and Critical* (1860)—the third part of the third edition of the *Philosophy*. His interest in Plato did not come out of the blue. Whewell had started his career as an orthodox follower of Bacon, became a Kantian idealist of sorts in the 1830s, and from the 1840s on was more and more attracted to Platonism. It informed his *Elements of Morality* and *Plurality of Worlds*, which revived the Platonic theory of ideas. Following Richard Owen's application of Platonism to biology and paleontology, Whewell defended the view that all objects and laws of nature reflect "Archetypal Ideas" in the Divine Mind.[72] Since he never ceased to self-identify as a Baconian, it would be more accurate to say that Whewell made true Coleridge's remarkable claim that Bacon was "the British Plato."[73] The *Platonic Dialogues* also, and once again, bore witness to German influence on Whewell, who admitted to having derived all his views on Plato directly from the Plato scholar Joseph Socher. For instance, Whewell followed Socher in arguing—rather controversially—that the *Parmenides* was not a Platonic dialogue because Plato himself could not have written such a harsh attack on his own theory of ideas. Whewell's final publication, the unfinished article "Grote's Plato," appeared posthumously in April 1866. It showed a crucial feature of his mature philosophical thinking. No philosophy for the present time can be derived ready-made from the past. Neither can it be the creation of a single author. Instead, it will always be a product of the age itself. Whewell clearly saw it as his role to be the voice of his age, eventually echoing a present that was no longer there.

Whewell died after falling from a horse in 1866 at the age of seventy-one. Following the loss of his first wife in 1855, he had married Lady Affleck—the sister of his former pupil Robert Leslie Ellis—in 1858. She died on Saturday, April 1, 1865. Early the next morning, Whewell rose to the pulpit of Trinity College Chapel, allowing all to witness "the saddest of all sights, an old man's bereavement, and a strong man's tears."[74] After a few months of sorrow, he managed to write his articles on Auguste Comte and Plato. But his final thoughts and efforts were for Trinity. During the last years of his life, it had been his singular place of refuge. Whewell's bond to Trinity was both spiritual and tangible. In 1860 the first Master's Court, now Whewell's Court—a building opposite the Great Gate erected at Whewell's own expense—was completed. In his will Whewell also established and endowed a chair of international law that materialized his view, put forward in *Elements of Morality*, that international obligations between nation states marked the highest development of morality. On the last morning of his life, Whewell ordered his bedroom windows to be opened wide, allowing him to see the sun shine on the Great Court. "He smiled as he was reminded that he used to say that the sky never looked so blue as when it was framed by its walls and turrets."[75]

Whewell Scholarship: Past and Present

Any new volume on Whewell must position itself in relation to Menachem Fisch and Simon Schaffer's *William Whewell, a Composite Portrait*. Fisch and Schaffer opened their 1991 book with a reflection on how best to deal with the problem of Whewell's polymathy. Unlike previous studies, which they held to be typically limited in scope and compartmentalized in focus, their editorial strategy was to aim for a holistic treatment. "In a darkened Oxford pub we sketched out on a paper napkin a list of chapter headings such as 'Whewell the Teacher,' 'Whewell the Priest,' 'Whewell the Historian,' and so on. However, as this volume emerged, these carefully constructed compartments dissolved. . . . The dykes burst under a deluge of cross-reference and debate."[76] The present volume is organized exactly in such compartments. It brings together a group of scholars, each of whom contributes a chapter on one particular aspect of Whewell's polymathic oeuvre and career. This editorial decision was inspired by three considerations.

First, rather than seeking to oppose, let alone replace, the holistic accounts of Whewell in Fisch and Schaffer's volume—some of which have become classics—*William Whewell: Victorian Polymath* builds on, updates, and complements them. Whewell scholarship has developed significantly since 1991. It has undergone what might be described as simultaneous horizontal and vertical development: horizontal in that previously unexplored subjects in Whewell's publications have come to be studied, and vertical in that well-known topics and themes have been revisited through study and debate. Moreover, Whewell scholarship has been shaped by broader developments within the best historical research from the past thirty years or so. For instance, the history of the humanities, including the history of the book, has been emerging in interaction with the older fields of the history of science and the history of philosophy, together transforming the study of Victorian intellectual culture through conceptually rich, highly contextualized studies of science. Or, to give another example, Whewell often figures in accounts of neo-Kantian philosophy of science, so prominent today, in which he is often recognized as a founding father of the history of philosophy of science. The chapters in the present volume take stock of and add to the current state of Whewell scholarship, offering new, pithy, and authoritative starting points for research on Whewell's life, work, and times.

Second, the volume takes issue with Fisch and Schaffer's opinionated approach to Whewell's polymathy. According to them, it is "misleading to use the term 'polymath' for Whewell, since he was precisely in search of a means of synthesizing a vast range of allegedly disparate material."[77] There are several problems with this claim. It suggests that *polymath* is a monolithic term, whereas in fact it can be said to have several meanings and come in different kinds. Indeed, the attempt to synthesize knowledge defines one particular, "centripetal" type of polymath. Fisch and Schaffer, in abandoning the term *polymath*, essentially replaced it by that of *omni-* or *meta-scientist*, at the heart of which they situated his "historico-philosophical" project. It is true that metascience is key to understanding Whewell, but it is not a definition of who he was and what he did. After all, history and philosophy of science were his main occupation during one period of his career (the 1830s) and represent only a fraction of his large oeuvre. It is too reductive to read metascience into his work from the 1820s and to extrapolate it to that from the 1840s to the 1860s. Whewell's metascience offers one possible solution to the riddle of pinpointing to what he owed his status as one of the leading men of science of the Victorian era. But others are needed as well that take into account, for instance, that Whewell addressed many different audiences: not just men of science but also Cambridge students and the general English reader. Taken together, rather than abandoning the idea that Whewell was a polymath, it is arguably more fruitful to recognize that his polymathy was complex, many-sided, changing, and not free from internal tensions. This, in brief, is what *William Whewell: Victorian Polymath* brings clearly into view through studies of Whewell's many and interlinked polymathic interests.

Last but not least, there is the issue of Whewell's place in the early Victorian landscape. During the revival of Whewell scholarship in the 1970s, Whewell was approached either rather narrowly as a philosopher of science or as a leading member of an expansive "Cambridge Network."

Since the publication of Fisch and Schaffer's *William Whewell* and Richard Yeo's *Defining Science*, the focus has been predominantly on Whewell as a meta-scientist. This shift in Whewell scholarship went hand in hand with a broader criticism of Susan Faye Cannon's Cambridge Network as somewhat of a fictitious entity—though it has lived on, in another form, in Laura Snyder's *Philosophical Breakfast Club*.[78] It neglected individual differences between key members in favor of an emphasis on collective agreement on a shared project to transform the whole of British science. Cannon, for instance, tended to regard Whewell's and Herschel's view of science as the view of science of all members of the Cambridge Network. Although it was recognized that there were several subgroups, she not just glossed over fundamental disagreements between Whewell and Herschel but also ignored their distance, in certain crucial respects, from each other and from someone like Charles Babbage. William Ashworth's recent *Trinity Circle* instead pits Whewell squarely against Babbage, while bringing into view other, lesser-known allies of Whewell's religious-scientific-moral project, such as Sedgwick, Connop Thirlwall, Julius Hare, and, by extension, Robert Leslie Ellis, John Grote, and Thomas Rawson Birks.[79] At the same time, it is clear that Whewell, Herschel, and Babbage remained on friendly terms, and there remained commonalities between them that set them apart from other scientific reformers, whether it was William Hamilton, David Brewster, or Henry Brougham.

All this presents at least two significant challenges for future scholarship. First, "Morrell's Challenge": to zoom in and explore British science in the early Victorian period in terms of the "singularity" or "individualism" of figures like Whewell, Herschel, and Babbage.[80] Rather surprisingly, very little work in this regard has been done for Herschel and Babbage.[81] Moreover, Fisch and Schaffer's claim notwithstanding, the same holds for Whewell, albeit to a lesser degree. By 1991 the number of compartments used to study Whewell's oeuvre and career was fairly limited. Looking at their *William Whewell*, at least eight out of a total of thirteen chapters dealt purely with the history and philosophy of science. The present volume seeks to redress this situation, opening up more windows to Whewell's polymathy, also beyond his metascience. Second is "Cannon's Challenge": to zoom out and reshuffle the individual pieces to recompose a new big picture of British science in the (first half of the) nineteenth century. Once the well-rounded accounts of all aspects of pivotal figures (organizations, societies, groups, etc.), rather than some aspects of only a few of them, are available, scholars of nineteenth-century Britain should be in a better position to draw large-scale comparisons, unearth broader developments, and flesh out new major themes, contexts, and geographies. For instance, important work is currently being done to study the efforts of the Society for the Diffusion of

Useful Knowledge, founded in 1826. It will be key to bring the results to bear on ongoing research on other visions of science—whether Babbage's rational-mechanical utopianism or Whewell's Anglican conservativism or that of the Royal Society or the BAAS—and to see how these connect to those coming after them in the late Victorian era.

It is important to emphasize that, however comprehensive, the present volume is not exhaustive. Among the topics not included are Whewell's contributions on English hexameters; his poetry; his work as editor of Mackintosh, Butler, Grotius, and Plato; and his diary-keeping. Furthermore, with the digitization and further cataloging underway of the Whewell Papers held at Trinity College Library, Cambridge, it is very well possible that new Whewell material will be found that has not been taken into account. The hope is that this volume will inspire future work that will make up for its shortcomings.

The issue of Whewell's polymathy is perhaps most strongly felt when structuring a book about it. There is no ideal solution to capturing it, and the roughly chronological rather than thematic structure adopted here surely has drawbacks. Indeed, the chronology is, and can, only be quasi-chronological. For only very few of Whewell's myriad activities come with clear-cut start and end dates. Some were pursued infrequently (such as his contributions on political economy) or with greatly varying intensity over time (such as work on his textbooks, which climaxed in the 1820s but continued through the 1840s, and the delivery of sermons). Others concern themes running, sometimes explicitly and at other times implicitly, through his entire oeuvre—such as his views on politics and on women.[82] The chosen order of chapters hopefully gives a lively sense of the nature and development of Whewell's polymathy and naturally bring out certain thematic clusters. This holds, for instance, for the chapters dealing with his thinking on the history and philosophy of science, which tellingly find their place at the heart of the book.

Throughout the book, letters to or from Whewell included in Todhunter's *William Whewell* or Douglas's *Life of Whewell* are quoted from those books, as these can be readily consulted. In all other cases, full reference is provided to archive, collection, and item number. An overview of Whewell's works can be found in the "List of Whewell's Published Works," which is published as an online supplement to this book.

CHAPTER 1

Whewell's Early Life and Education

CHRISTOPHER STRAY

This chapter aims to give an account of the education William Whewell received from his early boyhood in Lancashire until his graduation from Cambridge in 1816. This part of Whewell's life has received much less attention from historians than his later years, which have understandably been the major focus of interest for the historians of science and philosophy who have written about his work. No biography has been attempted since Janet Stair Douglas's appeared in 1881, and even this was by design not comprehensive in its coverage: the author's intention was to cover the "domestic and academic" side of Whewell's life, as opposed to the "literary and scientific" side previously surveyed by Isaac Todhunter; and even this bipartite division was the survivor of an original plan to publish in three parts: scientific, academic, and domestic.[1] The nearest approach to a comprehensive survey of his work is Fisch and Schaffer's *William Whewell: A Composite Portrait*, but its editors acknowledge in their preface that the volume "has skirted Whewell's massive accomplishments in the area of literary and moral sciences."[2] The only attempt to look at Whewell's early life in that volume was made in the first three pages of Harvey Becher's contribution.[3] Whewell's schooling is dealt with in three sentences and two footnotes; only one of the four schools he attended is mentioned, and its name is consistently misspelled.[4] Fisch and Schaffer's book reflects, both in its high quality and in its limited coverage, the state of historiography in its time, when the history of science and mathematics was a specialist field more highly organized and developed than that of the humanities. This led not only to a lopsided coverage of intellectual and academic life, but also to a lack of interaction between the study of science and of the humanities.[5]

A reassessment of Whewell's early life has to take account of two difficulties. The first is a simple lack of evidence in some areas, as, for example, his time in the two local schools he attended before entering Lancaster Royal Grammar School in September 1808. The second is the accumulation of anecdotes about different aspects of Whewell's life, of the kind that tend to be attached to famous individuals. For example, the story is told that when Queen Victoria visited Cambridge in 1843, she saw pieces of paper floating in the River Cam and asked Whewell, who was with her, what they were. He replied that they were notices forbidding bathing.[6] This story concerns Whewell the master of Trinity, the autocratic figure who in another riverine anecdote told an undergraduate standing on a bridge next to Trinity looking at the view that the bridge was "a place of transit, not of lounge."[7] Other anecdotes will be discussed in the chronological account given below, leaving aside those that, like the examples given above, clearly belong to his post-undergraduate life.[8]

Whewell's access to education followed a pattern seen in other cases in the late eighteenth and early nineteenth centuries, such as that of the Greek scholar Richard Porson four decades earlier: born to humble parents, talent-spotted, and then supported by local clergy and gentry. In Porson's case this took him to Eton; Whewell went to a local school, which helps to account for the difference in the stories told about them. Whewell went to Cambridge from a northern grammar school and has been portrayed as a rough diamond whose manners were polished only at university; Porson's time at Eton ensured that he was polished before he arrived there.

Birth and Family

William Whewell, the eldest of the seven children of John and Elizabeth (Betty) Whewell, was born on May 24, 1794, on Brock Street, Lancaster.[9] His father, a master carpenter, had his workshop either there or in the next street (Lucy Street).[10] John Whewell's occupation has been variously described:[11] Janet Douglas called him a master carpenter; more recently he has been described as a "carpenter and builder."[12] In his obituary of Whewell in *Macmillan's Magazine*, W. G. Clark stated that John Whewell was "a joiner, and not, as has a been stated, a blacksmith."[13] The last word could go to John Whewell's own last words: in his will, dated April 23, 1816, he described himself as a "joiner and house carpenter." Elizabeth Whewell (née Bennison) was described by Douglas as a woman of "powerful mind and considerable culture"; this may have been taken from Whewell's obituary by Clark, who claimed that Whewell's "intellectual strength came from his mother's side, though she never attempted any literary task beyond the humble one of contributing annually enigmas and charades to the Lady's Diary."[14]

Thomas Bond's School

As a young boy Whewell was sent first to a school in Mary Street (a turning off Brock Street) run by Thomas Bond. Bond was a member of the High Street Independent Chapel, a local church founded in 1774 by breakaway members of another chapel that had moved from Presbytarianism to Unitarianism. The dissentient members, while seeing Unitarianism as a step too far, also wanted to avoid what were seen as dull sermons given by Anglican ministers.[15] In 1796 the congregation joined the Sunday school movement begun by Richard Raikes in Gloucester in 1780, setting up a Sunday school in the building where Bond ran his school on weekdays.[16] The Sunday school was set up by William Alexander, a lay preacher at the chapel; he had come to Lancaster to work as a carpenter in the shipyards and apparently worked there with John Whewell.[17]

The Blue Coat School and Lancaster Royal Grammar School

In her biography Douglas wrote that Whewell was sent to the "Blue School," but does not explain that it was one of a large number of charity schools founded from the mid-sixteenth century to educate the children of the poor. Their full name was "Blue Coat schools."[18] There were two in Lancaster, the boys' school being held in the chapel of Penny's Hospital.[19] The chapel was very small, and pupils were probably no more than fifteen in number.[20] Very little is known of Whewell's experience there, though the curriculum would have been fairly narrow, with an emphasis on basic skills: in the terminology of the time, to "read, write and cast accounts."[21]

It is not known when Whewell first went to the Blue Coat school, but he presumably left it in 1808, when he was admitted to Lancaster Royal Grammar School.[22] One of his fellow pupils at the Grammar School was the comparative anatomist and paleontologist Richard Owen, admitted in September 1809, who sent a memoir of Whewell's early life to his biographer Janet Stair Douglas.[23] In it he recalled that the local parish priest and headmaster of the Grammar School, Joseph Rowley, who was Owen's godfather, lived on Brock Street, next door to Owen's mother, and met the teenaged Whewell when his father was mending a fence between the two properties.[24] John Whewell had gone home for his midday meal; Rowley found William Whewell in his garden, talked with him, and was struck by his knowledge and intelligence. William was about to be apprenticed to his father, who thought his son already knew more about parts of his business than he himself did.[25] Rowley urged him to send William to the grammar school, for which he would waive the usual fees as well as buying the boy's books. John Whewell agreed, and William entered the school on September 11, 1808.[26]

The Lancaster Royal Grammar School was one of hundreds of such schools in England founded by endowments left in the wills of charitable citizens.[27] Many of them were founded in the sixteenth century, but the grammar school's foundation was earlier; its records had been destroyed by fire, so its actual foundation date is unknown, but the first mention of it in the town records was in 1495. According to an 1818 survey, the school had room for 120 boys, but its average attendance was 70: 25 taught by the headmaster and 45 by the usher (second master) George Morland. The school at that point had a library of about three hundred books.[28] In 1802 a set of regulations was published, stipulating that both Latin and English were to be taught "grammatically"; history and geography were also taught, and a writing master, who charged fees, also offered arithmetic.[29] The rote learning of classical texts was a common element of schooling in this period, tested daily or weekly by the recital of memorized texts. Whewell annoyed his schoolfellows by reciting longer passages than they had memorized, thus presumably encouraging the master to set longer extracts next time. Owen remembered that his form-mates told Whewell, "Now Whewell, if you say more than twenty lines of Virgil today, we'll wallop you." Whewell would not oblige, and two boys tried to wallop him, but were themselves walloped. Schoolboy notions of fairness decreed that no more than two boys could confront him at one time, but more candidates could not be found.[30] While he was still at Heversham his strength had been remarked on: he had often carried one of the younger pupils up to the school on his back.[31] In adult life Whewell retained his powerful physique and was told by a champion boxer that he was a great loss to the ring; he intervened physically on several occasions in university ceremonial gatherings when dons were mobbed by undergraduates. Charles Bristed, who was at Trinity in 1840–1845, recalled that "the anecdote was often told, and not altogether repudiated by him, how in his younger days, about the time of his ordination, a pugilist in whose company he accidentally was travelling, audibly lamented that such lusty thews and sinews should be thrown away on a parson."[32] In 1856, during a controversial contested election for a university MP, he refused to hide from a combative undergraduate crowd, but walked along Trinity Street with a prize fighter on either side.[33]

Owen recalled a particularly trying incident for Whewell in which he pronounced the common abbreviation "viz." (for "videlicet," "namely, that is to say," equivalent to "sc." [scilicet]), as it was spelled, and not in full. Owen stressed the embarrassment Whewell had in being sent to the smallest boy in the school, to be put right about "viz." (he was several years older than Owen). Whewell was on good terms with his headmaster and benefactor Joseph Rowley and maintained a long and friendly correspondence with George Morland.[34]

Whewell was supported at the grammar school by Joseph Rowley, as we have seen. But the school had no university scholarships, and so in 1810 he transferred to Heversham Grammar School, which did. In November 1809 Thomas Satterthwaite of Lancaster wrote to his cousin John William Whittaker:

> We have a boy in this town of the name of Whewell who is quite a prodigy of learning. He is a carpenter's son. Mr Wilson of Dalham Tower has been interested to put him upon establishment at Heversham which will give him 40 a year for College expenses. He is to go to your College Trinity I think. We have had him to tea. He is fond of drawing—is long—he has taught himself mathematics algebra &c &c—learning is no labour to him. Mr Rowley who attends Thomas one hour every evening has the merit of cherishing this Genius.[35]

Heversham Grammar School

Heversham is a village in Westmorland (now part of the county of Cumbria), fifteen miles north of Lancaster and five miles south of Kendal. Heversham Grammar School was founded by Edward Wilson in 1613 for instruction in English, Latin, and Greek.[36] Wilson's family later provided scholarships for poor boys, including one to Trinity College, Cambridge, of about £50 per annum for four years.[37] The preferment was in the hands of Wilson's descendants, in Whewell's time Daniel Wilson of Dallam Tower. It was a small school, its pupils numbering forty-five in 1819;[38] there was a single master, and in 1818 its boarding fees were 25 guineas per annum.

The school's connections with Trinity College went back some way; the college owned land in Heversham, having been given the rectory by Queen Mary.[39] John Hudson, born in Haverbrack, a hamlet near Heversham, and a pupil at the school, entered Trinity as a sizar in 1793 and was senior wrangler in 1797. He was elected to a fellowship in 1798, and was assistant tutor 1805–1807 and tutor 1817–1825, before becoming vicar of Kendal in 1815. In 1810, after leaving Lancaster Royal Grammar School, Whewell went to see him. They met at the Bridge Inn, and Hudson, after a brief examination, declared that he would be one of the first six wranglers.[40] The master of Heversham School when Whewell arrived was Thomas Strickland, who died in September 1811. Strickland's replacement, Joseph Fawcett, did not arrive until the following Easter, and in the interim the seventeen-year-old Whewell was put in charge of the school. The future master of Trinity thus became a master thirty years earlier, while still in his teens. At this point he made contact with John Gough of Kendal, a blind mathematician who became well known as a coach of intending Cambridge undergraduates.[41] Whewell told his father in 1812 that Gough was "a very extraordinary person . . . of very eminent note in classics, mathematics, botany, and chem-

Figure 1.1: The earliest portrait of William Whewell, by James Lonsdale, 1825. *Source:* TC Oils P 207, Trinity College Library, Cambridge. Reproduced with kind permission from the master and fellows of Trinity College, Cambridge.

istry."[42] This suggests that Gough played a role in developing Whewell's polymathy.

Trinity College, Cambridge

In September 1811 Whewell traveled to Cambridge with John Hudson; on the twenty-third he was examined by the master, William Lort Mansel, and four fellows, and formally admitted to Trinity. His admission certificate reads, "Guglielmus Whewell examinatus et approbatus, admissus est

Sisator." The last word indicates that he was admitted as a sizar.[43] The term was used to refer to a poor boy who paid lower fees than the standard rates applied to a paying undergraduate (pensioner).[44]

How poor was Whewell? Laura Snyder stated that his father did not leave him much in his will. This has little detail, but contains a declaration that his estate was worth less than £200; it also states that he owed his sister-in-law Alice Lyon £150, a debt that he passed on to his son. On the other hand, John owned property in Lucy, Mary, and George Streets that he rented out. If he died poor in 1816, this may have been due in part to his generosity to his son, who in 1813 referred in a letter to his father to "the great expenses which you now sustain" and in a letter to his aunt in 1815 to his having been "such an expensive business to my father for so long."[45] Whewell's father sent him £50 in January 1815 and another £30 in September.[46]

One of the four fellows who admitted Whewell was John Hudson, who became his tutor. By the time Whewell was admitted, the college had two fixed tutorial "sides," the two tutors, John Hudson and Thomas Young, dividing the undergraduate population between them and each employing several assistant tutors as lecturers. After a marked increase in undergraduate numbers in 1800–1820, a third side was set up in 1822, and in 1823 Whewell became tutor of the side Hudson had run from 1807 to 1815.

Whewell began to reside in Cambridge only in October 1812, when he matriculated at the university.[47] His finances were bolstered by a subscription in Lancaster sponsored by his old headmaster Joseph Rowley.[48] His progress through Trinity was marked by success at every stage. In the college examinations of his first two years he came top of the first class.[49] In his second year he won the Chancellor's prize for an English poem, the subject set being "Boadicea," and a college declamation prize on "Caesar and Brutus," and he went on to gain a scholarship. In January 1816 he sat the Senate House Examination, the university's only degree examination, and came out as second wrangler, beaten to the senior wranglership by George Jacob of Gonville and Caius College, a man who was inferior as a mathematician but superior as an examinee.[50] It may be that Whewell's incipient polymathy held him back, as was the case with Augustus De Morgan (4th wrangler, 1827) who had at some time been Whewell's pupil.[51] He was elected a fellow in 1817, and appointed assistant tutor in 1818 and tutor in 1823.

Some accounts of Whewell's career have claimed that this rapid academic rise was accompanied by personal difficulties: "These academic achievements were accompanied by personal trials associated with the passage from Lancaster to Cambridge. For a time Whewell was not only very much alone, but also seen as unusual. His manners and speech were considered rude or rustic."[52] Whewell did indeed eat some meals alone in his lodgings, but that was true of most Trinity freshmen: rising admission numbers in the

1810s meant that only a minority of undergraduates could be housed in the college.[53] He told his father just after his arrival that "I shall dine and sup in hall, and have my bread and butter etc from the buttery."[54] On October 23 lectures began, and he would then have met the other thirty to forty members of his tutorial side in the classical and mathematical lecture rooms. Most freshman would have formed acquaintances mostly within their own college, but Whewell already had a connection through his fellow Lancastrian Thomas Satterthwaite with the latter's cousin John Whittaker, who had entered St. John's in 1810.[55] Through Whittaker, who was thirteenth wrangler in 1814, he became friendly in the spring of 1813 with other Johnians, including John Herschel (senior wrangler, 1813), Fearon Fallows (3rd wrangler, 1813), and Richard Gwatkin (senior wrangler, 1814).

The claim that Whewell's speech was considered "rude or rustic" rests largely on the story told in his obituary in *The Athenaeum*: "He arrived at Cambridge, they say, a raw and unformed northern youth. The following story may or may not be true, as much of one as of the other; but it shows the opinion entertained by his comrades of the forcible and elegant English writer, as he afterwards was. It is said that, lounging at the College gate, he saw a herd of swine driven by, and soliloquized as follows: 'They're a hard thing to drive—very—when there's many of them—is a pig.'"[56] In his *Defining Science*, Richard Yeo refers to Whewell's "social demeanour," and adds that it "transgressed the conventions of Cambridge life. An early manifestation of this was his English pronunciation and traces of north-country dialect."[57] Yeo then quotes from the pig-driving anecdote in support of his opinion, but it should be noted that he does not mention pronunciation and that the alleged remark has no dialectal specificity. The only evidence of a regional accent I know of comes from a later period, probably the 1850s: learning that country clergymen took out volumes of sermons from the university library but did not return them, "Whewell exclaimed, 'And a very good thing too, *warthless* sermons,' . . . in the strong Lancashire accent that he fell into when excited."[58] As for dialect, the only Lancashire vocabulary used in Whewell's correspondence comes in a letter of 1829 in which he describes a hut in the Swiss Alps in which he was staying as having a "shippen," a word helpfully glossed by Isaac Todhunter as "a Lancashire word for a cow-house."[59] The picture so easily painted of a rough rustic youth overlooks the fact that Whewell had been a sickly child and spent a lot of time reading, rather than playing with other children.[60] The earliest letters of his that we have are certainly written in a good, clear English style.

How did Whewell pronounce his own surname? In his obituary, Harvey Goodwin wrote that "a controversy used to exist in Cambridge as to the proper pronunciation of Whewell's name. He was described in a newspaper article as a man whose name was more easy to whistle than to spell; and in practice the pronunciation was somewhat various, some saying You-ell, others Woo-ell, or perhaps rather Whoo-ell. On a public occasion, when he recited his own name, I remember that his own pronunciation correspond- ed nearly to the last of these three, which therefore I presume may be re- garded as the correct rendering of the name."[61] But how does one pronounce "Whoo-ell"? Pronouncing dictionaries of the late eighteenth century rec- ommended pronouncing "wh" as "hw," and this suggests the pronunciation "hwoo-well."[62] The initial aspiration would explain the reference to whistling; Whewell's unpopularity as master of Trinity and as vice-chancellor prompt- ed undergraduates to give him the irreverent nickname "Billy Whistle." The *Saturday* reviewer of Douglas's biography wrote, "We well remember the de- gree-day of January 1843, when penny whistles sold for a shilling, so great was the demand for them, and the indignant undergraduate received the Vice-Chancellor with a concert the reverse of respectful."[63] The university registrar John Romilly recorded in his diary for that day, January 21, 1843, that "the V.C. was abundantly made conscious that he is most unpopular . . . a long sustained groan was kept up all the time he walked up the Senate House—from time to time the youngsters called out '3 groans for Whistle' . . . at other times they kept up whistling for half an hour together."[64] This ev- idence of rudeness in the 1840s, when his appointment as master of Trinity and then as vice-chancellor arguably went to Whewell's head, should not be read back to the 1810s. He could certainly be very rude to his friends: Airy wrote his wife in 1845 that "it is only those who have *well* gone through the ordeals of quarrels with him and almost insults from him, like Sheepshanks and me, that thoroughly appreciate the good that is in him."[65]

Whewell's transition from raw provincial boy to high-achieving don has often been portrayed; the available evidence needs to be sifted to sep- arate what can be known from what fits neatly into stereotyped narratives. William Clark stated in his obituary of Whewell that "there are those still who remember Whewell as he first appeared in Cambridge, a tall, ungainly youth, with grey worsted stockings and country-made shoes. But he soon became known in the college as the most promising man of his year."[66] A very similar account was given of Richard Watson, later bishop of Llandaff, an earlier alumnus of Heversham Grammar School.[67] When he arrived in Cambridge in 1754, it was noted that "he wore constantly coarse mottled Westmorland, and stockings of blue yarn . . . spoke his provincial dialect with so faithful a twang, that even his eminent attainments could not for some years secure him from laughter."[68] He was apparently sneered at by richer undergraduates, who called him "the Westmorland phenomenon."[69] In 1784 the Yorkshireman Richard Ramsden and John Bell of Westmorland were removed from their scholars' duty of reading grace in Hall "on account of their personal appearance and uncouth dialect."[70]

The Athenaeum obituary, after giving the pig-driving anecdote, continues: "But against this story it must be said that in 1814, when an undergraduate of two years' standing, the young man won the Chancellor's prize for an English poem. The first story may be—to his honour—as true as the second: for Whewell had great rapidity of acquisition, and a tremendous memory."[71] This account sets up a problem, the gap between what Whewell was at first and what he later became, and neatly solves it—he was a fast learner with an outstanding memory. The obituarist ascribes the pig-driving anecdote to Whewell's contemporaries ("comrades"), while admitting that it may not be true. But we can ask other questions about the anecdote. First of all, was it the kind of story that takes on a life of its own, as such anecdotes sometimes do? Other versions can be found, the earliest I have located dating from 1855: "Perhaps . . . legislators . . . will leave off the attempt to make the people good by compulsion, and will acknowledge a difficulty which resolves itself into the ejaculation of the troubled drover to the Cambridge student: 'hard things to drive—pigs—for one man—so many on 'em—very.'"[72]

Several aspects of this anecdote have not been noticed in the literature on Whewell. First of all, the reference to pigs may reflect the traditional rivalry with St. John's, the college next door to Trinity, whose members were widely known in Cambridge as "pigs" or "hogs." A herd driven past the front gate of Trinity north up Trinity Street could well have been seen as destined for St. John's. John Browne, an ex-pupil of James Tate who was an undergraduate at St. John's, wrote to his old headmaster in the year Whewell was admitted at Trinity. His letter opened: "Grunts extraordinary . . . communicated by post, April 16, 1811." It quoted thirteen Johnian toasts, beginning with "St John's for ever! May she always save her bacon!"[73] Whewell's alleged comment, then, if actually delivered, may have belonged to a long history of porcine references by Trinity men to their neighbor and rival.

Another aspect to be taken into account is that the pig-driving anecdote was seen at some point as a parody of the style of the Greek historian Thucydides, one that seems to have been well known in nineteenth-century Cambridge. In a letter of 1916 to Macaulay's nephew George Otto Trevelyan, Henry Jackson, Regius Professor of Greek and vice-master of Trinity, referred to "the old Cambridge parody, which I have always attributed to Augustus Vansittart . . . 'Awkward animals to drive is a pig, one man many of them very.'"[74] Vansittart, who died in 1882, had been a fellow of Trinity since 1848; he was known for his witty speeches at the college's annual Audit Feast. Jackson entered Trinity in 1858, so could have heard Vansittart's speeches at any time in the next two decades.

A final point to be made about the pig-driving anecdote is that it cannot be assumed that Whewell's comment, if he ever made it, was an unreflective response to what he saw. It is quite possible that a version of the comment was already in circulation, and that he was in effect quoting a stereotyped "peasant" remark.

Yeo states that some of Whewell's obituaries mention that his "social demeanour . . . contravened some of the conventions of Cambridge life." The evidence here is mixed. In *The Athenaeum*, J. W. Clark commented that "Dr Whewell was not popular. His manners had not the polish only acquired in early life. His spirit was sometimes boisterous and overbearing; he was occasionally uncivil. But he had a kind heart." William Clark, on the other hand, stated that "there were some who feared that the new Master would be imperious and overbearing, but their fears were dissipated by the result."[75]

In this chapter I have traced the progress of William Whewell's education at school and university. I have also attempted to trace the emergence of Whewell as a man, and to consider the ways in which he became the subject of partial accounts in his lifetime and biographizing in the nineteenth century and later.

CHAPTER 2

Whewell and Cambridge Mathematics

TONY CRILLY

From an early age William Whewell proved himself in mathematics, and this led him to Cambridge University and Trinity College, institutions that would govern his life. But for mathematics he would not have had the career he had. His earliest studies at Cambridge brought him into association with Charles Babbage, George Peacock, John Herschel, and the Analytical Society. Staying on, he went his own way and became a major designer of the mathematical curriculum—more specifically of its famous Mathematical Tripos, the cornerstone of a Cambridge education.[1]

In the passage of Whewell's early life, his father, a master carpenter, was advised by the local Anglican minister (and headmaster of Lancaster Royal Grammar School), Joseph Rowley, that his son would benefit from an academic education. Rowley was particularly impressed by his ability in arithmetic.[2] After a primary education in Lancaster the boy was transferred to the grammar school at Heversham in Westmorland to qualify for their £50 closed exhibition to Trinity College.[3] In 1809 Trinity don John Hudson, who had been a pupil at Heversham himself, interviewed young Whewell and forecast he would be "among the first six wranglers" at Cambridge.[4] Hudson should know; he himself had jumped through the hoops of the Mathematical Tripos in 1797 and emerged with the "Cambridge double" as the senior wrangler and the winner of the first Smith's Prize (the SW and 1SP).[5]

Whewell spent two years at Heversham, but for several months in the summer of 1812 he was tutored in mathematics by John Gough of Kendal.[6] He described the blind mathematician as "a very extraordinary person" and he would have been impressed by his polymathy, an attitude to knowledge that infused his own academic life.[7] Whewell was in good hands; Gough's pupils had included John Dalton, while both Richard Dawes and Thomas Gaskin followed him to Cambridge.

From an early age Whewell had a mathematical pedigree, for Gough had been a resident pupil of John Slee—a mathematical master of Mungrisdale in Cumberland and a man grounded in the principles of calculus. Studying with Gough, Whewell recalled, "I reviewed algebra, trigonometry, and other branches, which, as I had before gone over them, did not take up much time, and am now reading conic sections, mechanics and fluxions [calculus]."[8]

Undergraduate Days

Whewell began his first term at Cambridge in October 1812 as a subsizar, a rank below a proper sizar who received endowments from the college.[9] For a serious student, as Whewell undoubtedly was, the academic goal was to succeed in the famed Senate House Examination at the end of the ten-term course. Success in this once-and-for-all examination meant to be included in the wrangler class with the ultimate prize being the senior wranglership, the celebrity at the pinnacle of the whole competition.[10]

Mathematics permeated studies at Cambridge.[11] In 1822 a Classical Tripos was introduced, but to be allowed entry to this degree, students had first to achieve an honors degree standard in mathematics (or be the son of a peer). By 1851 degrees in moral sciences and natural sciences were introduced, but even then students still had to achieve a first-class pass in mathematics at the ordinary degree level.[12]

The Cambridge course Whewell studied as an undergraduate had been molded over the previous century, and existing textbooks reflected this progression. Euclid's *Elements* was highly valued and Isaac Newton's *Principia* (1687) revered. Following Newton, fluxions were taught from a geometrical perspective and a reliance placed on intuition gained from geometric figures—synthetic geometry.[13] James Wood (SW 1792) produced *Elements of Algebra* (1795) (based on Newton), *Principles of Mechanics* (1796), and *Elements of Optics* (1798). Samuel Vince (SW 1775) published *Elements of the Conic Sections* (1781), *A Treatise on Practical Astronomy* (1790), *The Principles of Fluxions* (1795), and *Principles of Hydrostatics* (1796). All these texts, ones Whewell would study, adopted this synthetic geometric style.

Settling into Cambridge, Whewell displayed a gregarious nature. He made friends easily, and several of them became lifelong. As well as Babbage and Herschel (SW 1813), among them were Richard Sheepshanks (10W 1816), Richard Jones (BA 1816), and Hugh James Rose (14W 1817). In Richard Gwatkin (SW 1814) from St. John's College, Whewell found someone to discuss mathematical ideas.[14] At Cambridge Whewell was mixing with people outside his social class, notably Babbage, the son of a wealthy

banker, and Herschel, son of the royal astronomer. Herschel was much admired, and on their first meeting Whewell described him as "a most profound mathematician and an excellent general scholar."[15] Whewell's own academic ability was clear, and in 1815 he was elected a scholar at Trinity. He
was on the first rung of the academic ladder set out for those with ambition.

Whewell's Tripos examinations in the Senate House took place in January 1816. Predictions of the degree results were not always realized and the
award of the Cambridge double, the senior wrangler (SW) and First Smith's
Prizeman (SP1), went to Edward Jacob from Gonville and Caius College;
Whewell came second in both competitions.[16] Whewell had been thought
the favorite by his friends, but in the scramble for marks put down his failure (as he felt it) due to a slow writing speed. Explaining the outcome to his
father, he wrote of Jacob as "a very pleasant as well as a clever man, and I had
as soon be beaten by him as by anybody else."[17] It is notable that as Whewell
matured, he disparaged the Mathematical Tripos as a competition.

The Analytical Society

In the 1810s mathematics teaching at Cambridge was in the doldrums. Little had changed over the previous century, and texts such as those by Wood
and Vince reflected this stagnation. There was little recognition of mathematics as a developing subject, let alone recognition of the rapid developments taking place on the Continent. The business of Cambridge was the
training of Anglican clerics, and mathematics was a way of explaining the
status quo and reinforcing Christian faith.[18] For most, teachers and students, mathematics was seen as a fixed body of knowledge—"permanent" as
Whewell would come to call it—rooted in ancient Greece and reaching its
climax with Newton.

There had been some earlier reform movement at the beginning of the
nineteenth century, propelled by Robert Woodhouse.[19] However, the main
beacon for change was the formation of the short-lived Analytical Society. Its main drivers were the young men Babbage, Herschel, and Peacock,
when Babbage and Herschel were twenty years old and Peacock twenty-one.
Herschel was elected president at its initial meeting held on May 7, 1812.[20]
Among members there was much enthusiasm for Continental mathematics,
and, as the name indicates, the objective of the society was the promotion
of the analytical style of mathematics (symbolic and algebraic in nature, and
above all, abstract), as opposed to the existing Cambridge synthetic style.
The separation of these distinct attitudes was felt most keenly in the way
calculus was traditionally presented at Cambridge. Newtonian fluxions (derivatives) involved motion, studied through geometry. Moreover, Cambridge
remained faithful to Newton's notation, the dotage ($\dot{x}$) rather than the Continental d-notation of Leibniz (dx/dt).

Babbage and Herschel were the radicals of the society, who might in today's terms be described as the movers and shakers of the membership. They were also the principal writers of the society's *Memoirs*, a single volume forged over the summer of 1812 when they were in constant communication with each other. They were joined by Peacock in translating the *Elementary Treatise on the Differential and Integral Calculus* (1816) by Silvestre Lacroix.[21] Their youthful enthusiasm was clear, but in racing ahead, they failed to recognize that the new viewpoints would be too steep for most. One member, the aristocratic Edward Ffrench Bromhead, warned Babbage in 1813 that "not one mathematician in 10^∞ can understand [analytics]."[22]

While Whewell would become such an important figure in guiding the future of the Mathematical Tripos, he evidently played no part in the formation of the Analytical Society. At the time he was just seventeen, studying with Gough and yet to go up to Trinity.[23] When at Cambridge he acknowledged the society's promotion of Continental analysis and for some years kept abreast of its activities, even briefly considering annotating a planned translation of Lacroix's "Application of Algebra to Geometry."[24] But soon his main focus would be to promote "mixed mathematics" (applying mathematics to physical problems) in the curriculum.

At Trinity many conservative dons mounted opposition to the "new mathematics." Daniel Mitford Peacock (no relation to George) was a conspicuous opponent of Continental analysis and the introduction of Lacroix's texts. This Peacock had been the SW and SP1 of 1791 and was a Trinity College fellow. He was critical of Lacroix for his abstractness, so for this view he would have found sympathy from Whewell. He warned against its use in the Mathematical Tripos: "Academical education should be strictly confined to subjects of real utility."[25] The eminent natural philosopher Thomas Young was another voice resisting the incoming "analytics."

While opposition was supported by the conservative elements in Cambridge, what about the young students exposed to the fervor of the impending analytic revolution? One Cambridge student wrote of his reading in his third year (1817): "Having already gone through Vince, Dealtry, and the 'small Lacroix,' it only remained for me to read the three huge quartos [of the large *Traité* by Lacroix]. This was a task not easily performed. I contrived to get through a very considerable proportion of it."[26] Another, James Challis (a future SW and SP1) was in his first year at Trinity and remembered his mathematical experience: "In 1822, when I commenced the study of the Differential Calculus, the introduction of the analytical method of reasoning was, as I well remember, still regarded with jealousy by the older mathematicians of the University, from the apprehension that it was not susceptible of the rigid reasoning they considered to be characteristic of the geometrical method, especially as exemplified in Newton's *Principia*."[27]

Wary of appearing a radical, Whewell did not share the dominating passion for pure analytics. He could not embrace the derivative as a coefficient in a Taylor expansion (the Lagrangian formal definition). What was needed in applications was the derivative based on the limit concept, as this was how it occurred in practical situations. He believed that George Peacock, in his role as moderator for the Senate House Examination in 1817, had taken the wrong step by introducing calculus abstractly: "He has stripped his analysis of its applications and turned it naked upon them."[28]

Later, writing in 1831, Whewell was mindful of his mission to preserve the spirit of the Mathematical Tripos as he had experienced it himself, and he downplayed the influence of the Analytical Society, pointing to the lack of connection between writers and readers (and printers) of its sole publication: "Those who are acquainted with English mathematical works, will probably recollect a very remarkable quarto, 'Memoirs of the Analytical Society, vol. 1" which seventeen years ago, issued from the University Press of Cambridge. In this publication, the extraordinary complexity and symmetry of the symbolical combinations sorely puzzled the yet undisciplined compositors of that day and led unmathematical readers to the conviction that the whole was a wanton combination of signs, left to find a meaning for themselves, like the Javanese character of Princess Caraboo."[29] When in 1831 a Scottish colleague was looking for advice on the mathematics required for research in the physical sciences, Whewell could not recommend Lacroix's *Traité* without qualification: "You would find a great deal there which would be of no use and would occupy much time. . . . The requisite preparation for this [research] appears to me to be some elementary treatise on Differential and Integral Calculus."[30]

Teacher of Mathematics

To stay in Cambridge following his first degree, Whewell needed to fund himself, and so he took private pupils and led reading parties during the Long Vacation, writing to Herschel on one occasion that he was "grinding down several specimens of senior sophs."[31] The next stage in his academic career was preparation for the Fellowship Examinations held in the autumn of each year, and succeeding in this, he was awarded a fellowship in October 1817. He had thus reached the first rung of the Trinity Foundation, a junior fellow in a hierarchal organization of some sixty fellows, the master in charge supported by eight senior fellows. In 1818 he was appointed in an official teaching role as an assistant tutor—a post that involved giving college lectures on mathematics. Though detached from the key objectives of the Analytical Society, Whewell revealed himself as a reformer, but his plans involved the application of mathematics to the problems of "natural philosophy." At the outset of his new appointment, he wrote to Herschel of his intention:

[I am] taking it for granted that you still retain some interest for your old plan of reforming the mathematics of the university. I have it now more in my power to further this laudable object by the situation I have taken of assistant tutor (i.e., Mathematical Lecturer) here. Whatever may be the disadvantages of the office this is one of its advantages. I shall have a permanent and official interest in getting the men forwards—I shall have the opportunity of directing their reading—and I shall write books (good ones of course) and be able to put them in circulation. By such powers wisely but discreetly much may be done. I have the first volume of a system of mechanics, containing statics and a little dynamics, ready for the press.[32]

On the same day he wrote to Rose, thinking, as he frequently did about his own future and how it might turn out. Commenting on Robert Woodhouse's *Physical Astronomy*, recently published, he wrote, "It is like his other books . . . executed in no very neat manner but still good metal—so that at worst it may be melted down and coined over again. It will have no doubt make its way into the Senate House [Mathematical Tripos examination]— especially as we have Gwatkin & Peacock as moderators [in 1819]. . . . If I go on here [at Cambridge] I shall have no doubt become a worthy successor to James Wood."[33]

Whewell's first book was the very successful *Elementary Treatise on Mechanics* (1819). It opens: "The peculiar objects of the following work are, to establish the theory of Mechanics upon principles proved with as much rigour as the nature of the subject admits; and to present the more elementary parts of the science in such a form that the student may be enabled to pass from them to its more difficult problems, and particularly to those higher applications of it in the system of the world which have given so much splendour to the progress of mathematics in modern times."[34] It was a weighty tome consisting of some 370 quarto pages. Whewell judged that statics was the more firmly established subject, and this received his greater attention; it had been studied by Aristotle, while dynamics arrived in the wake of Galileo. In Whewell's treatment, he used calculus (for example, with derivatives expressing the curvature of a curve, integrals in measuring volume), and he employed the Leibnitzian *d*-notation. He referred the reader to Lacroix's *Elementary Treatise*, and in preparing the way to "higher applications," the reader was introduced to elliptic integrals in dealing with pendulum problems.

Whewell's book bridged the gap between Newton's *Principia* and analytical processes. He actually used the term *new mathematics*. His text effectively displaced the earlier texts written by Wood and Vince. Peacock had already introduced some Continental notation inspired by Lacroix into the Senate House Examination, and there was a place for a readable elementary work in mixed mathematics. Whewell's text fulfilled this requirement.

In his writings Whewell showed he was up to date. In taking in the latest research of Eaton Hodgkinson on suspension bridges, for example, he investigated the shape of curves suspending a roadway, using such examples to convince students that calculus had real-world applications. "When we suppose the curve to be continuous," he wrote, "we must suppose δs and δy [lengths] to be indefinitely small; in which case the ratios of such quantities are the differential coefficients."[35] This shorthand reasoning was a typical pragmatic response at the time to the practical problem at hand.

Whewell wanted to make the idea of the derivative intuitive, whereas the analysts on the Continent desired pure definitions. In applications the ratio of infinitesimal amounts naturally pointed the way to the derivative as a limit. Amid practical examples, for instance, Whewell was content with describing limits in vague terms. He later addressed the definition of a limit problem in his *Doctrine of Limits* (1838), but presenting it in a geometrical language, he struggled. As a result, this text did not achieve popularity.[36]

Whewell followed the *Elementary Treatise on Mechanics* (1819) by *A Treatise on Dynamics* (1823). In this work he praised Pierre-Simon Laplace's use of calculus in the *Mécanique celeste*, for it showed "the application of analytics which might otherwise be considered as merely subjects of mathematical curiosity."[37] In focusing on mathematics to solve physical problems, he continually stressed the usefulness of mathematics and made clear his opposition to mathematical abstractions. These basic texts of 1819 and 1823 went through many editions, and as they did, they steadily reduced the mathematical demands on its readers. By the fourth edition of *The Elementary Treatise on Mechanics* (1833), Whewell had all but given up on proselytizing on the values of calculus. Those who wanted advanced theory would have to consult a companion volume he had separately authored.[38]

In 1823 Whewell was promoted to tutor, where he was now in charge of teaching on one of the "sides" at Trinity.[39] His authority was bolstered two years later when he was ordained as a priest in the Anglican Church. In his paternalistic role as tutor, he was in loco parentis and conscious that he had a responsibility to the students who sat in his classes, not all of whom would be wranglers. This gave him an interest in pedagogy, and as a result, many of his writings voice this concern. The tutor role suited him. It embedded him into the life of a don, and it was financially attractive, giving him security for the next sixteen years. He made use of it by writing pamphlets and textbooks and conducting a scientific research program. He took his turn in administration and in 1820 and 1828 was a moderator for the Mathematical Tripos and an examiner in 1829.

George Biddell Airy was a pivotal figure in the transition to the analytical methodology at Cambridge. He gained a scholarship to Trinity in 1820 and enrolled there as a sizar, a commonality of background with Whewell's own. They became close friends and worked together on several experimental projects. As a student at Cambridge, Airy carried all that was set before him and in 1823 emerged as SW and SP1.[40] In 1826 he was elected Lucasian Professor of Mathematics; two years later he was appointed to the Plumian Chair of Astronomy, a position with a higher salary. Like Whewell he promoted the application of mathematics, done through the analysis of differential equations.[41] He gave lectures in Cambridge and published various *Tracts* that were of use to undergraduates in their understanding of calculus in action.[42]

Reforming Cambridge Mathematics

A sound mathematician he undoubtedly was, but Whewell was not content to be what he rather scathingly referred to as a "mere mathematician." Over the course of the 1820s–1830s, he published numerous papers on subjects to which he applied mathematics, including mineralogy and political economy. He also presented a British Association for the Advancement of Science (BAAS) report on recent research on electricity, magnetism, and heat, and was involved in the founding of its new statistics section (Section F).

Upon his resignation of the chair of mineralogy, Whewell applied for the Lowndean Chair of Astronomy and Geometry in 1836, to which his friend George Peacock was appointed. During the 1830s he encouraged the expansion of mixed mathematics, both for ordinary and honors degree students. As he wrote in a letter to James Forbes in May 1836: "We are at present trying to introduce some mixed mathematics (Mechanics, Hydrostatics, Optics) into the general examination for degrees, as I recommended."[43] Somewhat later Whewell was instrumental in gaining entry for heat, electricity, magnetism, and the wave theory of light as examinable subjects. This continued involvement with mixed mathematics informed his thinking in the way he approached the philosophy of science.[44]

Around that same time Whewell decided to step back from day-to-day involvement in college business. He retired from being a tutor in 1839 and a year later, when he was thinking of taking up a college living, wrote to a friend: "I have done what I could, or at least in a few months shall have done all I can, to improve the mathematical studies of the college and of the university."[45] From the time he was elected to the Knightbridge Chair of Moral Philosophy, philosophy came to dominate his thinking, his metascientific work, though mathematics was never far away. Book 2 of volume 1 of *The Philosophy of the Inductive Sciences* (1840) contained chapters on the "Pure Sciences" (geometry, arithmetic, algebra), for instance. A few years later, when composing his *Elements of Morality, Including Polity* (1845), Whewell brought his mathematical lens to bear once more, controversially claiming for the "elements of morality a foundation philosophically similar to that of the elements of Euclid."[46]

In the summer of 1841 Whewell again contemplated leaving Cambridge, where he admitted to having spent "the thirty best years of one's life."[47] He was also facing up to the annual BAAS meeting at the end of July, where, as he wrote to a friend, he was "to preside at this great, ugly meeting at Plymouth" and give the Presidential Address.[48] In this address there is scant attention made to mathematics, except to tell the audience that government funds had allowed his friend Airy, now the astronomer royal, to employ a dozen computers (i.e., human calculators) for the calculations in astronomy, and he informed the audience of his continued support for the statistics Section F.

Quite suddenly, at forty-six years of age, Whewell's prospects took a change in direction. The year 1841 was a decisive one: in November he took possession of the Lodge at Trinity as its new master, and so began his dictatorial regime. With members of the Trinity Foundation, and with students alike, there was no doubting who was the master of Trinity.[49]

The university was expanding. During his time as tutor, Whewell had witnessed the doubling of student numbers graduating in the Mathematical Tripos. In the year he graduated (1816), 48 students had graduated with an honors degree (19 in the wrangler class), while in the first cohort he witnessed as master of Trinity (1842), 114 graduated with honors (38 in the wrangler class). The balance between pure and applied mathematics in the curriculum had shifted too, mainly due to Whewell's influence. In the examinations of January 1842, for example, exactly half the questions were concerned with applications of mathematics and the other half with pure mathematics. In the contest of 1842 Arthur Cayley, who became a champion of the analytic method, graduated as SW and SP1.

As master of Trinity, Whewell returned to thinking about the education of undergraduates, a process he had started in the mid-1830s with *Thoughts on the Study of Mathematics* (1835) and *On the Principles of English University Education* (1838). He focused on the two extremes that had been identifiable in the Cambridge student population for many years, the poetical man as opposed to the mathematical man: "Your poetical or critical man you educate by educing his reasoning powers through the discipline of mathematics. . . . And in like manner the spontaneous mathematician is educated by educing his imagination and philological faculties. . . . Observe, I do not want much mathematics from our classical men; but a man who either cannot or will not understand Euclid, is a man whom we lose nothing by not keeping among us."[50] A glimpse of his mindset is the occasion he reminded his niece that Euclid would be relevant to her: "I was amused with your account of your legal studies; but, seriously speaking, a little discipline in the precision of law language and the application of legal principles is a good element in education; just as Euclid is."[51]

The young American Charles Bristed, who arrived at Trinity in October 1840, hated mathematics but found himself confronted with it: "For in Trinity mathematics are not the sine qua non, though imperious Whewell is doing his worst to make them so."[52] Whewell had his own concerns and saw the expansion of analytics in the Mathematical Tripos threatening the basis of a liberal education, to educate the whole man, as had always been done at Trinity. In his efforts to achieve this ideal he continued to write textbooks and pedagogic tracts, and he wielded his political power in the university to maintain it.[53]

Whewell retained a belief that some physical subjects should be treated geometrically, though he confessed he had not always been consistent: "So far as the analytical method has superseded the geometrical, I am obliged to say (though I believe that I myself, by College Lectures, may have formerly contributed to bring about such a change), the result has been very unfortunate." He recommended that some texts written in the geometrical mode, such as Wood's *Elements of Optics* (1798), should continue to be read. He also continued to condemn the abstract analysis of analytics as "general methods which express all problems alike, but actually solve none." He opined that had English mathematicians developed Newton's procedures, "the geometrical method of treating the problem of three bodies might have had its triumphs to point to, as well as the analytical."[54] Whewell in effect became a mathematician for the eighteenth century who championed Newton and the synthetic geometry in opposition to analytical geometry.

Though in his *Elementary Treatise on Mechanics* (1819) Whewell had not adopted the abstract formulation of calculus as portrayed by Joseph-Louis Lagrange, he had stressed the power of analysis. But over time his views changed, and by the 1830s and 1840s he was advancing the importance of the geometrical method in undergraduate teaching and lauding textbooks based on synthetic geometry as the way forward.

Whewell wrote to Herschel in 1845, "Cambridge is, and I hope will continue [to be], the principal school of *English* mathematics." He was still—or, in fact, once again—worried that the most active students, on whom the future reputation of Cambridge would rest, were being encouraged to study "the last [latest] supposed improvements, contained in memoirs, journals, and pamphlets" rather than the standard works of mathematical literature.[55] Beside his own textbooks, the works gaining Whewell's seal of approval for inclusion in the standard Mathematical Tripos course included the *Tracts* written by Airy. Recently published papers were specifically excluded, "not admitting them into our scheme till some time has elapsed, and the mathematical world has given them its sanction."[56] Whewell gave his blessing only to contributions that had gained the status of "permanent" knowledge. In this grouping he was not against including work by the giants of mathematics, such as Leonhard Euler, Lagrange, Laplace, and Gaspard Monge, but he could only recommend them for the most advanced students.[57]

During his first vice-chancellorship of the university in 1843, Whewell proposed the setting up a board of mathematical studies to oversee the content of the curriculum of the Mathematical Tripos, and in 1848 one was established. The range of examinable subjects was widening, and students were unable to discover what was likely to be asked by the moderators and examiners. In addition, the Mathematical Tripos competition being praised as a level playing field was open to question. It was perceived by Whewell that there was the need to bring stability to the tripos, and in 1849 and 1850 the board recommended some of the more difficult subjects be discontinued. Out went elliptic integrals, Laplace coefficients, theories of electricity, magnetism, heat, capillary action, and the figure of the earth. Some parts of the lunar and planetary theories were also omitted. The curtailment of these subjects from the tripos lasted until 1872.[58]

Whewell also proposed reforms of educational provision, but some of these ultimately failed. Through college statutes in 1844 he sought to limit the system of private tuition, but this did not happen. By 1845 he was looking back to the lost golden age of his youth, seeing value in the old curriculum. In it was "a great store of beautiful examples of mathematical logic and mathematical ingenuity"—of *permanent* knowledge.[59]

Whewell as Mathematician

Whewell's early abilities in mathematics gained his entry to Cambridge and a path in life that would otherwise have been unavailable. But in maturity he did not see himself as only a mathematician, in contrast to George Peacock, whose research concerned almost entirely the foundations of pure mathematics. Whewell advocated the utility of mathematics and had firm ideas on its teaching to Cambridge undergraduates.

Whewell's educational prospectus concentrated on classics and mathematics. The latter consisted of geometry and mixed mathematics, and his textbooks reflected his pedagogic interests.[60] If mathematical education had been an established field, he would have been a rising star. To his credit, he understood the golden rule for teaching mathematics, that most students learn from the particular to the general and from the concrete to the abstract and not the reverse.

For Cambridge students of the 1840s, particularly those at Trinity College, there was a spirit of research engendered by the creation of the *Cambridge (and Dublin) Mathematical Journal* (edited by Duncan F. Gregory, Robert Leslie Ellis, and William Thomson).[61] In character its birth is similar to the founding of the Analytical Society in 1812. It too was started by young students dissatisfied with the attitude that mathematics was there for the sake of the Mathematical Tripos handed down unchanging as if on tablets of stone. Their manifesto of 1837 published in the preface to the first volume of the *Journal* begins: "Our primary object, then, is to supply a means of publication of original papers. But we conceive that our *Journal* may likewise be rendered useful in another way—by publishing abstracts of important and interesting papers that have appeared in the Memoirs of foreign Academies, and in works not easily accessible to the generality of students. We hope in this way to keep our readers, as it were, on a level with the progressive state of Mathematical science."[62] The word *progressive* was directly opposed to Whewell's *permanent*, and these words were not the sort of thing the inhabitant of the Master's Lodge would wish to hear.

For this 1840s generation of leading young men, mathematics escaped the boundaries set by merely viewing it as a component of a liberal education; for them it ceased to be a fixed body of knowledge anchored to the past, but became a moving field of inquiry in which workers published their research in European journals.

Whewell ultimately lost the battle of maintaining the Mathematical Tripos as he had known it as a student, and *analysis* became the new watchword in the 1830s and beyond. In his role as teacher between 1816 and 1839, he had insisted that mathematics should be applied to the world, and this belief and the textbooks he produced shaped the Mathematical Tripos as it evolved.

Whewell's academic reputation did not survive his demise. Harvey Goodwin, a member of the 1840s generation, pointed out his shortcomings as a mathematician, as a writer of books, and as a lecturer. He acknowledged that "Whewell was our great Cambridge man. As Master of Trinity he was a prominent feature of the University till the day of his death. He handed on the lamp; and though his books may become antiquated, the direction given to the scientific and philosophical thought by Whewell's writings may have an influence upon men's minds deep and permanent, and not to be adequately measured by the size of his printed works."[63]

Today's assessment of Whewell's legacy in the field of mathematics would diverge from Goodwin's. In the 1840s and 1850s Whewell was a member of a generation of teachers of mixed mathematics paving the way for later generations and the flowering of such notables as William Thomson, George Gabriel Stokes, John Couch Adams, and James Clerk Maxwell. In this role he used his position as a formidable university politician, possessed of great physical and intellectual energies, to maintain his essentially conservative views.

CHAPTER 3

Whewell
and Mechanics

BEN MARSDEN

William Whewell wrote and revised textbooks on mechanics and dynamics from the beginnings of his academic career in the 1810s to the 1840s. These textbooks were at first designed for the use of students at the University of Cambridge, but were increasingly used by students at other educational establishments, by scientific practitioners, and by technical experts. They were perused by British engineers studying in academic spaces from the late 1830s, and by 1859 Whewell's views on mechanical pedagogy had reached China. Isaac Todhunter devoted a chapter of his *Life and Letters* to Whewell's books on mechanics, and we are indebted to Harvey Becher's and Elizabeth Garner's work in this regard.[1] No assessment of Whewell's textbooks on mechanics and dynamics, however, takes fully into account recent research in the history of "science and the book," the history of engineering pedagogy, and Whewell's extended pedagogical networks, including his connections with British engineering elites. This chapter offers such a reassessment, arguing for a closer engagement by Whewell with practical engineers and their academic mediators than hitherto suggested.

The *Elementary Treatise on Mechanics*

Whewell had shown his mathematical skill while at school.[2] He won an exhibition at Trinity College, Cambridge, and his student life there, beginning in October 1812, prepared him to be a mathematical tutor—and, relatedly, a mathematical author. His broad intellectual abilities brought a string of prizes, though he graduated as second, not first, wrangler in January 1816, and he was second, not first, Smith's Prizeman. He forged friendships with John Herschel and others associated with the Analytical Society (f. 1811) and its advocacy of Continental calculus and algebraic analysis.[3] From 1817 Whewell was tutoring mathematics students, studying for Trinity's fellowship exam, and discussing with Hugh James Rose a proposed review to be published by J. Deighton & Sons in Cambridge.[4] This firm, and its successor J. & J. J. Deighton, became the chief publisher of the pedagogic works of Cambridge professors in a "textbook revolution."[5] Whewell's election as a fellow at Trinity in October 1817 provided a platform for literary, mathematical, and intellectual projects.

Although Whewell's interests were wide, mathematics teaching dominated his period as a fellow until and beyond 1823, when he was promoted from assistant tutor to tutor. In October 1817, deciding against the proposed review, he noted, "I have got several projects for mathematical works. I think I could at least be useful but I have not put any of them in a tangible shape yet."[6] In March 1818 Whewell was teaching a growing number of private pupils;[7] guided by that experience he was also writing a mechanics textbook, having already been inspired by the history of mechanics. In June 1818 he wrote to Herschel about Simon Stevinus, voicing an enthusiasm later elaborated in *The History of the Inductive Sciences* (book 6, chapter 1). Whewell gave Herschel an update on the textbook: "I have nearly finished my Statics, and feel tempted to publish it by itself; for Dynamics, to treat it as I am going on, working out all kinds of problems, will take up much time. I shall say nothing about the metaphysics of the subject just as yet."[8]

Whewell would eventually fulfill the ambitions of his younger self, producing numerous textbooks in mechanics and dynamics, and the history and philosophy of the inductive sciences that represented the culmination of his studies. In August 1818 Whewell admitted to Richard Jones the scope of his plans. Tutoring seven pupils left him comfortably off enough, with time to dream of "undertakings metaphysical, philological, mathematical." But in the meantime he wrote, "Nothing prospers but Mechanics." He wanted to get this "afloat as soon as possible" since he had been offered, and had accepted, a mathematical lectureship. He imagined a "hope of doing some good," regretting that he would be obliged "to talk with some moderation" of James Wood's then-favored *Principles of Mechanics* (1809).[9]

Whewell recognized and wanted to use his institutional leverage in Cambridge. He wrote to Herschel in November 1818 about his plans for "reforming the mathematics of the university":

> I have it now more in my power to further this laudable object by the situation I have taken of assistant tutor (i.e. Mathematical Lecturer) here. . . . I shall have a permanent and official interest in getting the men forwards—I shall have an opportunity of directing their reading—and I shall write books . . . and be able to put them in circulation. . . . I have the first volume of a system of

mechanics, containing statics and a little dynamics, ready for the press. It consists in a great measure of a classification of problems. To serve as an elementary book introductory to the higher applications of mechanics is one object, and another is to establish the science on simple and satisfactory principles, which I do not think has been done.[10]

In March 1819 Whewell received the proofs of his *Elementary Treatise* and "agreed with Deighton for copyright for three guineas a sheet for first edition and two guineas a sheet for succeeding editions, with three guineas a sheet for additional matter."[11] When Whewell and his friend Richard Sheepshanks were shipwrecked traveling to Paris, sheets of the "mechanics" went under with everything else. But Whewell worked on his treatise back in Cambridge, expecting it to be out by October 1819. "Fortune" had shone on him by making him moderator for the Mathematical Tripos examination in 1820 and thus in good position to secure the treatise's success.[12] The appearance of *An Elementary Treatise on Mechanics* in November 1819 marked, in Todhunter's terms, the start of Whewell's "long course of authorship."[13] Publication coincided with the start of the university session. Whewell began a habit of sending out copies to friends, rivals, and influencers. Julius Charles Hare was complimentary, but Herschel had his doubts.[14] On December 1, 1819, he carped that Whewell had made too many "concessions to the cramming system . . . and that the work would have been productive of more extensive good . . . had you conformed a little more to the taste of the age and a little less to that of the University." The gifted junior Whewell had failed to do the bidding of his intellectual elders in the Analytical Society.[15]

Todhunter devoted the second chapter of his homage to Whewell to a bibliographic analysis of the "Publications Relating to Mechanics."[16] Whewell's repeated revision of his works was not "fortunate," Todhunter opined, seeming as it did to "shew a want of stability in the author which shakes the faith of his readers." Whewell had "in his treatises on *Mechanics* exemplified the . . . process of rearrangement with unsatisfactory results."[17] Elizabeth Garner, more generously, presented Whewell's aim to be drawing Cambridge students "into the study of mathematics through the consideration of physical problems," though his views of which physical principles mattered, and what mathematics students needed, changed, leading him away from his early enthusiasm for Continental analysis to attack aspects of Continental mathematics, and to see the virtues of traditional geometry for a liberal university education.[18] Neither author focuses on Whewell's dalliance with engineering mathematics and his conversations with engineers and their mediators—a crucial feature that emerged over the years.

The first edition of the *Elementary Treatise* was an octavo volume with an extensive preface, 346 printed pages, and fifteen folded leaves of plates. The book was manufactured in Cambridge by John Smith, "Printer to the University," like all of Whewell's subsequent textbooks in mechanics and dynamics. The expected publisher, J. Deighton & Sons, partnered with G. & W. B. Whittaker, and Whewell would continue to use this company and its successors as his London outlets.[19] The book carried the teasing subtitle: *Vol. I Containing Statics and Part of Dynamics.* There was, as we shall see, no clear sequel.

Todhunter considered the first edition of the *Elementary Treatise* as a book that, with Peacock's and Herschel's publications, "introduced the continental mathematics . . . to replace the system of fluxions which had so long prevailed at Cambridge." Although these books helped to reform mathematics teaching in Cambridge, and Whewell adopted Continental symbols, Herschel had his doubts about Whewell's commitment. Nevertheless, Whewell's book, according to Todhunter, provided "copious" accounts of elementary and "the higher parts" of statics, and used the differential and integral calculus "very freely." In dynamics, it treated "only the elementary parts." Whewell's long introduction discussed the logical order of the book, the relationship between definitions and empirical issues, and Dugald Stewart's ideas on mathematics—a topic that had long agitated him. The Royal Astronomical Society quickly praised the work for its logic, accuracy, and being "in advance" of contemporaries in its treatment of bodies in contact, laws of motion, and composition of forces.[20]

On April 13, 1820, when Whewell was elected a fellow of the Royal Society of London, the campaign in his favor had been built on his authorship of the *Elementary Treatise*. When thanking him for a copy of the work, Herschel told Whewell that he had recommended his candidature to President Joseph Banks.[21] Whewell's election certificate repeated his Cambridge credentials, his skill in "various branches of Natural Philosophy," and the fact that he was "the author of a treatise of Mechanics."[22] Whewell's active fellowship increased his profile and influence within the organized community of scientific practitioners.

Whewell revised his *Elementary Treatise* on the basis of his experience as an author and pedagogue teaching from his own textbook. A fifty-page *Syllabus of an Elementary Treatise of Mechanics* (1821)[23] repeated the "most simple and important" propositions of the *Elementary Treatise*, referred to the book for full demonstrations, and made additions and alternations.[24] This *Syllabus* offered navigational help for local consumers like William John Speed, a sizar at Trinity College from March 1820.[25] Interleaved blank pages allowed students to make notes or practice proofs. The *Syllabus* also gathered material for the first of many revisions of the original.

A Treatise on Dynamics and Mechanical Revisions

Whewell partnered his book on mechanics with one on "dynamics," or the investigation of "the relation of the time, space, velocity, and force, when bodies are in motion under any circumstances."[26] Although he had projected a dynamic successor to the "Vol. I" that was the *Elementary Treatise*, in September 1822 he told Jones, "My book [on dynamics] is swelling out larger than I expected. As I do not think it will gain much by coming under the protection of the former volume [the *Elementary Treatise*], I intend to print it as a separate work, instead of calling it Volume II." Whewell meanwhile meditated on writing a "History of the Metaphysics of Mechanics, though as yet it is only intention."[27]

In the New Year, with financial support from the syndics of the University Press, the *Treatise on Dynamics* (1823) appeared. The book had swollen to 403 pages, with 144 geometric diagrams on six folded leaves. Whewell claimed there was barely in English a scientific treatise on dynamics beyond that of Cresswell's translation of Venturoli. His own work developed demonstrations and collected mechanical problems "selected from the works of the best mathematicians," as a mark of "proper respect" to authors such as Colin Maclaurin, the Bernoullis, Leonhard Euler, Thomas Simpson, Joseph-Louis Lagrange, and Pierre-Simon Laplace.[28] It extended, but muted, the Analytical Society's campaign, bringing the "simplicity and evidence of the mathematics of a century ago" with the "generality and rapidity of modern [French Continental] analysis."[29] Sharing respect for Newton and the French, Whewell included almost all the propositions of the *Principia*'s (1687) first two books and echoed Newton's topical arrangement. He indicated that this *Treatise* included a toolbox of mathematical techniques.[30] Whewell was particularly obliged to "the excellent Treatise on Mechanics of Mr. Poisson."[31] Though he never revised the *Treatise on Dynamics* in print, two later books would be designated the first and second parts of "a new edition."[32]

With the *Treatise on Dynamics* completed, Whewell revised the *Elementary Treatise*, as earlier editions had sold out and in response to factors linking Cambridge to wider worlds. Whewell hoped to motivate mathematics students with real-world problems and to show engineers the value of mathematical mechanics. In November 1824 Whewell's friend Henry Wilkinson expected the "new [second] edition of the first volume of *Mechanics*" to be out before Christmas.[33] When it appeared, and in good time, Whewell suggested the considerable changes, especially in statics, made it a new and separate work, certainly no longer "Vol. I."[34] Jones protested, "Your first literary Babe will think you an unnatural Papa for not letting your youngest darling call it Brother."[35] A third edition of the *Elementary Treatise* (1828)[36] included thirty pages of "improvements and additions" including topics of engineering interest, such as friction and the "connexion of pressure and impact." Whewell had included material on the "elasticity and compression of solid materials," and would "gladly have given a section on the strength and fracture of beams, had there been any mode of considering the subject, which combined simplicity with a correspondence to facts. The common theory [deriving from Galileo] . . . is manifestly and completely false."[37] Here were opportunities for men of science and engineers to come together to do better.

Engaging with Readers beyond Cambridge

Whewell's early books might have been intended for Cambridge students, but he was concerned that popular audiences, incautious scientific readers, and theory-friendly engineers might prefer, and be misled by, accessible yet intellectually inferior works. He kept an eye on rival works on mechanics and censored their transgressions.[38] In May 1830 Whewell persuaded David Brewster to publish some cutting observations on the maverick Dionysius Lardner's book on mechanics, coauthored with Henry Kater as part of the *Cabinet Cyclopaedia*. This book gave "slighting notice" to Newton's propositions, and claimed that Newton's three laws had "little or no utility."[39] Whewell showed the absurdity of the claim by elaborating a thought experiment in which bodies tended to move more slowly, the longer they continued in motion.[40] The "purity of our scientific logic," itself "essential to the respectability of science," was in danger of being "corrupted" by a work with "a prospect of extensive circulation." If the "more general diffusion" of science was to be by contempt for sound reasoning, "the advantages of the spread of knowledge will be grievously diminished."[41]

Brewster's protégé James David Forbes was one who actively sought guidance from Whewell on proper scientific reading in the summer of 1831.[42] At first Whewell offered Forbes advice;[43] later, when Forbes became professor of natural philosophy at Edinburgh, they regularly reflected as (near) equals on the needs of Cambridge and Edinburgh students, often returning to discuss Whewell textbooks.[44] Forbes's forte was experimental natural philosophy, typical of the Scottish universities,[45] but he advocated a thorough grounding in theoretical mechanics, praised Whewell's works, and, eventually and repeatedly, suggested how they might be improved.

Whewell, Lardner, and Forbes all wanted to see mechanics, within the context of natural philosophy, made practically useful. After the formation of the British Association for the Advancement of Science (BAAS, f. 1831), with its annual meetings often in provincial industrial centers, Whewell, as a respected "gentleman of science," had greater opportunities to fraternize with practical men. Successful civil engineers might court his patronage; he saw them as outliers in a scientific community in danger of fracture, worth courting in return. In April 1833, for example, George Birkbeck, mastermind of the London Mechanics' Institution, thanked Whewell for sending a copy of the recently completed *Astronomy and General Physics* (1833), thus ensuring it would be available to the practical men frequenting the institution.[46]

Most consequential for the mechanics treatises was Whewell's engagement with the "philosophical engineers" seeking to deploy natural philosophy and its methods in their practice. In June 1832 the strength-of-materials experimenter Eaton Hodgkinson sent Whewell essays that might supply deficiencies Whewell had identified in the preface to his *Elementary Treatise* (1828).[47] Hodgkinson asked Whewell, then vice president of the BAAS, if he might present at its Oxford meeting. He had approached Whewell only after being praised by practical mathematicians Olinthus Gregory and Peter Barlow. Hodgkinson's "On the Strength and Best Forms of Iron Beams" (1832) duly appeared in the "Arts" category, and the author sat with Whewell on the Committee of Section A of the BAAS. This continuing relationship showed that philosophical engineers like Hodgkinson responded to opportunities revealed by Whewell, reading the *Elementary Treatise* as a developing project compatible with practical mechanics.

"An Easy Book" on Mechanics and Grappling with Newton's *Dynamics*

In the summer and autumn of 1832 Whewell diverted from the *Elementary Treatise* to produce three distinct works on mechanics and dynamics. The first, a popular, mathematically unchallenging book, was a corrective to the errors of Lardner and his ilk. Whewell wrote to Wilkinson, "I am on the point . . . of publishing another *Mechanics*, which is to be an easy book explaining the matter about as fully as I know how, and containing as much mathematics as Wood does. It is an employment at which I sometimes grumble, to have to write so many elementary books,—a very difficult and ungrateful office; but when you have to lecture and instruct about these things, you have the defects . . . of existing books so strongly and repeatedly forced upon you that it is difficult to abstain from trying to remedy them."[48] By October 1832 Whewell had published *The First Principles of Mechanics* (1832), an "easy book" of about 120 pages.[49] With its Continental approaches and calculus, the *Elementary Treatise* was not for the mathematically fainthearted. In his *First Principles* Whewell had "wish[ed] to put the elementary portions of Mechanics in a form in which they might be extensively studied in Universities."[50] A popular book, enticing readers with historical and practical illustrations, catered to the less mathematically adept Cambridge students, beginners at other universities, popular audiences and, possibly, engineers. Whewell characteristically indicated his sources (including Arnott, Herschel, and Babbage) and also his motivations: logical exposition remained, even in an accessible work with little abstruse mathematics.

This partly historical work showed Whewell limbering up for more sustained treatments and also reflected an established passion. From his earliest days as an author, Whewell had charted the historical development of scientific ideas. His understanding, and representation, of mechanics was bound up with his fascination with the history of the inductive sciences at large. During the construction of the first edition of the *Elementary Treatise*, Whewell's friends knew he was working historically. Julius Charles Hare, receiving a copy of the book in December 1819, had asked whether Whewell intended "to keep your history of the science [of mechanics] for the next volume, or to publish it separately, or to keep it back to form a part of a history of mathematics from the creation of the world down to the year of the Lord 1836."[51] The *First Principles of Mechanics*, with its "historical illustrations," began to fulfill Hare's expectations.

The book was not, in truth, an "easy" one: its definitions terse, its examples abstruse, its classical illustrations unfathomable to untutored students. On Whewell's circulation list were people like Edward Hill, a fellow at Christ Church, Oxford, who became an examiner in mathematics for Oxford University. A critique of the *First Principles* in the *Westminster Review*, although supportive of Whewell's mathematical and historical projects, mocked the conceit that everyday practical traditions should be retrospectively presented under the logical gaze of the organizing mathematician.[52] Another atypical recipient, University College London mathematics professor and Cambridge alumnus Augustus De Morgan, reflected, "I think it [the book] highly calculated to do good: especially among the lower species of Wranglers. However might it not be useful to enter a little into what becomes of the motion lost by friction and other resistances, so as to shew that we have no reason to believe in the absolute loss of momentum?"[53]

Whewell left the *First Principles* unrevised, for once, and the ample development of the history of mechanics would wait for the *History*, overshooting by one year Hare's estimated date of completion. Todhunter speculates that the *History* grew from Whewell's original plans to write "merely" a history of astronomy or of mechanics, and that what Todhunter saw as weaknesses (lack of unity, the downplaying of deductive reasoning) resulted from that starting point in Whewell's historical work.[54]

In 1832 Whewell also brought out two books on dynamics. The first was the sixty-four-page *An Introduction to Dynamics* (1832),[55] which explained the laws of motion and dealt with the first three sections of Newton's *Principia*. Whewell's alignment of his works on dynamics with the *Principia* caused some alarm. The Rev. H. Wilkinson had been "disturbed" that Whewell might have aimed to supersede the "reading of Newton's text," but Whewell insisted that only mathematicians actually read Newton and those who tried to get Newton's ideas were often misled by "manuscripts generally illogically constructed."[56] Whewell remained concerned that the market was flooded with popular books, light in mathematics, that did violence to the logic of science. Whewell's two-part *Treatise on Dynamics* (1832 and 1834) was not light in mathematics, and Newton was center stage. The first part, completed in April 1832, bore the title *On the Free Motion of Points and on Universal Gravitation.*[57]

Modifying the *Elementary Treatise* to Please University Students and Practical Engineers

Despite the proliferation of Whewell's textbooks, key supporters remained unsatisfied. In the spring of 1833 Forbes complained that his hope to mix "pure demonstration with experiment and collateral illustration" in teaching would be wasted without an adequate mechanics textbook: "Your mechanics [the *Elementary Treatise* of 1828] has appeared to me by far the best book," but it was too long, too difficult, and in statics, too complete. Whewell should slim down the mathematics, add propositions from the first three sections of Newton's *Principia* (via *An Introduction to Dynamics*) and include the mathematical theory of hydrostatics. Competitor volumes by Thomas Jackson and John Leslie might be "a little repulsive" and "incredibly bad," respectively, but they were superior to Whewell in ease of use and coverage.[58] Whewell recognized the need but balked at the proposal, concerned at the "sacrifice of time," delayed by his preferred "fastidious . . . style," and bogged down in the writing of the second part of the new *Treatise on Dynamics* (not published until 1834).[59] Furthermore, he was increasingly engaged in the endeavors that culminated in the *History* and the *Philosophy of the Inductive Sciences* (1840). On October 6, 1834 he wrote to Jones, "The history of mechanics and astronomy is so important and instructive, that it must be tolerably full. It has never been written according to any philosophical view. . . . I expect too that by means of it I shall be able to stop the mouths of all gainsayers of my philosophy."[60]

After worrying he might have to write the proposed mechanics textbook himself, Forbes discovered in July 1833 that Whewell was going to revise his *Elementary Treatise* (1828) by dividing it into two, presumably as Forbes had suggested.[61] He told Whewell, "I cannot tell you what a weight is off my mind since you agreed to modify your mechanics. I am so fully convinced that no inexperienced person [like himself] can undertake to write a good systematic work." Forbes had marked in the *Elementary Treatise* (1828) the parts he thought should be included in the first volume of the two suggested by Whewell, and he hoped it would be cheaply available by October 1833.[62]

Before that time, Whewell used the Cambridge BAAS meeting to reconnect with one engineer keen to add mathematical rigor to his practice while reciprocally providing further impetus for Whewell to rethink his

Elementary Treatise. Whewell and Herschel both fraternized with Hodgkinson, whose efforts to enhance engineering through systematic large-scale experiment supported by theoretical investigation were increasingly well known. On September 7, 1833, Hodgkinson and Whewell corresponded about Hodgkinson's paper on the effects of impact on bars as determined by engineering experiment. Hodgkinson explained his method of "seeking by an approximate mode for the inertia of the bar," since a "strict determination" (by mathematics) was "above my feeble powers." Hodgkinson was keen to read more about impact, and although he worried about his "want of . . . mathematical knowledge" he intended to read the new edition of Poisson's treatise on mechanics.[63]

Whewell grasped the opportunity to augment the practical dimensions of his treatment of mechanics. Hodgkinson allowed Whewell to use some of his work on beams and chain bridges in the next edition of the *Elementary Treatise*, and he recommended a paper by the engineer Edward Dixon, whose brother was connected with the new Liverpool & Manchester Railway (LMR), on the performance of steam engines. Hodgkinson offered to provide as much information as Whewell wanted, indicating that Dixon's paper included a definition of "horsepower" applied to engines, argued for the use of engines rather than horses on railways, tackled friction, and treated the power of engines to carry loads up inclines.[64]

Whewell was keen to give timely examples to satisfy multiple audiences. Encouraged by Forbes and Hodgkinson, he spent the autumn of 1833 dividing the *Elementary Treatise* with an eye on students with varying mathematical abilities, whether at university or in engineering practice. The fourth edition of the *Elementary Treatise* (1833), was slimmed down (as Forbes had suggested) to comprise the "absolutely elementary portion only." What it lost in higher mathematics, it gained in engineering relevance.[65] It contained a demonstration of the "parallelogram of forces," conjured up by Whewell nearly simultaneously with Poisson in the latter's new *Traité de mécanique* (1833). New sections on the "friction of bodies in motion" and the "principle of work" reflected suggestions by De Morgan, Hodgkinson, and Forbes that Whewell engage with the engineers.[66] Yet investigations on the "forms of bridges" and on "species of the elastic curve," though of engineering relevance and included in the third edition, were removed in the fourth as insufficiently elementary. Whewell had shed all the material that required analytical geometry and differential calculus, the material that had partially aligned the original publication of 1819 with the Analytical Society's endeavors.

Those parts assuming higher mathematical knowledge were not altogether jettisoned. By November 1833 Whewell had prepared a "supplement" to the shortened fourth edition of the *Elementary Treatise*.[67] Whewell's *Analytical Statics*,[68] despite its title, was no systematic treatise, and rather than focusing on mathematical abstractions, it attempted practical utility. Some additional material (including a proof of the general principle of virtual velocities) benefited from Poisson's treatise. Whewell also introduced "propositions which have a bearing upon practical applications of mechanical knowledge." He turned to Davies Gilbert and Hodgkinson (on suspension bridges) and, to transcend the Galilean theories complained about in 1828, to Barlow and especially Hodgkinson (on strength of materials). Whewell respectfully deployed Hodgkinson's memoirs brought to his attention a year before.[69]

As part of this mechanical reshuffling, he published a pamphlet, *Additions, in the Fourth Edition of an Elementary Treatise of Mechanics* (1833). Instead of buying a new edition, students and tutors could work from the third and refer to these extras.[70] The new topics were: "the friction of bodies in motion" (sliding, rubbing, of wheels, the power of traction), with examples from roads, railways, and canals; "the measure of the power of mechanical agents and of work done by machines" (with passages on Stephenson's *Rocket* locomotive then at work on the LMR); and the introduction of engineers' terms, including *duty* (further specified by Whewell as "theoretical" and "practical" duty), with discussions of the power of water, air, elastic bodies (including springs), steam (including condensing and high-pressure steam engines), and of "men and horses."

For the next two years Whewell tried, at least, to focus on completing his outstanding mechanical writing projects. The second part of the *Treatise on Dynamics*, on "points constrained and resisted," came out in October 1834.[71] Forbes suggested that the sale of Whewell's mechanics books alone had firmly established a connection between Edinburgh and the favored Cambridge publisher Deighton, who should send copies of every new book published, "and several of such books as yours."[72] The *Elementary Treatise* on mechanics was out of print by January 1836, and Forbes urged, "You must give us a new one."[73] Whewell obliged, announcing at the end of May, "My Mechanics, new edition, will be ready in a week or two."[74] One lucky owner would be Thomas Graves, an Irish mathematician who had studied with William Rowan Hamilton at Trinity College, Dublin, in the mid-1820s.[75]

The Mechanical Euclid

Whewell had gradually reduced his expectation that Cambridge students, and other readers of his elementary pedagogic works, should master analysis and geometry; he had lately attempted to create popular introductory works on mechanics and dynamics that exemplified logical thinking even where higher mathematics was relegated to the background. His *Mechanical Euclid* (1837) attempted to combine these features of his developing approach to mathematical education.[76]

Whewell's title implied "a coherent system of exact reasoning . . . for which Euclid's name is become a synonym." He had constructed a "system of Mechanics" that could "hardly fail to be of use in that disciplinal employment of Mathematics in which Euclid's *Elements of Geometry* have hitherto most deservedly held their place without a rival." The University of Cambridge had appointed the elements of mechanics and hydrostatics as essential for the examinations "for the usual degrees," and here Whewell produced the requisite "manual." To be sure, he had been a member of the syndicate that recommended the curricular changes subsequently adopted by the senate; but the book was his own, not the university's. By treating more than the published "list of Propositions," it was future-proofed in case the university should extend "this line of examination."[77]

In the *Mechanical Euclid's* preface, Whewell admitted his motivation for writing the *First Principles of Mechanics*, but he now considered "the effect of mixed Mathematics as a discipline is . . . likely to be far better answered by the more rigorous scheme of Mechanics which the University has sanctioned, than by such a treatment of the subject as was then presented; and the Historical Illustrations . . . are given much more completely in the History [*of the Inductive Sciences*]." The historical excursions in the *First Principles* had been "superseded"; the "practical illustrations," though redeployed in his books of 1833, might, he teased, be "incorporated in some future publication in an improved form."[78] Again Whewell responded simultaneously to Cambridge students and to engineers.

The *Mechanical Euclid* dedicates twenty pages to "algebra" (addition, subtraction, multiplication, division, fractions, proportion, and progressions arithmetical and geometric) and five to "geometry." The "algebra" had been taken "with little alteration" from the "well-established Treatise of Dr Wood."[79] There follows "mechanics" (including the lever, composition of forces, "wheel and axle," and pullies), and "hydrostatics" (with propositions moving from general physical issues, to specific machines and instruments, including the siphon, air-pump, and thermometer). Mimicking Euclid, "definitions" and "axioms" preceded propositions in each section. The third book treats "the laws of motion" (Newton's three laws, with gravity mentioned). The Cambridge market alone ensured the book would go through five editions, the last being in 1849.

As to the work's reception, Thomas Turton, bishop of Ely, reflected on his presentation copy: "Your remarks upon mathematical reasoning and Induction . . . contain a great deal of very important matter; and I am glad to see the attention of men directed to the grounds of science. The producing of results is not all that is wanted in academical education."[80] An assistant tutor at Trinity College, however, dared to suggest that doubts could persist, even on Whewell's home turf, about this attempt to restructure mechanics and to discipline mathematical reasoning. John Moore Heath told Whewell, "Your manner of treating of geometry I think admirable, but your self-evident mechanics stagger me as yet I confess." Similarly, "everybody knows practically that all bodies have a centre of gravity; but not axiomatically."[81]

The *Mechanics of Engineering* as "the Only Possible Course for the Present Time"

Whewell's relationship with Robert Willis informed a different way of presenting mechanics to students in and beyond Cambridge.[82] Willis excelled as a practical experimenter before the Cambridge Philosophical Society and supported the university's "voluntary science" curriculum.[83] As Jacksonian Professor of Natural and Experimental Philosophy he taught mechanics in Cambridge using specially created apparatus, suited to the display of "kinematics" (the geometry of machine parts, ignoring forces). By the late 1830s Whewell and Willis contemplated yet another redivision of mechanics, catering to multiple audiences. A short and accessible version of the *Elementary Treatise* would be accompanied by a collection of engineering examples and a book setting out Willis's kinematics.

Stimuli for this redistribution included the growth of academic engineering classes in Britain and the approaches made by the Institution of Civil Engineers to Whewell, Willis, and George Biddell Airy, all of whom were made honorary members in 1837.[84] Cambridge alumni were key to this rapprochement between theory and practice. Hodgkinson eventually became a university engineering professor. Forbes advocated academic engineering, so long as those engineers teaching "practical mechanics" did not usurp the professors of mathematics and natural philosophy, who had intellectual ownership of theoretical mechanics. Indeed, Forbes supported an experiment in engineering education at the new University of Durham, which from about 1837 offered select gentleman engineers a multiyear course.

Forbes even examined the Durham engineers, and it was Whewell he turned to in February 1840 for advice on their question papers.[85] Whewell supported this new venture to extend his mechanics beyond Cambridge to the engineering community. In September 1840 he was preparing a revision of the *Elementary Treatise* and wanted "to fit the book more for practical men, and to make it correspond with a work which Willis is very soon to publish on *Kinematics*, or *Pure Machinery*, as I have called the subject."[86] By October Whewell set about "entirely modifying . . . my *Mechanics* . . . to make it a book for civil engineers. It is the only possible course for the present time, on grounds of the highest principle, as well as convenience."[87] Forbes was keen to welcome engineers into university classrooms to teach them mathematics according to Whewell's plans, but he was less keen on engineers masquerading as university professors of theoretical mechanics, and he resisted his former student Lewis D. B. Gordon's 1840 appointment as an engineering professor at the University of Glasgow.[88]

Whewell pursued his splitting of mechanics to serve Cambridge students and engineers alike. In January 1841, as he worked on a new edition of the *Elementary Treatise*, he was considering relocating the engineering topics to a separate book.[89] The revisions to create an abridged sixth edition were straightforward. Expanding the engineering material into an independent book during the academic term took longer. In March Whewell wrote, "You see how it has been sweated down into a little book [of 150 pages] by the process to which it has been subjected. I am printing my Engineering [*Mechanics of Engineering* (1841)] and Willis is printing a book about Mechanism [*Principles of Mechanism* (1841)], so we shall between us do something towards giving a form to that science."[90]

The two fellows collaborated with mutual admiration: Whewell told his friend Jones that Willis's work was "excellent," and noted that the *Mechanics of Engineering* had a "good chapter on the Measure of Moving Power, borrowed from the French Engineers and improved. Willis and I agreed to call the Measure of Moving Power 'Labouring Force'" (rather than *travail*).[91] Whewell promoted "engineering" into his title, and though he called on Deighton to publish the book, he also recruited John W. Parker (copublisher of the 1837 *Mechanical Euclid*) to provide an outlet on the Strand, close to King's College London (KCL), with its large cohort of engineers. Whewell's and Willis's literary projects responded to a surge in the university and college training of engineers.[92] Neither Whewell's recent election as master of Trinity College nor his marriage were enough to inhibit reviewers from joking, though, about "the progeny of these two fellows."[93] Detractors, in the *Mechanics' Magazine* and elsewhere, carped that, by aiming at high theory and at engineering practice, Whewell and Willis had succeeded in doing neither well.

The publication of the *Mechanics of Engineering* stimulated an exchange between Whewell and a key academic engineer, the Cambridge wrangler, mechanical author, and KCL professor Henry Moseley. Moseley developed his *Illustrations of Mechanics* (2nd edition, 1841) into a substantial *Mechanical Principles of Engineering and Architecture* (1843), which drew from French scholarship, including Jean-Victor Poncelet's *Mecánique industrielle* (1839). The *Mechanical Principles* extended Moseley's lectures to KCL's engineers and cannot have ignored Whewell's efforts. Letters from Moseley in October 1842 detailed their discussions. Moseley's book rivaled Whewell's, and reviewers tended to treat it more kindly.[94]

In the seventh and final edition of the *Elementary Treatise* (1847), Whewell reverted to earlier organizational schemes.[95] Even he admitted that the strategic division effected in the sixth edition had been too much for readers to stomach. Commentators might have noted that it showed the strain between Cambridge and practically oriented centers of learning in London, Glasgow, Dublin, and elsewhere. Whewell thus reinstated sections removed from the *Elementary Treatise* to restore its former usefulness. Nevertheless, he maintained that the shortened sixth edition, the *Mechanics of Engineering*, and Willis's book together represented a valuable treatise on mechanics.[96] For more than twenty years, as pedagogic and practical environments, the liberal education, technical provision, and foreign pedagogic provision had been in flux, Whewell had striven to make his textbooks relevant to Cambridge students and to engineers.[97]

Simplistic views about the polymathic Whewell, portraying him as an early follower of an Analytical Society agenda and later as a "meta-scientist" relatively aloof from scientific (and engineering) practice, need revision. Whewell stepped back from the analytical fundamentalism of his early textbooks partly to accommodate students' capacities, in and beyond Cambridge. Whewell worked with figures like Forbes to optimize the distribution of material in his textbooks, whether of Newtonian laws or practical illustrations. Whewell's posture is shown in this chapter to be closer and more responsive to practical mechanics than scholars have imagined (closer, perhaps, to the posture of the carpenter's son). In particular, Whewell stood in a long, constructive, and two-way engagement with mediators of practical mechanics like Hodgkinson, even before the creation of British engineering chairs. This becomes apparent, for instance, from the inclusion of practical engineering matters and examples in his books on mechanics, which climaxed in his collaboration with Willis. Whewell appears as ahead of the curve in the training of gentleman engineers, rather than, retrospectively and from a distance, hoping to take control of it.

CHAPTER 4

Whewell and Architecture

EDWARD GILLIN

William Whewell's name is inseparable from Trinity College. As an undergraduate from 1812, a fellow from 1817, and master from 1841, he lived amid the medieval splendor of King's College Chapel and Cambridge's collegiate architecture. Following a fatal fall from his horse, Whewell died in the city and was buried in Trinity's chapel: from his tomb to his college's Whewell's Court, he still remains part of Cambridge's built environment. As the university's center of undergraduate mathematics, Trinity was an important intellectual context for his career, but equally important was the architectural setting it provided. From within these largely Gothic surroundings, Whewell engaged critically with architectural history, determined to establish the discipline as a model of observational science, comparable to botany or geology. Throughout the 1820s and 1830s he developed a body of terms to describe the various elements of medieval ecclesiastic buildings and categorized Gothic structures according to their composite parts into clearly defined historical periods, giving a chronology of architectural development. By analyzing medieval construction techniques and decorative elements, he traced how the architects and builders of the Middle Ages cultivated new skills and knowledge for suspending weight and managing pressure. In what he identified as an age devoid of scientific progress, Whewell sought to account for the construction of Europe's increasingly elevated and lofty churches and cathedrals, explaining the principles behind such feats through the mechanical sciences. As with the history of the natural and experimental sciences, he portrayed the history of architecture as a story of progress toward clear laws and established scientific rules through epochs, discoveries, and the fundamental idea of upward growth.

The Carpenter's Apprentice

Architecture was in Whewell's blood. As the son of a master carpenter, he had been expected to follow his father, John, and become an apprentice carpenter, until the local schoolmaster persuaded Whewell's parents to send him to Lancaster Royal Grammar School.[1] However, Whewell was to remain interested in construction throughout his subsequent career, regularly undertaking architectural tours. In 1823 he explored the buildings of Normandy and Picardy before spending several months in Germany, rigorously recording details over style and construction (figure 4.1). Cumberland followed in 1824, before Germany in 1825, and the Netherlands and Germany again in 1829. Wherever Whewell traveled, he carried a copy of Thomas Rickman's *An Attempt to Discriminate the Styles of English Architecture* (1817), which proved exceptionally influential to Whewell's early architectural writings. A Quaker antiquary who later designed a series of Gothic buildings, including the New Court of St. John's College, Cambridge, Rickman's treatise systematically categorized English Gothic architecture, providing an early scholarly model for architectural history. In deconstructing the various elements of medieval churches, in terms of buttresses, ribs, pinnacles, vaults, and window tracery, he provided a chronology of stylistic development, dividing ecclesiastical buildings into four periods: "Norman," "Early English," "Decorated English," and "Perpendicular English."[2] Whewell read Rickman's treatise in 1818 and, hugely impressed, became friends with its author. In 1830, a year after he lectured on "Pointed" architecture to the Cambridge Philosophical Society, Whewell toured Devon and Cornwall with Rickman, before again traveling to Picardy and Normandy in 1832. Whewell later confessed that, before Rickman's work, architects had not understood the Gothic "and did not discriminate one age from another, nor the style of churches from that of houses. Mr Rickman had not yet enlightened us and given us eyes."[3] By 1834 Whewell still referred to Rickman's book as his "usual travelling companion."[4]

In 1830 Whewell published his own architectural treatise that sought to refine Rickman's analysis and extend it to Continental Europe. Whewell's *Architectural Notes on German Churches, with Remarks on the Origin of Gothic Architecture* (1830) represented what he hoped would establish architectural history as a model observational science, based on clear general laws. Written from the notes of his German travels, Whewell's treatise outlined "the progress of ecclesiastical architecture in Germany," setting out a "theory and system" of how the region's churches had developed over time.[5] As in his study of nature, Whewell's Anglican theology informed both the topic and the manner of his study: believing the medieval churches of the Rhine to be the most developed examples of German Gothic, Whewell attributed par-

Figure 4.1: Drawing of the Basilique Royale de Saint-Denis, by William Whewell. *Source*: R.6.12, Sketchbook No. 6 D.1823, TCL. Reproduced with kind permission from the master and fellows of Trinity College, Cambridge.

ticular moral value to these constructions as expressions of Christian faith. Likewise, his intention to determine the laws of medieval ecclesiastical architecture mirrored his endeavors to reveal the laws of nature as evidence of a Creator.[6]

Drawing on Rickman's terminology, Whewell explained that English medieval architecture could be divided into three distinct ages, these being "Early English," "Decorated," and "Perpendicular." Early English Gothic dated from the late twelfth to the late thirteenth centuries, having evolved from the "Romanesque" style of the Norman between 1145 and 1190, as French builders introduced increasingly delicate and refined elements. Also known as "First Pointed," this early style was characterized by stronger walls, vaulted roofs, arched ribs, rib vaulting, and the use of buttresses to reinforce walls against horizontally and downward-acting pressure from the weight of a building's roof. Eventually, these evolved into flying buttresses, conveying the thrust of a nave's roof, pressing on a church's walls, outward over the roof of the aisles. Stone pinnacles added weight to these wall-supporting buttresses. Over time, Whewell observed, the Early English became increasingly unified, resulting in the first great stone cathedrals, such as Lincoln, Salisbury, the east end of Canterbury, and the great transept of York Minster. Following this period, from the late thirteenth to the late fourteenth centuries, came what Whewell and Rickman termed Decorated Gothic, marked by an expanded ornamentation of most architectural elements. As this style grew aesthetically richer, taking influence from the French Rayonnant period, buttresses became more commonplace, as did sculpture and fan vaulting, with the first example of this beautiful element employed in the cloisters of Gloucester Cathedral in 1373. Lavish Lierne vaulting, also known as "stellar" vaulting, bound roofs together, while elaborate tracery combined with intensely stained glass to produce visually spectacular windows. Leading examples from this period included the choirs of Exeter and Ely cathedrals, as well as York Minster's nave (figure 4.2). Finally, Whewell dated the Perpendicular Gothic to the late thirteenth through the mid-sixteenth centuries. Here medieval builders emphasized the vertical lines of their churches and cathedrals with higher walls and larger windows, and reduced the structure around these spaces to create thin supporting piers. Tracery grew more geometric, ornament more lavish, and towers like that at York Minster were preferred to spires. King's College Chapel at Cambridge provided Whewell with a familiar example of this style. In general, English architecture appeared to lag behind that of France and Germany, Whewell speculated, probably because its builders had not wanted to abandon "the beautiful simplicity and sobriety of the Early English, even for the rich and elegant complexity of the succeeding style."[7] Following the Perpendicular, Whewell claimed that the Gothic had declined, with France transitioning

Figure 4.2: Sketch of a clerestory window in the choir of Ely Cathedral, by William Whewell. *Source*: R.6.12, Sketchbook No. 3 A.1822, TCL. Reproduced with kind permission from the master and fellows of Trinity College, Cambridge.

to the Burgundian, England going Tudor, and the Netherlands adopting "the *Belgian* style" of the "magnificent town-houses" of Ghent, Louvain, and Bruges.[8]

Unlike Rickman's work, which remained largely descriptive, Whewell's *Architectural Notes* explained historical change in architecture, tracing the development of ecclesiastic Gothic. He wanted to account for why medieval buildings looked the way that they did and, through the methodical collection of facts, provide principles that would allow an observer to know the age of a medieval church simply from its style and construction features; this was about changing how students of architecture looked at buildings and was, effectively, a visual epistemology of ecclesiastical structures.[9] From this empirical study of stylistic elements, Whewell sought to find "consistency and uniformity in the several buildings of the same epoch" so that scholars would agree on a church's age and chronological development through visual observation alone.[10] For Germany Whewell identified similar trends to those Rickman had outlined respecting English churches, but with a slightly different chronology of the Romanesque transition into "Early German." The Decorated appeared to have developed in Germany soon after the start of the construction of Cologne in 1248. The development of geometrical

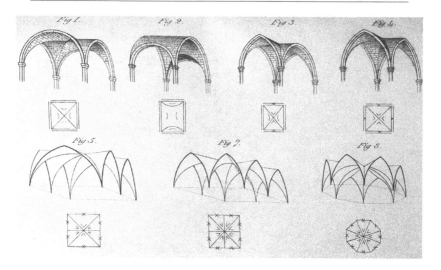

Figure 4.3: Image of the development of increasingly pointed and elevated arches, portrayed in William Whewell's *Architectural Notes on German Churches*, plate 1. *Source*: Photo by the author, 2021.

► **Figure 4.4**: Study of arches inside Notre-Dame de Bayeux, by William Whewell. *Source*: R.6.12, Sketchbook No. 6 D.1823, TCL. Reproduced with kind permission from the master and fellows of Trinity College, Cambridge.

tracery, as exhibited in Cologne, along with the introduction of buttresses, which brought order to the disorder of the early German style, inaugurated what Whewell defined as the "Complete Gothic." There was, Whewell contended, moral value to these high structures, pointing upward to heaven, rather than the low temples "of heathen antiquity." Buttresses, pinnacles, and woven tracery suspended the weight of the vast arches and roofs of churches and cathedrals, these being "the strong arms which the Christian architects learnt to make powerful and obedient."[11]

Whewell's account of architectural development was marked by "epochs," "preludes," and "sequels" to discovery. While the fundamental principle of Grecian architecture had been the sustaining of horizontal masses on vertical props, the dominating idea within the Gothic was the pointed arch in which the weight of a building's roof was carried down through its supporting walls and directed outward toward their lower portions. Whewell explained how this was the manner in which the old system of architecture, derived from the Classical styles, was finally converted into one of a different and opposite kind. According to this view "all the other changes which are found in company with the newly-adopted pointed arch, may be considered as the natural manifestations of the new character impressed upon art. The features and details of the later architecture were brought out more and

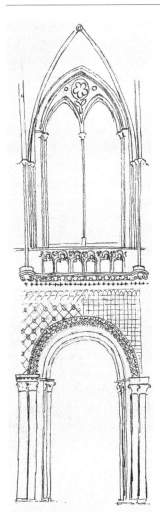

more completely, in proportion as the *idea*, or internal principle of unity and harmony in the newer works, became clear and single, like that which had pervaded the buildings of antiquity: the characteristic forms of the one being horizontal, reposing, definite; of the other vertical, aspiring, indefinite."[12] As the fundamental principle of the Gothic style, the pointed arch completed the transition from the predominating horizontal lines of Grecian and Roman buildings, still prominent in the Romanesque, to the upward projection of medieval ecclesiastical structures (figures 4.3 and 4.4).

Over time the Gothic had emerged into an increasingly coherent system. Whewell explained how the post-Romanesque development of Gothic architecture represented an expansion of "unity and harmony," with new principles of predominating vertical lines through shafts and buttresses "applied with as extensive a command of science and skill, as great a power of overcoming the difficulties and effecting the ends of the art, as had ever been attained by Greek or Roman artists."[13] All of this was made possible thanks to the development of the pointed arch. Whewell claimed the cause of this element was organic, originally deriving from single vaults running longitudinally down the sides of a church, which then progressed to a series of vaulted spaces running perpendicular to the building's length. The pointed arch constituted what Whewell called an "architectural revolution," allowing greater expanses of space, with the semicircular Romanesque arch abandoned. Buttresses contributed to this idea of upward growth, with the introduction of the flying buttress marking what Whewell defined as "the complete Gothic style."[14] As Whewell put it, the "inventor of this exemplification of architectural and mechanical skill must be considered as having done, for the advancement of Gothic architecture, far more than the inventor of the pointed arch: or rather as having given the means of executing in their full extent, those wonderful works of which the pointed arch contained the first imperfect rudiment and suggestion."[15] The builders and masons of this period "delighted in lifting to immense height in the air the most gigantic and magnificent clerestories, enclosed by enormous areas of transparent wall. In this way are constructed and suspended the magical structures of Amiens, Strasburg, Freyberg, Cologne, and in the neighbourhood of the latter, the exquisite abbey-church of Attenberg."[16] Through the idea of upward growth, the pointed arch and flying buttress represented epochs in the progress of ecclesiastical architecture.

As much as Whewell was taking Rickman to a new level of philosophical analysis, this remained a shared intellectual project, with the two scholars remaining in communication throughout the 1830s. When, for instance, Whewell attended the Edinburgh meeting of the British Association for the Advancement of Science in 1834, he took the opportunity to examine Scottish architecture. Keen to share his "Scotch Architectural Observations" with Rickman, he described how Scotland had "clear well-formed examples of Norman, of Early English, and of Decorated, exhibiting very slight differences from the English examples of the same style. For instance, Iona, Dryburgh and Kelso are good *light* Norman; Holyrood, Iona, Glasgow, part of Elgin, Dumblane, and Dunkeld are Early English." Effectively, Whewell's Scottish observations represented the testing of the architectural laws he had established for English and German Gothic. It seemed to him that the main divergence from England's churches was the more extensive use of geometrical tracery. As in France and Germany, the formation of "a complete Gothic style is almost always accompanied by geometrical tracery."[17] Together, Rickman and Whewell were mutually engaged in the cataloging and systematizing of Gothic architecture throughout Europe, from Scotland and England to France and Germany. Architectural tours and the collection of observations were, effectively, empirical studies for validating and refining the general laws of medieval ecclesiastical buildings.

The Science of Architecture

Whewell published his treatise on German churches at an auspicious moment for ecclesiastical architecture. Recognizing that there was a chronic shortage of church accommodation within the urban industrial centers that had expanded so rapidly since the late eighteenth century, Parliament passed the Church Building Act in 1818. Pledging an initial £1million for new churches, this initiated an unprecedented program of church building across the British Isles; by the early 1850s the State had funded 612 new places of worship at a cost of over £6 million. For economy, each church's budget was capped at £20,000, which meant that Neoclassical structures were often impractical, with the cost of stone for a temple's colonnades leaving little cash for the rest of the building. The preferred style for post-1818 churches was, therefore, Gothic, employing cheap brick and minimal ornament. However, with little Gothic ecclesiastical architecture built since the sixteenth century, few architects possessed the experience to design aesthetically pleasing medieval structures.[18] Within this context the fashioning of new architectural knowledge took on considerable urgency, with the historical study of medieval ecclesiastical Gothic crucial to recapturing lost building techniques and mechanical principles. Whewell's and Rickman's interventions were, therefore, timely. From 1835, however, it was Robert Willis who would provide new direction to this refashioning of architectural knowledge on a more empirical foundation. After studying at Gonville and Cauis College between 1822 and 1827, Willis became Cambridge's Jacksonian Professor of Natural and Experimental Philosophy in 1837, developing the science of "mechanism."[19] His *Principles of Mechanism* (1841) was intended to be of value to both students and engineers, delivering a study of movement in machines. Willis was equally concerned with questions of construction and cultivating a new discipline of architectural analysis. As Whewell had brought historical explanation to Rickman's classification, so Willis brought an emphasis on the mechanical techniques of Gothic construction to the study of architectural history.[20]

After honeymooning in France, Italy, and Germany between 1832 and 1833, Willis published his observations from this tour in 1835. *Remarks on the Architecture of the Middle Ages, Especially of Italy* built on Whewell's German study by distinguishing between "decorative," or the apparent, and "mechanical," or actual, forms of construction to deduce "the system of Middle Age Architecture."[21] Treating buildings as puzzles of these decorative and mechanical elements, Willis analyzed the constituent parts of Gothic structures in a similar manner to his examination of the various parts of a machine. He asserted that any building would be unsatisfactory to the eye unless "the weights appear to be duly supported," so a complete style, be it Classical or Gothic, had to have a decorative element representing the construction, even if this "apparent frame" was different to "the real one." To analyze architecture, therefore involved differentiating between "how the weights are really supported, and how they seem to be supported."[22] In the Classical structures of Rome and Greece, all horizontal weight was sustained through perpendicular props, this being the mechanical construction. However, in the Gothic the arch and vault complicated this management of pressure. Gothic builders had, Willis explained, "wisely adapted their decoration to the exact direction of the resisting forces required by the vaulting structure."[23] The complete Gothic style harmonized all these ornamental and mechanical elements within a harmonizing structure, with moldings appearing to sustain weight and "proportioning props" giving cathedrals a feeling of "lightness."[24] Willis would go on to publish an examination of vaulting in 1842, exploring the geometrical principles through which medi-

Figure 4.5: The interior of Amiens Cathedral. *Source*: Photo by the author, 2019.

eval masons achieved these complex systems of weight suspension, before publishing a series of studies focusing on the buildings of various cities and towns over the next thirty years.

Whewell responded to Willis's *Remarks* with a new edition of his *Architectural Notes*, published later that same year. This revised text included a new preface written largely in reference to *Remarks*, adopting Willis's language and analysis. Whewell embraced Willis's interpretation of Gothic churches as systems of decorative and mechanical construction, with horizontal weight distributed through piers and pillars and diagonal pressure managed through buttresses and pinnacles, giving unity and combining "into a harmonious whole."[25] Through such examination, he believed that Willis had effectively confirmed architectural history's status as a science. Whewell concluded his revised architectural treatise with the notes of a tour of Picardy and Normandy that he had made with Rickman in 1832 with the aim of

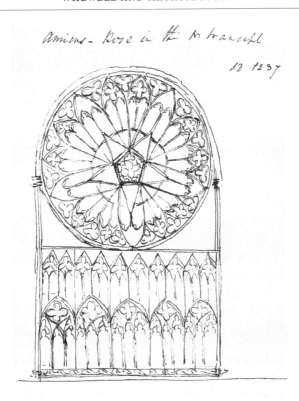

Figure 4.6: William Whewell, sketch of the rose window in Amiens Cathedral. *Source*: R.6.12, Sketchbook No. 4 B.1822–3, TCL. Reproduced with kind permission from the master and fellows of Trinity College, Cambridge.

extending principles drawn from English and German churches to French ecclesiastical buildings. It was an industrious trip, with the two scholars averaging five churches a day. On one occasion the expedition almost met with disaster when Whewell's and Rickman's intensive note-taking at Norrey church, between Caen and Bayeux, had panicked the local authorities into calling out the national guard and having the suspicious English tourists arrested. Both had been particularly impressed with the church of St. Wulfran at Abbeville, of which Whewell asserted "that the English never attained the art of making their Perpendicular fronts so rich and effective as this French one."[26] But it was the cathedrals of Amiens and Beauvais that Whewell and Rickman thought best displayed the French medieval builders' love of "inordinate height." The interior of Amiens was "one of the most magnificent spectacles that architectural skill can ever have produced," far more lofty than any English nave (figure 4.5). Especially distinctive was the height-to-breadth proportion, being almost double that typical in English churches, resulting in slim and graceful interiors that Whewell felt inspired "devotional feeling with the upward lines which the eye traces to their concourse apparently in another region." Whewell's treatment of Gothic architecture was rarely, if ever, emotional, but he undoubtedly ascribed a certain moral value to such buildings as testament to God's greatness and valuable for understanding Christianity. In terms of classification, Amiens appeared to be similar to Early English, but with an abundance of what Rickman termed "*geometrical*" tracery, consisting of circles, trefoils, and "figures which may easily be made by the compasses"[27] (figure 4.6).

On returning from France, Whewell reflected that "we have classified to our own content the styles of church-building which we found there, and I think Rickman will probably illuminate the world upon the subject as soon as he can find time to put his notions on paper."[28] But it was Whewell, rather than Rickman, who was to publish the notes of their French tour within his 1835 and 1842 editions of *Architectural Notes*. It was through his historical work on the development of scientific knowledge that architecture was to take on increased significance for the Cambridge polymath. Following his successful 1833 Bridgewater Treatise, Whewell devoted his efforts to his three-volume *History of the Inductive Sciences* (1837) and his two-volume *Philosophy of the Inductive Sciences* (1840). It was in the first of these extensive philosophical works that Whewell positioned his architectural work within the broader history of scientific progress, marked by moments of imagination and inductive observation leading to the discovery of truths. While the ancients' "clear and distinct general laws" allowed the "inductive sciences" to flourish, Whewell argued that the sciences had stagnated during the late Roman period.[29] Architecture offered a prime example of "this indistinctiveness of ideas," with Roman architects merely imitating Grecian colonnades and pediments. Here Whewell's analysis again mobilized Willis's terminology, as he explained how "all architecture, to possess genuine beauty, must be mechanically consistent. The decorative members must represent a structure which has in it a principle of support and stability. Thus a Grecian colonnade was a straight horizontal beam, resting on vertical props; and their pediment imitated a frame like a roof, where oppositely-inclined beams support each other. These forms of building were, therefore, proper models of art, because they implied supporting forces."[30] Yet in merely copying Classical principles, Whewell argued that late Roman buildings "were destitute of their mechanical truth, belonged to the decline of art; and showed that men had lost the idea of force, and retained only that of shape."[31] Whewell claimed that science continued this decline in the Middle Ages that followed. However, through "practical architecture" and "architectural treatises," he traced "the progress of scientific ideas," believing these to represent the "prelude to the period of discovery" of the sixteenth-century Renaissance.[32]

The formation of the Gothic style marked what Whewell believed to be an advancement of science that would be critical for the later development of philosophical knowledge. He explained how, from the twelfth century, "the idea of true mechanical relations in an edifice had been revived in men's minds.... The notion of support and stability in the decorative construction again became conspicuous and universal in the forms of building. The eye which, looking for beauty in definite and significant relations of parts, is never satisfied except the weights appear to be duly supported, was again gratified."[33] This subdivision of weight was pleasing to eye, bringing the arch and vault together as medieval masons cultivated and applied "the idea of mechanical pressure and support," and Whewell claimed it was the "possession of this idea, as a principle of art" that led "to its speculative development as the foundation of a science; and thus architecture prepared the way for mechanics."[34] The next three hundred years saw what Whewell believed to be the cultivation of mechanical science through the construction of Europe's great cathedrals. Along with Gothic buildings, architectural treatises marked the second area of medieval scientific progress. Whewell claimed that it was impossible to look on twelfth- to fifteenth-century French, German, and British architecture without recognizing that it was a closely connected "system" in which builders exchanged knowledge with each other. He was certain that this diffusion of Gothic elements across Europe was evidence of a sophisticated and interconnected network of masters and scholars in existence throughout the Middle Ages. Such a communication of knowledge was "proved beyond dispute by the great series of European cathedrals and churches, so nearly identical in their general arrangements, and in their particular details." The question then was whether this "system of instruction" had ever "been committed to writing."[35] There were, Whewell confirmed, no known treatises from "the Gothic masters," but in place of books, a firm verbal tradition of knowledge formation and exchange had been in place. He concluded that architecture therefore represented the one area of scientific progress to characterize the "Dark Ages": architects and engineers had precipitated a new era of discovery and scientific enlightenment. It is clear then, that by 1837, architectural history featured prominently within Whewell's broader understanding of the history of science.

The Master Carpenter

Together Whewell, Willis, and Rickman instigated a broad expansion in the study of architecture. Across Europe this cultivation of architectural knowledge through empirical analysis shaped a Gothic revival in church building, as well as for new domestic and public buildings. In France it was architect Eugène Viollet-le-Duc, responsible for a string of new churches and restoration projects including the rebuilding of Notre-Dame de Paris, who took up Whewell's and Willis's scientific treatment of ecclesiastical buildings. As Viollet-le-Duc drew on Willis's *Remarks*, translated into French in 1842, so Willis reproduced the French scholar's sketches of Notre-Dame de Reims for British audiences.[36] In Britain it was Augustus Pugin who did the most to extend this architectural analysis, having read both Whewell and Willis before developing his own account of the Gothic as a fundamentally moral art form. Pugin's *Contrasts* (1836), taught that architecture physically embodied a society's religious values, with medieval churches representing the three doctrines of Christianity. Their cross-plan layout symbolized Christ's sacrifice on the cross and man's redemption, while their vertical lines, lofty height, pointed arches, and spire all invoked the resurrection of the dead. Likewise, triangular arches, tracery, and tri-part divided naves embodied the Trinity of Father, Son, and Holy Spirit.[37] Catholic theology and architecture were, for Pugin, perfectly combined. The scientific treatment of ecclesiastical architecture that Whewell had done so much to cultivate, through Pugin's elaborations took on passionate religious value and found material shape in the churches of leading architects like George Gilbert Scott, William Butterfield, George Bodley, and George Edmund Street.

After Pugin, the art critic John Ruskin developed this architectural analysis even further, exerting a profound influence on architects and public taste between 1838 and the 1880s. Having read Willis's treatise in 1846, Ruskin, in his *Seven Lamps of Architecture* (1849) and *The Stones of Venice* (1851–1853), promoted his belief that buildings could be read. The twelfth- to fifteenth-century architecture of England, France, and Germany was especially important to Ruskin's understanding of the Gothic, focusing much research on the cathedrals of Lincoln, Wells, Salisbury, Amiens, Reims, Chartres, Bayeux, Rouen, and Bourges.[38] Of course this emotive approach to architecture, in which a building's aesthetic worth was defined by its supposed moral value, was contrary to Whewell's objective architectural analysis. While they enjoyed an amicable relationship, with Willis, Whewell, and Ruskin visiting Ely Cathedral together in 1851, Ruskin was unaware that Whewell had been the author of an anonymous review, published a year earlier, savaging his *Seven Lamps*.[39] Whewell accused the art critic of failing to establish clear architectural principles and a coherent analysis, producing what was instead a highly emotive, unscholarly work. Yet Ruskin's conviction that the Gothic was something that actively influenced congregations and endured well into the twentieth century. Written in the twilight of his career, Ruskin's *The Bible of Amiens* (1881) united the scholarly treatment of the Gothic within a religious framework.[40] In particular, this work caught the attention of Marcel Proust, who embraced the idea that churches communicated sentiments to those inhabiting them in his description of the church at Combray in *Du côté de chez Swann* (1913).

Proust had, in fact, translated Ruskin's *Bible of Amiens* into French during the early 1900s.[41]

From Pugin and Viollet-le-Duc to Ruskin and Proust, there was a shared consensus that the nature of ecclesiastical Gothic architecture and religious belief were fundamentally connected. And through the writings of natural philosophers like Willis and Whewell, a growing scholarship surrounded the construction of medieval churches and cathedrals. However, Whewell's architectural works were also influential with practicing architects. Having been made an honorary member of the Institute of British Architects at some point between 1835 and 1836, and then earning election to the Société Française pour la Conservation des Monuments in 1840, Whewell was particularly prominent within the formation of Cambridge's Camden Society, of which he was a member. Established in 1839 to return the Church of England and its churches to the medieval splendor of the Middle Ages, both theologically and aesthetically, the Camden Society founded its own scholarly journal in 1841, *The Ecclesiologist*, and became the Ecclesiological Society in 1845.[42] While echoing many of the principles of the Oxford Movement, especially that the Anglican Church drew its authority from the Apostolic Succession of the medieval Roman Catholic Church, the society paid considerable attention to the role of architecture within this return to Catholic ritual and aesthetic. Promoting a radically High Church approach to church building, it published recommendations for church architects, specifying the importance of a well-defined chancel raised at least two steps above the floor of the nave and kept for the clergy alone, the separation of the congregation from the altar by a rood screen, stone rather than brick, and the tripartite division of naves to symbolize the Holy Trinity. Whewell might have lacked the extreme Anglo-Catholic tendencies of some of his fellow members, but presumably endorsed this refashioning of church building as a science, based on established laws drawn from the study of architectural history.[43]

Throughout the 1840s and 1850s Whewell continued his architectural tours, industriously observing church buildings and working to produce a catalog of all the various elements of ecclesiastical construction according to country and historic period, although he never completed this project. Nevertheless, in 1863 he delivered his most complete account of architecture to the Royal Institute of British Architects (RIBA). Although recognizing that other scholars had advanced the study of architectural history beyond his own observations of the 1830s, Whewell proceeded to define architecture's artistic character, distinguishing between the imitative *"fine* arts," including sculpture, painting, and music, and the "constructive" arts. As Classical architecture imitated "in a structure of masonry the form of a constructive frame-work of stone," it was essentially a fine art. Yet architecture was also governed by "the idea . . . of a constructive frame-work," and therefore was also "constructive" in character. The development of a style represented the harmonizing of various elements, such as columns, masonry, vaulting, and ribs. Whewell explained how an interior like that of Salisbury Cathedral represented "to the mind indeed a construction, in which collected pillars support collected pier-arches, and vaulting shafts support vaults."[44] Architecture was, Whewell continued, most comparable to music. While a musical strain needed rhythm, divided into bars, or equal intervals of time, a colonnade in a temple required equal divisions of space by columns. This musical-architectural comparison was essentially an analogy of time and space. As the ear detected musical harmony, so the eye identified the architectural harmony. It was not, Whewell asserted, "too fanciful to say that the aisles of a cathedral may be compared to the parts of a psalm tune—the treble, alto, tenor and bass. The concord or discord of those parts in the tune, as the bars succeed each other, are a sort of *transverse* relation between *parallel* strains of melody—much as the agreements and disagreements of the piers, shafts, vaulting spaces, triforium arches and windows are a transverse relation in the building."[45] Whewell's RIBA audience concurred with his claims. Having listened throughout, George Gilbert Scott praised Whewell's combining of architecture with "a perfect knowledge of philosophy."[46]

Whewell's architectural contribution was not, however, purely philosophical. Having evaded life as a carpenter, he nonetheless engaged in the design and restoration of several buildings in his position as master of Trinity College. Along with the education of undergraduates, his duties entailed the management of the college's extensive medieval and early modern buildings, which included the Great Court and Trinity's college chapel. On his appointment in 1841, Whewell already harbored architectural ambitions, accepting £300 from Alexander Beresford Hope to recover the "antique character to the old court" and renovate the Master's Lodge by restoring its oriel and mullioned windows as they were before Master Richard Bentley's eighteenth-century alterations.[47] A graduate of Trinity, Hope combined his career as a reactionary Conservative MP with a penchant for architecture. Phenomenally wealthy, Hope was a founding member of the Camden Society and supervised the construction of Butterfield's All Saints Church on Margaret Street in London, which became a celebrated beacon of High Church Gothic architecture. With Hope's funding in hand, Whewell invited the Gothic specialist Anthony Salvin to the Master's Lodge in January 1842. Having trained under John Nash during the 1820s, Salvin was known for his mastery of the Tudor and Gothic styles thanks to several castle restoration projects, and in November 1841 he was made an honorary member of the Camden Society.[48] Whewell and Salvin subsequently examined the Master's Lodge, attempting "to discover traces

Figure 4.7: The façade of the Master's Lodge at Trinity College, Cambridge. *Source*: Photo by the author, 2021.

of the oriel which formerly existed as part of the front of the Lodge. We found the foundation of the wall of the oriel immediately below the surface of the ground. The plan was semicircular, the diameter of the semicircle 13 feet and 7 inches, exactly opposite to the oriel which exists in towards the garden."[49] Throughout August and September Hope and Whewell directed workmen in restoring the lodge's façade, including its windows and the oriel, before Hope donated an additional £1,000 to further enhance the dignity of the building (figure 4.7).

In 1843 Whewell corresponded with Salvin on remodeling and rebuilding the chimneys and dormers of the Great Court, again thanks to Hope's financing.[50] Two years later Whewell paid for the construction of a new wooden ceiling, which Willis designed, for Trinity's Great Gate. But his most significant architectural projects would come in his later collaboration with Salvin during the 1850s and 1860s. Though not built until the 1870s and to William Milner Fawcett's designs, Salvin's initial selection to design a new physical laboratory for the university confirmed his growing reputation in Cambridge, and during the 1850s he continued his collaboration with Whewell to expand Trinity College. Whewell wanted to establish a chair of international law, believing that the development of the science of mo-

Figure 4.8: Salvin's Gothic Whewell's Court on Trinity Street, Cambridge. *Source*: Photo by the author, 2021.

rality depended on an increased recognition of international obligation. He therefore proposed the construction of a new court of student accommodation, opposite the Great Gate on Trinity Street, with the rents financing the new chair. At his own expense Whewell purchased the freehold of the Sun Inn, which occupied the location, for £7,000. In 1850 he proposed to gift this to the college, but Trinity's senior fellows refused the offer. Undeterred, Whewell commissioned Salvin to design a new Gothic accommodation that was built between 1859 and 1860 and eventually became Whewell's Court (figure 4.8). Including twenty-five sets of rooms and costing £10,000, this private property soon became part of the college. In 1865 Whewell purchased adjacent land and charged Salvin with adding a second new court, completed by 1868. Whewell spared no expense, with his total architectural endowment to the college valued at around £100,000 when Trinity took full possession of these assets. In their Gothic style these constructions followed Whewell's conviction that "whenever new buildings connected with the College were erected, that their architectural character should improve the appearance of the town."[51]

Whewell's architectural enterprise was not limited to his college. In 1855 his wife, Cordelia, died and was buried in nearby Mill Road Cemetery. In 1851 George Gilbert Scott had submitted plans for a new chapel for this cemetery that had initially been deemed too expensive. Whewell now donated a substantial sum for the erection of a church complete with a tower and spire, followed by a later gift for financing its internal fittings. Constructed between 1856 and 1858, Mill Road Church represented a collaboration between Scott and Whewell. Reviewing Cambridge's latest Gothic edifice on its completion, *The Ecclesiologist* praised Whewell's recommended "experiment" of a spire and tower, as well as his innovation in avoiding heavy external buttresses.[52] Both through his writings and building projects, Whewell contributed directly to Cambridge's built environment.

"Ruskin was not an architect and he never designed a building."[53] With this statement historian Geoffrey Tyack opened his study of John Ruskin's profound influence on nineteenth-century architecture. Though less renowned for his artistic insights, Whewell could boast a considerable architectural influence as well and, unlike Ruskin, could actually claim to have orchestrated the construction of several buildings. Originally destined for a career as a master carpenter, Whewell had instead become a college master with considerable architectural responsibility and patronage. With Rickman and Willis he transformed medieval Cambridge into a center of Gothic investigation. While Whewell's university career and direct contribution to the city's built environment were significant, to understand his place in architectural history we have to see his work in its national and international contexts. He had founded an empirical approach to architectural study that produced the knowledge to inform the construction of new Gothic buildings. He remained prominent within a growing network of architects and scholars who looked to fashion understandings of medieval construction that would result in increasingly authentic historical edifices. Whewell was in this way part of a broad cultivation of architectural knowledge that shaped a wave of church building and architectural restoration across the British Isles, France, and rest of Europe, as well as new secular works, including railway stations, domestic homes, town halls, commercial properties, and parliaments. This "Gothic Revival" was, in fact, a very global project. Throughout Europe's vast colonial empires, the builders of new religious and government buildings took inspiration from these recent empirical understandings of medieval architecture. In Britain's overseas territories, architects designed and built in this ecclesiastical fashion, from Bombay, where Scott's university buildings and the fantastical Victoria Railway Terminus were central to the Gothic rebuilding of the city, to medieval parliament buildings in Barbados and New Zealand, to great cathedrals and churches throughout

Africa, India, Australia, China, Singapore, Hong Kong, the West Indies, and the Pacific. Whewell's efforts in the 1830s and 1840s constituted an early initiative within an architectural tradition that emphasized historical authenticity and changed the built environment around the world. While Ruskin and Pugin allowed religious feeling and emotion to shape their writings, Whewell's self-avowedly scientific approach was predicated on a belief that architectural history had to be disciplined by a defined language of terms, established chronology, and systematic cataloging of constructive and decorative elements. But in fashioning this architectural knowledge as an observation-based science, Whewell came to see ecclesiastical architecture as a crucial part within the broader history and philosophy of science. In Whewell's grand vision of scientific progress through time, architecture had provided a rare area of philosophical advance during the Middle Ages, presenting a prelude to the Renaissance and stimulating new understandings within the mechanical sciences.

Perhaps most intriguingly, however, is the architectural work that Whewell never completed. Buried deep in his archives, now held in the Wren Library at Trinity College, Cambridge, is an unfinished volume cataloging various architectural elements from around Europe. Various types of Gothic vaulting, pinnacles, column heads, ribs, window tracery, arches, and vaulting from specific buildings are all presented, showing the diversity of building styles employed within medieval churches and cathedrals. This manuscript is undated, but probably drawn up some time in the 1840s, after over two decades of architectural touring around Continental Europe and the British Isles. While it was only partially completed, nevertheless it embodies Whewell's architectural ambitions: to divide Gothic ecclesiastical buildings, according to their constituent parts, into a catalog, akin to the recording of botanical or geological samples.[54] It was an ambition never fully realized, but it represents the degree to which Whewell believed the study of architecture could be reduced to empirical observations and lawlike principles.

CHAPTER 5

Whewell and Political Economy

HARRO MAAS

William Whewell published several essays on political economy that are considered as pioneering forays into mathematical economics,[1] and a series of lectures to the Prince of Wales (1862) that only rehearsed received wisdom.[2] Yet his engagement with political economy was more profound than that and provides us with a perfect example of how problems of knowledge are entwined with problems of social order.[3] It is impossible to understand the importance of political economy for Whewell without considering his lifelong correspondence with his friend Richard Jones. Whewell was the driving force behind Jones's ambitious project to develop a scientific alternative to the political economy of David Ricardo that would explain the true principles of the production and distribution of rent, wages, and profit. After Jones's death Whewell secured the publication of his literary remains, for which he wrote a preface that articulated one of the strongest contemporary attacks on the method of Ricardian economics.[4]

Whewell considered that Ricardian economics departed from arbitrary definitions from which deductions were drawn with pernicious social and political consequences. His mathematical forays into economics served to prove the scientific inadequacy of Ricardian deductivism.[5] In contrast, a scientific approach to political economy should follow the inductive way from facts to theory, showing a more cheering message. Even though Whewell and Jones ran joint courses on political economy for most of the 1820s and 1830s, a full agreement of their views on induction never existed. Whewell, it is well known, moved in the 1830s from a more or less naive Baconian position to what Menachem Fisch labeled, for lack of a better term, a "post-Kantian" one that found its expression in the *Philosophy of the Inductive*

Sciences (1840) and for which Charles Sander Peirce would invent the term *abduction*.[6] Unsurprisingly, traces of this method existed before 1830, and they can be gathered from discussions on the subject arising from the correspondence between Whewell and Jones.

Cambridge resistance to Ricardo was aggravated only when Whewell and Jones found out that the Oriel Noetics, most notably Richard Whately and Nassau Senior, fundamentally agreed with Ricardian deductivism. The ensuing fight was an important impetus for Whewell and Jones to organize institutional support for their inductive approach to political economy, in the form of Section F of the British Association for the Advancement of Science (BAAS, f. 1833) and, somewhat later, the Statistical Society of London (f. 1834).

Jones became a member of the Tithe Commission in 1835 and no longer found the time to work on his project, and the papers presented at Section F (and later the Statistical Society of London) proved seriously disappointing in quality. As a result Whewell's interest in the subject quickly faded away. John Stuart Mill's essay on the definition and proper method of political economy, published in the *Westminster Review* in 1836,[7] seemed to settle the distinction between the science of political economy and its practical applications. In addition, his *Principles of Political Economy* (1848) gave just sufficient credit to the Cambridge men to iron out the differences between Whewell's and Jones's inductive approach.

Cambridge Opposition to Ricardo

Whewell and Jones had started their discussions on political economy during their student days at Cambridge. Thomas Robert Malthus's essay on population, published anonymously in 1798, caught the attention of the group of reading men splendidly brought to life in Laura Snyder's *Philosophical Breakfast Club* (2011).[8] Malthus's argument that population pressure would inevitably lead to social inequality and pauperism for the greatest part of mankind had been received with a sense of shock and horror by the British establishment that initially lumped Malthus's ideas together with the philosophical radicalism of Jeremy Bentham, James Mill, and David Ricardo—who were all openly hostile to the landed interests and revealed religion. In the preface to his book on rent (1831), Jones argued that Malthus's population principle had thrown a "gloom over the whole subject" by enforcing "an apparent inconsistency between the permanence of human happiness, and the natural action of the laws established by Providence." If nature was so constituted that its laws kept large parts of humanity in a state of inescapable misery, this implied a direct challenge to God's wisdom and goodness.[9] But Jones especially blamed the philosophical radicals for extending Malthus's "dismal system" to an inevitable conflict within the different layers of society itself.[10] The Ricardian theory of rent showed "essential contradictions and differences" to exist between the "fortunes and varying relative position of the different orders of society." A "feeling of dislike" had consequently crept into the public mind that had caused political economy, as a subject, to become "distrusted."[11] In contrast Whewell and Jones set out to show there was nothing in the nature of political economy that pushed humankind permanently into a state of want and conflict.

Shortly after the publication of Ricardo's *On the Principles of Political Economy and Taxation* (1817), Jones embarked on a project that aimed to show the true principles governing the production and distribution of wealth. Whewell was so convinced of the value of Jones's work that he almost literally bullied him into writing and publishing his "Euridice of Political Economy" to counter the theories of the "rotten, pseudo-political-economists" who were "driving tandem with one jack-ass before the other."[12] Instead of theorizing from first principles, Jones would follow the sustained path of induction used in the other sciences. As we will see, just what such an inductive path looked like would become a source of contention between the two men.

To support Jones's project, Whewell sent him as many tracts and other printed materials as he could lay his hands on in Cambridge and on his travels to the Continent, materials that were not available to a country vicar in Sussex. But time and again Jones found himself indisposed to work and write, either because of his hypochondriac state of mind and health or because of any other inhibition Jones found in his way. Whewell also had to maneuver carefully so as not to trespass into Jones's domain of expertise. For instance, he left unpublished five sermons prepared in February 1827, and did not even deliver the fifth that explicitly dealt with the relation of political economy to natural theology.[13] In that fifth sermon Whewell drew together a discussion of Ricardo's theory of rent and Malthus's population principle, warning that both were based on premature deductions that led to an underestimation of the bounty of nature, and an overestimation of the conflict between social classes. The speculations of these economists thus undermined Christian faith by preventing the general public from seeing "God's government" in economic matters.[14]

Both men harshly criticized Ricardo's theory of rent, especially for its disruptive social consequences. Ricardo distinguished between rent as the payment to the landlord "for the use of the original and indestructible powers of the soil" and profit as a revenue on capital, introducing a definition that differed from its use in "popular language" in which rent referred to "whatever is annually paid by the farmer to his landlord."[15] From Ricardo's definition it followed irresistibly that the landed interests were opposed to those of the capitalists (the farmers) whose margin of profit was squeezed between the rents of the landlords and the subsistence wages of the day laborers. Ricardo's approving citation of David Buchanan that "the landlord gains . . . at the expense of the community at large" captured Whewell and Jones's worries.[16] To John Ramsay McCulloch's observation that Ricardo did not use the term *rent* in the "ordinary and vulgar sense of the word," Whewell replied that he left it to the reader "to decide for himself which subject of inquiry is better worth his notice,—the rents that are actually paid in *every* country, or the Ricardian rents, which are *not* those actually paid in *any* country."[17]

The only part of Jones's project published during his lifetime was *An Essay on the Distribution of Wealth, and on the Source of Taxation* (1831). When Whewell wrote to Jones about his intention to publish a paper read to the Cambridge Philosophical Society in which he algebraically showed the deficiencies of the Ricardian theory of rent, Jones warned him to clarify that his mathematical deductions proceeded on *their* assumptions and definitions and that he at no point subscribed to the enemies' views. For Jones, no mechanisms or laws could be imposed on society that did not emerge from a careful study of historical facts. Whewell fundamentally agreed, because one could only see the providential order in nature and society from the gradual discovery of its principles, not from their stipulation.

By comparing different social arrangements, Jones showed that rent did not universally result from Ricardo's principle of diminishing returns on marginal lands but depended on the specific mode of production in place. Jones questioned the relevance of Ricardo's theory of rent even for those nations that could be classified as having the (Ricardian) farmer's rent system (i.e., England and the Netherlands). Ricardo's conclusions were valid, for example, only under the artificial assumptions that farmers did not, or were not able to, improve the quality of the soil, or to improve yields by improving on farming techniques—assumptions contradicted by the available evidence. Jones's conclusion was that no society, neither now nor at any point in history, could be identified where the landed interests were opposed to those of the other classes in society. His more "cheering, because more comprehensive" conclusion was that "the same providence which has knit together the affections and sympathies of mankind . . . has, in perfect consistency with its own purposes, so arranged the economic laws which determine the social condition of the various classes of communities of men, as to make the permanent and progressive prosperity of each, essentially dependent on the common advance of all."[18]

The Oriel Noetics on Political Economy

By the time of the publication of Jones's book in February 1831, Whewell and Jones had made an uneasy peace with Malthus. Malthus, a Cambridge man himself, agreed with them that the proper method of political economy was the inductive, and that the definitions and generalizations of political economy should follow from and not precede observation. For the subsequent editions of his essay on population, Malthus himself had collected a vast array of observations in support of his population law, and in correspondence with Ricardo had defended the inductive method against Ricardo's deductivism.[19] Like Whewell and Jones, Malthus considered political economy a deeply moral subject that should not limit the vision of the political economist to man's selfish motives. But he did not want to dismiss Ricardo's theory out of hand, as he felt the Cambridge men were doing, even when he agreed that Jones's book convincingly showed its limited applicability.[20]

One can imagine Whewell's and Jones's horror when they found out that the Oxford economists Richard Whately and Nassau Senior—both religious men—fundamentally agreed with the Ricardians on questions of method. Whately was archbishop of Dublin and Senior was Whately's predecessor as Drummond Professor of Political Economy at Oxford and chair of the Royal Commission for the Revision of the Poor Laws. With Edward Copleston, provost of Oriel College, they formed the so-called Oriel Noetics (after the Greek *noèticos*—roughly, "intellectual power"), a group of speculative thinkers that used the Oriel Common Room to question "nearly every conceivable practice of church and state" so that the place, as Ralph Pomeroy quotes a contemporary observer, "*stunk* of logic."[21] Neither Whewell nor Jones scorned debate, but they considered the Noetics' speculative approach to political economy as just "mad."[22] The ensuing fight with the Oxford men over the appropriate method of political economy, and the proper relation between theory and observation, would incite Jones and Whewell to create an institutional home for their inductive approach.[23]

Jones discovered the Ricardian tenets of the Oxford Noetics while his own book was in press. In his appendix to Whately's *Elements of Logic* (1826), Senior explained that the proper "foundations of political economy [are] a few general propositions—deduced from observation or from consciousness & generally admitted as soon as stated."[24] Enraged, Jones asked Cambridge to stop the printing press and insert an additional page to state that his own book was *not* based on abstract definitions without grounding in historical observation. "On a subject like this, to attempt to draw conclusions from definitions, is almost a sure step towards error. A dissertation, however, on the use and abuse of definitions would be out of place here.... If any reader ... is really puzzled to know what we are observing together [in the book], I shall be sorry: but I am quite sure that I should do him no real service by presenting him in the outset with a definition to reason from."[25] For Jones, the Oxford men were exemplars of the "anticipators" whom Francis Bacon had opposed as the "father of errors & destroyer of the sciences."[26] But Jones also realized that it was much easier to accuse Ricardo—a Jew who converted to Unitarianism—of a faulty method and a blind spot for God's hand in nature and society than the Oxford men, with their strong ties to the establishment and the Anglican Church. To counter the Noetics, Jones realized, Whewell and he had "a hard battle to fight."[27]

Almost from the day of publication of Ricardo's *Principles*, Whewell and Jones could have foreseen Oxford divergence from their views. In a public letter of 1819 to Robert Peel on the Poor Laws, Copleston had referred approvingly to Ricardo, who had "well established . . . the fact . . . that a rise in wages is a diminution not of rent, but of profits." Copleston continued by claiming that "whatever is theoretically true we may be sure is really acted upon by the interested party, however unconscious he may be of the abstract principle, and even unable to comprehend it."[28] In his inaugural lecture for the Drummond professorship, delivered in December 1826, Senior voiced the same predilection for theory over facts, which he based on the common sense of mankind. The pure theory of political economy articulated propositions on the basis of a few axioms and definitions that were considered true as soon as they were stated. The theory was to be distinguished from its practical applications, or the "art" of political economy, which depended on a manifold of factors extraneous to political economy and involved the judgment of practical men.[29]

The Noetics' recourse to "common sense" as the basis for the truth of the pure theory of economics followed from their acceptance, with a twist, of Dugald Stewart's distinction between the inductive processes in the physical and moral sciences.[30] Stewart had explained how political economists like Adam Smith had discovered their laws not by way of the natural philosopher—that is, by observation with the eye—but by way of the philosopher of mind: inner reflection. Thus Adam Smith "indulged in theory" by simply exposing the "common sense which guides mankind in their private concern." Stewart contrasted this procedure with the "collection of facts" by political arithmeticians such as William Petty and others who could "never afford any important information" because a heap of facts did not say anything about the "combination of circumstances whereby the effect is modified."[31]

Whately concurred. Never afraid to take on a fight, he added a ninth lecture to his own *Introductory Lectures* of 1831 to argue against the reliance of the Cambridge men on scientific inductions from historical facts. Quoting from Senior's introductory lecture, and in explicit opposition to Whewell and Jones, Whately affirmed that "it is in fact as impossible to avoid being a practical Economist, as to avoid being a practical Logician."[32] A "naive collector" of facts would just be "heaping . . . materials together . . . into some, probably faulty, system." He would thus easily overlook the common experiences available to everyone from which the true principles of the subject could be inferred. "Yet such is History!"[33] Instead, "the Logical, not the Physical investigation" should have priority in political economy.[34]

Whately specified that such an investigation made sense by probing common sense. There was no subject of study in which "paradoxical theory" was more important than "in matters concerning Political Economy." That someone could be "a benefactor to the community by building himself a splendid palace" was a *"paradoxical* truth" that might make *"sense,"* even though it was not *"common-sense."*[35] The purpose of political economy was to clarify such paradoxes by showing how private interests played out in the marketplace. This also motivated Whately's preferred name for the discipline, "catallactics," the science of exchange. Whately considered such clarification "as [a] sort of continuation of Paley's 'Natural Theology'"—the work that Whewell sought to remove from the Cambridge curriculum.[36] William Paley had shown "contrivance" to be central to natural theology. Similarly, the political economist was to show that there was not only "contrivance" in the "structure of organized bodies" of the natural world but that "an attentive study of the constitution of Society, would bring to light a no less admirable apparatus of divinely-wise contrivances, directed no less to beneficial ends."[37] The Oxford men thus shared Whewell's and Jones's belief in the providential order of society. But their providential order was not that of the Tory belief in the values of the existing social order; instead, it was a social order based on the utilitarian belief in the virtues of commerce, trade and industry.

Whately was convinced that man's rational ability to trade in a commercial environment made him "perhaps the most wonderfully contrived specimen of divine Wisdom."[38] Referring to Bacon, he argued that political economy did not "dispense with common-sense" but used it "more profitably," just like "the adoption of Arabic numerals and of the Algebraic symbols, does not supersede calculation, but extends its sphere."[39] Instead of conceiving of the political economist as inferring laws from a collection of historical facts, Whately saw him as an observer of what the political economist Alfred Marshall would later call the ordinary business of life. This position defined the political economist as a theorist. According to Whately "the looker-on often sees more of the game than the players. Now the looker-on is precisely (in Greek θεωρὸς) the *theorist.*"[40]

Whewell on Political Economy as an Inductive Science

Being called naive collectors of historical facts was of course more than the Cambridge men could bear. And nothing could be further from their position than to think of the observer as a theorist and to start a science with a set of definitions that defined man as "an animal that makes exchanges." When Whewell met Senior in the Athenaeum in London and confronted him with the fact that more than half of the world's wealth was not exchanged at all, Senior answered that in that case, as a political economist, he had nothing to do with it.

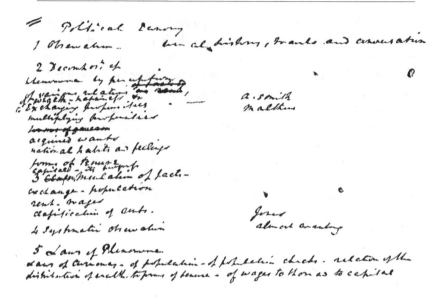

Figure 5.1: William Whewell's inductive scheme for political economy from his notebook on induction of the early 1830s. *Source*: R.18.17.15/41, TCL. Reproduced with kind permission from the master and fellows of Trinity College, Cambridge.

Whewell launched a "guarded attack" on the Oriel Noetics' school in a brief paper "On the Uses of Definitions," published in 1832 in the short-lived *Philological Museum*.[41] Drawing on Whately's distinction between geometry and the other sciences, Whewell argued that in geometry its definitions "*are* themselves the first principles of our reasonings" that were "independent of the external world." But in the empirical sciences, "exact definitions have been, not the causes, but the consequences of an advance in our knowledge" that "*followed* the settlement" of scientific disputes.[42] The same was true for political economy. "Voyages and travels, history and legislation, politics and statistics" were all needed to "arrive at general classifications." It would be only at that point that "definitions enable us to express our laws." Whewell continued that the "most celebrated teachers" of political economy (implicating Whately and Senior) ignored this clear answer to the question of how political economy should best be studied. This because they "scorn[ed] the prospect of collecting *their* principles by this slow and laborious process of observation and comparison. Their truths are to flow from the inexhaustible fountain of *definition without previous knowledge and classification of facts.*"[43]

Shortly after the publication of his 1831 book, Jones had encouraged Whewell to write a "good thing to the public" on induction. Such a text was to cover the moral as well as the natural sciences. He added a warning

Political economy

1. Observation		...history, tracts and conversation
2. Decomposition of phenomena by perception of various relations of wealth – happiness		
exchanging propensities	–	A. Smith
multiplying propensities		Malthus
acquired wants		
national habits as feelings		
forms of tenure		
capital – its progress		
3. Insulation of facts – exchange – population rent – wages		
classification of rents		Jones
4. Systematic observation		almost wanting
5. Laws of Phenomena		

Laws of currencies – of population – of population checks – relation of the distribution of wealth to forms of tenure – of wages to those as to capital

Figure 5.2: William Whewell's inductive scheme for political economy from one of his notebooks on induction of the early 1830s. *Source*: R.18.17.15/41, TCL.

against Whewell's metaphysical inclinations: "But mind if you insist on German phraseology or anything like it I wash my hands of the job." Jones made it clear that he did not want to have anything to do with "neological friends who speak high dutch & think smoke."[44] Around this time Whewell himself was still searching for the philosophical foundations of his new theory of induction. From his notebooks Menachem Fisch carefully reconstructed the path that would eventually lead to Whewell's three-volume *History of the Inductive Sciences* (1837) and the "moral" of that seminal work, the two-volume *Philosophy of the Inductive Sciences* (1840).[45] Fisch analyzed, for instance, the different schematic outlines Whewell drew to distinguish between the different types of progress in the sciences, but he left Whewell's draft scheme for induction in political economy undiscussed. This scheme did not find a way into Whewell's published work, indicating Whewell's inconclusive stance on the matter.

Figure 5.2 shows Whewell's rudimentary scheme for induction in political economy, one that combined the classification of historical facts with a search for "laws of phenomena." Smith and Malthus are listed as mere "decomposers" of phenomena, while Jones reached a higher stage with his classification of rents. Whewell perceived the significant merit of Jones's work in its classification of clearly distinct economic orderings. But Jones had not proceeded to their causal explanation. Whewell mentioned the "insulation of facts" as an important step in arriving at "laws of the phenomena," characteristic of any mature science. He listed various relations for which such laws might be derived, but we find no names of political economists who had done so fruitfully. Indeed, for steps 4 and 5 of the inductive process, practical examples are "almost wanting."

Whewell's struggles with the "mixed" character of political economy—that is, its combination of natural and mental phenomena—do not become clear from this scheme. In the early 1820s Whewell had written to Jones that he did not know of any "peculiar principles of observation or deduction employed in [political economy]—they may as well talk of the metaphysics of chemistry."[46] He had mentioned several examples of methods to arrive at laws of phenomena—tables, graphs, functions—that could be applied to political economy. For example, his former student John Lubbock had made a table of marriages and birth rates that showed the "dependence of the increase of population upon the number of marriages and the number of births per marriage, and vice versa." Lubbock thus made "many of Malthus's statistical details" more obvious by giving insight in the "mathematics of population."[47] In his first mathematical paper on economics, presented in 1829, Whewell had discussed a table that originated from two political arithmeticians, Gregory King and Sir Charles Davenant, showing the relation between shortages of supply and wheat prices.[48] Whewell considered that "it would be easy to find the law by which the increase in supply may depend on the defect and the square of the defect of supply, so as nearly to satisfy the above data."[49] In all these examples Whewell approvingly discussed methods of research in political economy in which quantification filled in the blanks between the collection of data and the establishment of causal claims, methods Jones was less sanguine about.

According to Whewell, when "the connexions" between phenomena had been "insulated," the "*form*" of the law could be determined and following on this, its cause. But for the "subjective sciences as opposed to the objective," it was hardly possible "to discover a *law of phenomena* but by a perception of its cause."[50] This difficulty was induced by the fact that the causes in the "subjective sciences" were "facts of consciousness," and there was no obvious way to observe, that is to measure, these facts in the same way as facts of "external objects"; indeed the very concept seemed to exclude the possibility of measurement. Even though many of the facts of political economy were "external (statistical and commercial details)," its classificatory concepts, like "wealth, property, exchange, relations of social life" were "mixed."[51] While the Oxford economists, following in Stewart's footsteps, claimed a separate and easy entry to a man's motives, Whewell remained at a loss how political economy might systematically make the step from the collection of data to the facts of consciousness. While Whewell kept exploring the possibilities of discovering "laws of phenomena" and their causes, Jones was satisfied with a systematic classification of facts. This divergence proved important in their efforts to institutionalize an inductive alternative to Ricardian deductivism.

Section F of the BAAS and the Statistical Society of London

The story of the establishment of Section F (statistics section) of the BAAS and of the Statistical Society of London has been told many times, even though the emphasis on the importance of the various actors involved differs. For sure, not only Whewell and Jones, but the entire "Cambridge Network" was of importance for the "irregular session" at Jones's premises at Caius College that was foundational for Section F during the meeting of the BAAS in 1833 in Cambridge. However this may be, Whewell and Jones discussed the matter in advance of the meeting and the presence of Adolphe Quetelet was then used by a group consisting of Whewell, Jones, Malthus, and Charles Babbage to persuade the BAAS to establish a separate section devoted to political economy and statistics.[52] Well documented are Whewell's conversations with Adam Sedgwick, then president of the BAAS, to alleviate the fears and objections of Section A—the mathematics and physics section—that some sort of "rabble rousing" club would be added to the BAAS. The new section was explicitly restricted in its research to the collection of facts that could be expressed in numbers and that might eventually lead to general laws. Indeed, as Sedgwick clarified the purposes and limitations of Section F to the BAAS, insofar as statistics dealt with "objects, whether of pure or mixed nature" that were "capable of being reduced to measurement and calculation," there was nothing against considering them "legitimate objects of our inquiry."[53] This restriction met with Whewell's full consent, who was as fearful of the intrusion of opinion in science as Sedgwick.

As an offspring of Section F, the Statistical Society of London was founded in 1834. Although roughly the same group of people was involved, Whewell was less and Jones more present in its founding. Jones originally designed the classification of subjects for the Statistical Society, which was accepted in modified form.[54] An inspection of its prospectus reveals an important difference in the plan of operation of the Statistical Society of London with Section F of the BAAS. While the BAAS document made mention of the establishment of laws, this reference had vanished in the prospectus of the Statistical Society. The society should "exclude all Opinions from its transactions and publications [and] confine its attention rigorously to facts—and, as far as may be found possible, to facts that can be stated numerically and arranged in tables."[55]

What followed was a dry enumeration of the classes of facts to which the inquiries of the Society should be limited. In his account of the history of the Statistical Society at the Jubilee meeting in 1884, Frederic Mouat argued that this strict limitation to the gathering of facts in tabulated form was motivated by Whewell's fear of the "daemon" of opinion and discord.[56] Whewell of course never excluded the possibility of using statistical facts as building blocks for the search for the laws of political economy. However, as Hilts has pointed out, Jones's emphasis on fact gathering made the Statistical Society an acceptable forum for widely different audiences, which, against his own intentions, included political economists like Senior and McCulloch. Neither Senior nor McCulloch denied the usefulness of statistical data, but as Senior wrote to Quetelet, he did "not consider the truths of political economy" to depend on "Statistical facts."[57] Indeed, in the early 1830s McCulloch himself was an important collector of statistical facts, but his statistical pursuits did not at all affect his zealous defense of Ricardian economics. The limits set by Jones to the work of the Statistical Society thus blocked the road to the establishment of general economic laws and stabilized the "curious separation" between theory and facts that is characteristic of Victorian political economy.[58]

A strong impetus in the demand for statistics came from businessmen, (local) administrators, and statesmen who needed them for practical policy purposes. These constituencies, as well as leading political economists and men of science, were all present at the formal founding of the Statistical Society in London, attended by some 250 people. When measured in these terms, it was an overwhelming success, and quite a contrast with the "irregular" founding of Section F by the small group around Whewell and Jones. Importantly, however, the agendas of these different audiences did not all align with the purposes of science as perceived by Whewell. On an abstract level it excluded "opinions" in the pursuits of the society, but the practical implication was a gulf between fact and theory, and a kind of statistics that was of immediate use for policy making.

The logo of the Statistical Society—selected by Charles Babbage—a loosely bounded wheat sheaf with the words *Aliis Exterendum* (to be threshed out by others), precisely expressed the self-imposed restriction to the collection and classification of statistical facts. There was no link to theory. At the BAAS nothing much came from the invitation in the founding act of Section F to use statistical data for the discovery of laws governing society. At the same time the papers presented at the Statistical Society and at Section F all too easily jumped to preconceived policy prescriptions—which, somewhat paradoxically, was precisely what someone like Whewell had hoped to avoid. But his failure to establish a viable counterweight against Ricardian economics can be traced to the different purposes of Section F and the Statistical Society, and to his own wavering on how political economy should proceed to inductively establish what he called the "laws of the phenomena."

Jones soon lost interest in the Statistical Society, certainly after he became completely overwhelmed by his work as one of the commissioners for the conversion of the tithes—traditionally a tax payment in kind from farmers to the clergy—to money payments. With Jones gone, Whewell abandoned all hope for an inductive reform of political economy. He became so annoyed with the papers produced for Section F that he eventually came to seriously regret its founding and even considered its abolition in the early 1840s, when he was president of the BAAS. In his opening address as president of Section F in 1860, even Nassau Senior expressed his disappointment and concern with the "unscientific character" of many of the papers presented.[59]

The Aftermath

In this chapter we have seen how Whewell's and Jones's defense of the inductive method in political economy was enmeshed with a fight against the image of society as a place of want and antagonism of the landed interests against the other classes. Historians of economics for a long time treated Whewell's and Jones's contributions as a footnote in a history of the progress of the science. They considered Whewell as a precursor of mathematical economics and Jones as a precursor of the historical school and a key influence on Karl Marx. As a result, it went unnoticed that Whewell intended his mathematical rendering of Ricardo as an *assault* on his theories. This did not mean that Whewell saw no future for the use of mathematics or, for that matter, of theorizing in economics. However, in line with his emerging theory of induction, it was crucial that work on the subject was based on the careful and gradual development of an empirical basis from which generalizations could be drawn. Whewell saw Jones's work as an important step in that direction and encouraged him to move forward, even though they came to disagree on what the end goal of an inductive science like political economy was.

Their fight with the Oxford economists Whately and Senior only increased the stakes to develop a theory of political economy that would not be limited a priori to a conception of man as a mere profit-seeking animal in the marketplace and a utilitarian vision on economic politics that turned particularly harshly on the poor. It was this image of man and its concomitant politics that Whewell and Jones hoped to counter with an ethical view of mankind in the social state. As Jones complained, the Oxford men "have got a new-fangled system of their own ready cut & dried which they would like to force down the throat of England & Ireland on their authority & without the slightest regard to local difficulties, reasonable modifications or any ones views or experience but their own."[60]

Whewell perceived political economy as a "mixed science," combining the "subjective" side of human motives with the "objective" side of external observation and measurable facts. Whewell had reproached the classical economists for jumping "from one or two trivial facts to the conclusion that every man will get as much money as he can. An axioma generallissimum,"[61] an approach defended by the Oxford economists on commonsense grounds that went back to Dugald Stewart. But in his own explorations of the inductive method in political economy he could not find a principle that would "colligate"—as he would express himself in his mature works on induction—the subjective and objective side of economics.

Whewell's hopes for a viable alternative to Ricardian economics increased when Jones became professor of political economy, first at King's College London and later at the colonial Haileybury College, as the successor of Thomas Robert Malthus. But these hopes soon vanished when Jones became completely embroiled in his work for the commission for the revision of the tithes. In a letter to John Herschel of December 4, 1836, Whewell once again expressed his frustrations about Jones's delays: "I am going to stay with him [Jones] in the Christmas vacation. The only misfortune is, that he is less and less likely to write the books he owes the world. He professes that he shall still do much in that way, but I confess I doubt it: and I doubt with grief, for in certain branches of Political Economy I am persuaded he is a long way ahead of anybody else, and might give the subject a grand shove onwards." But there were also disappointments on Jones's side. In Whewell's mature view on induction, observations, classifications, and taxonomies, important though they were, more and more became a preliminary step in the discovery of causal dependencies, rather than a goal in itself as they were for Jones. For Whewell, mental concepts, when properly approached, were the guiding principles in the discovery of the laws of nature. When Whewell published his *Philosophy*, Jones wrote to Herschel that they were the only true Baconians left.

Over the course of the 1830s the exchanges between Whewell and Jones on political economy became increasingly sporadic. When John Stuart Mill published his famous essay on the proper definition and method of political economy in his *Essays on Some Unsettled Questions of Political Economy* (1844), Jones wrote to Whewell that "young Mill has been publishing a paper to prove that a priori reasoning is not only good in Pol. Econ. but the *only* reasoning applicable to it. God help him and those this belief leads to trust in him his Papa and his school."[62] Mill's essay, originally written in 1831 and published in the *Westminster Review* in 1836, was a defense of the a priori method in political economy, in which Mill took up some of the themes already pursued by Senior in his introductory lecture of 1829. Mill argued just like Senior that the plethora of disturbing causes resulted in political economy being able to arrive at tendency laws only because its laws, though as valid as the laws of mechanics, could not be found in their purity in the statistical data. But instead of taking recourse in common sense, Mill argued that its first principles could be found by way of introspective experiments on man's motives for action. Mill also defended the economists' limited view on the motives guiding man's actions. "No economist" was of course so "absurd" as to embrace a view of man as guided by selfish motives only.[63] But it was in order to secure its laws that economists limited their field of vision, not because this vision exhausted the full meaning of mankind in the social state.

Only after Jones's death in 1855 did Whewell concede that "in Mental and Social Science, we are much less likely than in Physical Science, to obtain new truths by any process which can be distinctively termed *Induction*; and that in those sciences, what may be called *Deductions* from principles of thought and action of which we are already conscious . . . must have a large share; and I may add, that this observation of Mr. Mill appears to me to be important, and in its present connexion, new."[64] To this staggering statement by Whewell—coming after years of disdain for Ricardo and the Oriel Noetics—we may add Whewell's positive reception of Mill's *Principles*, in which he could find little to disagree with, as he wrote to Jones from Trinity's Master's Lodge.

Mill's *Principles of Political Economy* effectively papered over many of the differences between the Ricardians and the Oxford and Cambridge men, and conceived of political economy as a far wider subject than the restrictive definition of the science of "catallactics" had allowed. Mill acknowledged the importance of Jones's classification of rents, though only as a "copious repertory of valuable facts."[65] And it may well be that Mill's *Principles* motivated Whewell to write two new memoirs that explored, among others, Mill's rendering of Ricardo's theory of the comparative advantages of international trade. Read in 1850 to the Cambridge Philosophical Society, these papers show a new spark of Whewell's interest in political economy.[66] But by the end of the 1840s Whewell had in fact lost contact with the subject. In a letter to Jones in 1848, he mused about the days in which they had studied and discussed "Malthus and Ricardo" and wished those days might return. "I mean, I wish this, that I might have your help in getting hold of the subject now, as I had then. Besides, it was very pleasant—and would be so again. Adieu."[67]

CHAPTER 6

Whewell and Language

JAMES CLACKSON

Simon Schaffer, in a celebrated chapter in the 1991 collected portrait of William Whewell, drew attention to his use of metaphor and language as part of a larger scientific endeavor.[1] We read how Whewell tried to make scientific language "fit for public consumption" and how he saw language as a tool whose task was "to help *permanent* knowledge emerge," with accurate and shared definitions vital for ensuring scientific progress.[2] Schaffer stresses the importance Whewell placed on linguistic reform as a bulwark against the error and radical ideas that seemed to threaten Church and State.[3] Certainly Whewell's impact in shaping the language of science has been enormous; if he had done nothing other than coin the terms *scientist, anode, cathode, ion, uniformitarianism,* and *Eocene* (among others), he would still be justly celebrated. His two published papers specifically on language in the short-lived journal *Philological Museum,* "On English Adjectives" (1832) and "On the Use of Definitions" (1833), can be read in line with Schaffer's account of Whewell's wider concerns for the correct use of language for the proper understanding of new scientific knowledge. Whewell's 1828 *Essay on Mineralogical Classification and Nomenclature* further displays his careful attention to developing the right language for a classificatory system. Here, as elsewhere in his work, Whewell recommends restraint in the creation of new terms, rejecting the "unscrupulous recourse to Greek words and their combinations" with a preference for words that were already in general currency.[4] As the list above shows, Whewell was himself not above "unscrupulous recourse" to Greek, with some of his new creations causing distress to the classically minded.[5]

Whewell's view of language as an instrument at the service of other disciplines was in the tradition of eighteenth-century writers on language, including Samuel Johnson and Sir William Jones.[6] Like Johnson and Jones, however, Whewell was interested in language also for its own sake, not just as a handmaid to the sciences. He saw early on in his career that language shed light on questions of history and prehistory, and, through the short-lived Etymological Society (active in 1830–1832, and possibly earlier), he was involved in a collaborative effort to write a new dictionary of English that anticipated the work of the Philological Society of London in the creation of what is now known as the *Oxford English Dictionary*. Whewell further carried out his own researches into the structure of the English vocabulary and the differences between various derivational suffixes for noun and adjective formation. Most of the Etymological Society's work, done in collaboration, remains unpublished, being only briefly summarized in Whewell's own account in a letter sent to the Philological Society twenty years after the events.[7] The full range and depth of Whewell's studies on language can be seen from the papers held in Trinity College Library, Cambridge.[8]

The New Philology

Before embarking on a survey of Whewell's linguistic research, it is worth describing the intellectual context. The nineteenth century was a golden age of language science. Many modern accounts present the period as the beginning of a new era. Holger Pedersen, in a book tellingly titled *The Discovery of Language*, went so far as to claim that work on language in Europe before 1800 had advanced "but little beyond the knowledge of linguistics achieved by the Greeks and Romans."[9] The first encounters of eighteenth-century European scholars with Sanskrit, coupled with a growing acquaintance with the diversity of languages around the world, led to the recognition that all languages could not be directly traced back to a single language of Eden. Earlier scholars had noted the affinities of different languages within Europe, but the similarities between Sanskrit, Greek, and Latin were inescapable and famously led Sir William Jones in 1786 to declare to the Asiatick Society of Calcutta, in his "Third Anniversary Discourse, on the Hindus," that these languages were "sprung from some common source, which, perhaps, no longer exists."[10] Jones's statement, combined with his other writings on ancient Indian culture and his translation of Kālidāsa's play *Shakuntala* (1789), led to a wave of interest in Sanskrit, notably among Continental scholars, who benefited from French and German translations of Jones's work.

Knowledge of Sanskrit and examples of cognate words shared in Sanskrit and in Latin and Greek steadily increased in the first decades of the nineteenth century. Jones had also surmised that "Gothick" was in the same family, and the recognition of the relationship between Sanskrit and European languages proved to have the greatest impact among German-speaking scholars. The *Mithridates* of Johann Adelung, a compendium of all the languages then known, gave pride of place to Sanskrit and included the most extensive list of Sanskrit words then available outside India, with examples of comparable forms in the languages of Europe (and occasionally citations of Persian and Hebrew).[11] Friedrich Schlegel, taking his lead from Jones, published *Über die Sprache und die Weisheit der Indier: Ein Beitrag zur Begründung der Altertumskunde* (1808), which was to become a foundational text of the new philology.[12] Against a broader discussion of Indian philosophy and mysticism, Schlegel gave examples of correspondences between Sanskrit and European languages, although he took Sanskrit to be the origin of everything, without realizing, as Jones had done, that Sanskrit was a sibling to Latin and Greek rather than their parent. The young Franz Bopp, inspired in turn to learn Sanskrit by Schlegel, wrote the first lengthy account of the grammatical correspondences between Greek, Latin, Sanskrit, Persian, and the Germanic languages.[13] The language family to which they all belonged had already been christened Indo-Germanic or Indo-European—the British polymath Thomas Young introduced the latter term (which I shall henceforth use) in his review of Adelung's *Mithridates*.[14] The history and connections of the languages in the Germanic branch of Indo-European became much better understood thanks to the *Deutsche Grammatik* of Jacob Grimm (1819, swiftly followed by a second edition, the first volume of which was published in 1822).

By the 1820s and 1830s new work in comparative philology was thus principally associated with scholars in Germany. Despite the production of English grammars of Sanskrit by Colebrooke in 1805 and Wilkins in 1808, and a dictionary by Colebrooke in 1808, instruction in Sanskrit was largely limited to the British in India and to those at the East India Company College (later to become Haileybury School) in England.[15] Most of those at Oxford and Cambridge remained ignorant of the recent advances in philology. It is indicative that the two earliest accounts of the rise of comparative philology written in English in the late 1830s only mention Jones in passing, and not as the founding father of the discipline.[16] Instead, both look to German works, in particular Grimm's *Deutsche Grammatik*, singled out by John Donaldson as "by far the most important book of the kind which has made its appearance since the revival of letters."[17] In contrast, with the exception of James Prichard's work on Celtic,[18] British scholarship is seen as largely an embarrassment in comparison with the work done on the Continent: "England, we are sorry to say, has little to offer that will bear comparison with the performances of our continental neighbours, in regard either to comparative philology in general, or to Indian Scholarship in particular.[19] Donaldson records lectures given "before the University of Cambridge" by John Kemble in 1834 as among the first evidence for the diffusion of the discoveries made by Grimm and others in England. Kemble was, like Donaldson, a Trinity man and closely associated with Whewell as well. Whewell was a committee member for the Kemble Memorial, established after his premature death in 1857,[20] and the Whewell papers contain records of correspondence between Kemble and Whewell on etymological matters.[21]

In a letter to Jacob Grimm sent sometime in the spring of 1834, Kemble wrote that his Cambridge lectures were "eminently successful,"[22] but they appear to have been abandoned before they reached the promised number of twenty, with an audience of half a dozen or fewer at the end. An anonymous pamphlet, *The Anglo-Saxon Meteor*, recorded that these survivors consisted "chiefly of the unemployed and good natured fellows of his own College, who seemed desirous that a Trinity man should not be quite forsaken."[23] It is not unlikely that Whewell was one of those attending,[24] and his interest in the subject was certainly more than a demonstration of collegial support. Indeed, Kemble's lectures both sowed the seeds for the future growth of comparative philology in England (with Donaldson one of the earliest to flower) and started the process of cutting back the home-grown philological endeavors in the tradition of John Horne Tooke, the first volume of whose *Epea Pteroenta* (more usually known by its alternative title *Diversions of Purley*) was published in 1786, the same year as Jones's "Third Anniversary Discourse."

To many modern readers, the missteps and wrong turns taken by the likes of Jones, Schlegel, and Bopp are made up for by the fact that they were generally moving in the right direction, on paths that would eventually lead to the modern discipline of linguistics. The works of Horne Tooke and his followers, although seen as "scientific" at the time,[25] are now mostly cited as examples of fanciful etymologies and unbounded speculation.[26] Horne Tooke's work is framed as a dialogue, in the tradition of the philosophical dialogues of Plato (the title *Epea Pteroenta* is a Homeric formula meaning "winged words"). The study of language, or *grammar* to use Horne Tooke's term, is part of a larger philosophical enterprise. The springboard for Horne Tooke's etymological interests is the observation that the English word *if* derives from the Old English (Anglo-Saxon in his terms) *gif*, which he erroneously takes to be the imperative of the verb "give." This leads him to attempt to derive, first, all conjunctions from verbs or nouns and then prepositions and adverbs, and in the second volume, adjectives and participles. Thus the color terms *brown*, *yellow*, *green*, and *white* are taken by Horne Tooke to be participles of verbs meaning "burn," "set alight," "flourish," and "foam," respectively.[27] Horne Tooke supports his arguments through citations of words in a number of different languages, but the relationships among them are never clearly enunciated beyond the fact that Anglo-Saxon is the clear predecessor to English. This is not, however, his aim, which is to relate the study of language more closely to philosophy. Aarsleff cites William Hazlitt's words of praise, showing how enthusiastically Horne Tooke's linguistic analysis was greeted at the time: "Mr Tooke ... treats words as the chemists do substances; he separated out those which are compounded from those which are not decompoundable. He did not explain the obscure by the more obscure, but the difficult by the plain, the complex by the simple."[28]

"On English Adjectives"

Whewell's longest published article on language, "On English Adjectives," shows that he was well acquainted with Horne Tooke.[29] He refers to the list of adjectives from Latin roots that Horne Tooke gives in chapter 6 of part II of *Diversions of Purley*, but does not attempt to follow Horne Tooke's derivation of color terms from participles. Indeed, the article serves as a repudiation of Horne Tooke's basic thesis that all other parts of speech can be derived from nouns or verbs. Whewell divides up adjectives into two classes: first, adjectives of quality that are not derived from another word, such as *good, wise, bright, soft, red, sweet*, and *foul*. Whewell notes that some of the members of this class might be in turn be borrowed from Latin, including *long, large, chaste, grand, severe*, and *gay*. His second class contains adjectives that are derived from other words, such as *joyful* or *virtuous*; members of this class can also be loaned or native. Having made this division, Whewell is largely concerned with the second type of adjective, in particular those derived from nouns, which he terms *adjectives of relation*. Here one can see Whewell's concern with classificatory systems and the correct use of nomenclature coming to the fore. Although in some cases, such as *earthen, earthy*, and *earthly*, speakers of English have a variety of adjectives of relation at their disposal, for the most part there is a general poverty of adjective-forming suffixes. In consequence, the scholar who has need of an adjective corresponding to the substantive *taste* has no native English alternative to the unsuitable *tasty*. Latin *gustatory* is seen to be an appropriate word in connection with bodily taste, for example in the denomination of the gustatory nerve, but not for questions relating to taste as a faculty of mental judgment. Whewell accordingly considers various alternatives, such as *esthetical*, which "has not yet become an established English word; and we may express a doubt whether it deserves to be so."[30] Although the description of language is not Whewell's primary concern and he is by no means in thrall to Horne Tooke, it is striking to the modern reader that there is no appeal to the advances made in linguistic understanding through the works of Grimm and others. The framework in which Whewell operates is largely that of Horne Tooke and his followers, even though there is disagreement on the details.[31] The major advance on Horne Tooke is the clear separation out of loanwords from the native vocabulary of English.

Whewell's Letter to the Philological Society, 1852

Whewell's categorization of English adjectives, in particular his distinction between loan- and native words, corresponds to what we know of the activities of the Etymological Society. The single published account of its activities, Whewell's 1852 letter to the Philological Society, is vague about the dates and the members of the Etymological Society, stating that it was formed "at a period a little previous to the establishment of the Philological Society in London"[32] thereby giving only a *terminus ante quem* of 1842. Whewell further notes that some of their "speculations" were published in the *Philological Museum*, which lasted between 1831 and 1833; this accords with a letter he sent to Julius Hare in 1851, dating the society to "(I think) 1832."[33] The papers of the Etymological Society, included among Whewell's papers in the Wren Library, give no firmer indication of the dates of the society, but modern accounts have reckoned that Hare's departure from Trinity to a family living in Hurstmonceux in 1832 marked the end of its operations, which seems likely given that Hare was, in Whewell's words, "always the great workman among us."[34]

In Whewell's account the Etymological Society proceeded by designating "certain Classes of words, marked by some peculiarity in their relation or history," and each member of the society was assigned one of these classes "with the injunction to collect as many specimens as he could of the Class, and to produce them at the next or some succeeding meeting of the Society."[35] According to Whewell, the members of the society were aware that "our Classes were not philosophically framed, nor co-ordinated according to sound philological views. . . . We knew too that the assignment of any word to its place in one of our Classes was but one step in the deduction of its pedigree, and required other steps in order to complete the etymological story of the word."[36] Whewell then lists some of the Classes used by the Etymological Society:

Class I. Saxon-English words that are not German
Class II. German-French words
Class III. Non-Latin Italian words
Class IV. English words from Italian, Spanish etc.
Class V. Words derived from names of places or nations
Class VI. Words derived from names of persons
Class VII. Ecclesiastical words from Greek or Latin
Class VIII. Medical words from Greek and Latin
Class IX. Astrological and alchemical terms
Class X. Hawking terms
Class XI. Words implying ancient customs

Class XII. Bifurcating etymologies
Class XIII. False etymologies

Discussion of the final class that comprises what are now sometimes termed *folk etymologies*, such as *sparrow grass* for *asparagus*, occupies the largest part of the paper, and Whewell adds examples from other languages as well.[37] Whewell further mentions papers published in the *Philological Museum*, including his own on English adjectives, as a direct result of the speculations of the Etymological Society, and it may be significant that he also mentions Kemble's paper "On English Preterites,"[38] stating that its author allowed the Etymological Society to consider him a "fellow-labourer." Whewell finally mentions that the Etymological Society "had a grand, but I fear, hopeless, scheme of a new Etymological Dictionary of the English language; of which one main feature was to be that the three great divisions of our etymologies, Teutonic, Norman and Latin, were to be ranged under separate alphabets."[39]

Whewell's description of the work of the Etymological Society has generally been taken as indicative of an early flowering of English philological interest, in parallel to, if not always in step with, the development of the subject in Germany and elsewhere on the Continent. Among the other leading members of the Etymological Society, we have already mentioned Julius Hare. Hare's mother's sister had been married to William Jones, but he had died before Hare's birth. More importantly, Hare was fluent in German, having spent time in Weimar in his youth. Together with Connop Thirlwall, another fellow of Trinity who was also fluent in German, Hare edited the *Philological Museum*. Hare's collection of German books was judged by Crabb Robinson "the best collection of modern German books I have ever seen in England" in 1825, and included first editions of the works of Bopp and Grimm.[40] The Danish scholar Frederik Grundtvig, who visited Trinity College in 1831, also made a comparison of the intellectual activities there with what was happening on the Continent, writing home on June 26 that he found a more Germanic spirit at Trinity than anywhere else in England.[41]

The Etymology Society Papers

Among the surviving papers of the Etymological Society held in the Wren Library, there is no surviving record of any meeting of the society nor anything approaching a constitution or membership list. The papers present a rather different story than either that suggested by Whewell's 1852 account or what we might surmise from reading about Hare's German knowledge or Grundtvig's appreciation of the Germanic spirit. There is little corresponding to the thirteen classes into which Whewell says the members of the Etymological Society categorized words, and there is not much in the archive that can be directly linked to anything published in the *Philological Museum*.[42] From Whewell's letter to the Philological Society, it appears that the etymological dictionary was nothing more than a pipe dream, but among the papers in the Wren there survive what appear to be substantial remains of the enterprise. Whewell R.6.5/4/1b, in Whewell's hand, sets out rules for using the sheets of paper with pre-ruled columns designed for the work of the society: "It is proposed to insert in the following scheme the proximate sources of English words and the forms in other languages which illustrate the origin or history of such words."[43] The eleven columns reserved for different languages show a more complex categorization of origins than the "Teutonic, Norman and Latin" that Whewell later recalled. The rules set out the columns as follows: Celtic including Welsh (W) Breton (Br) Gaelic etc.; Scandinavian, with Danish (D) Swedish (S) Icelandic (I) Norwegian (N); German; English Dialects; Old French; Modern French; one column with Italian (I) Spanish (S) Portug. (P.); Low Latin; Latin; Greek; and Oriental "to contain Hebrew (H) Arabic (A) etc."[44] Another large piece of paper,[45] is ruled as a table with the English words to be discussed in the center, and a slightly different set of languages arranged in columns to each side, on the left Germanic, on the right "French not Latin"; "French from Latin"; "Italian Span(ish)"; "Latin"; "Greek" and "Hebrew," although this last column may contain other languages such as "Persic" (Persian). Another document assigns tasks to different members of the society,[46] so that Hare, for example, is allocated "etymological stories." Whewell also seems to apportion work to members of the society, with each receiving a letter, as reproduced below:[47]

A Hare

B Whewell

C Rose (i.e. Hugh James Rose, fellow of Trinity)

D Malden (i.e. Henry Malden, fellow of Trinity)

E Lodge (i.e. John Lodge, University Librarian)[48]

F is blank, as are *L, N, O* and all the letters after *Q.*

G Gwatkin (i.e. Richard Gwatkin, fellow of St John's)

H Coddington (i.e. Henry Coddington, fellow of Trinity)

I and J Kennedy (i.e. Benjamin Kennedy, fellow of St John's)

K Romily (i.e. Joseph Romilly, fellow of Trinity)

M Rose jr (i.e. Henry John Rose, fellow of St John's)

P Worsley (i.e. Thomas Worsley, fellow of Downing)

Q Jeremie (i.e. James Amiraux Jeremie, fellow of Trinity)[49]

Material for an Etymological Dictionary

Most surprising of all the papers in the Whewell archive is the existence of hundreds of pages of records for the letters *D, E,* and *F.* Here the English headwords are given in the leftmost column, followed by a full dictionary entry, including citations of literary passages to illustrate the meanings as well as etymologies. Words are grouped together, according to whether they derive from Latin,[50] Greek (in some cases including "Other Languages or Doubtful"),[51] Saxon,[52] or French.[53] Included with *F* there are thirty-two pages where all the headwords are Old English.[54] I give below a sample of the etymologies provided:

DAUGHTER Junius[55] from the Greek θυγατηρ. Skinner from the Lat. *dos puella enim sine dote vix elocari potest,* Wachter from Low Sax. *tygen gignere.*[56]

EBULLIATE Lat. *Bulla,* a bubble, which Vossius thinks may be from the Gk φλυ-ειν *fervere, ebullire,* to bubble up, to swell.[57]

FATHER The Gr πατηρ Lat. *pater,* Fr *pere,* It *padre,* Dut. *vader* Ger. *vater,* Sw *fadder,* A-S *fæder,* Goth *frad-rein sunt parentes* all which, Wachter thinks, must have had a common origin, either in the infantile cry, *pa, pa,* or in some Scythian word, dispersed by that people over the whole world. For the former Vossius decides.[58]

All the etymologies cited above are reproduced directly from Charles Richardson's *New Dictionary of the English Language.* Richardson's dictionary was first published as a two-volume work in 1835–1837, but the bulk of it had already appeared in print as installments in the *Encyclopaedia Metropolitana* (forming the opening section of the fourth and largest division of the work, *Miscellaneous and Lexicographical;* the volumes were published in alphabetical order).[59] The initial publication of the *Encyclopaedia Metropolitana* was superintended by the poet Samuel Taylor Coleridge, who promised in his "Preliminary Treatise on Method" that the fourth division would include "a Philosophical and Etymological Lexicon on the English Language, or the History of English Words,"[60] and Richardson was deputized to perform the task. As the entries above show, Richardson relied heavily on earlier etymological work by Skinner, Vossius, Junius, and Wachter, the most recent of which was Wachter's, published in 1737. The works of Horne Tooke are frequently cited, and Richardson avowed himself a disciple of Horne Tooke.[61] In his preface to the dictionary, Richardson shows familiarity with the works of Bopp and Adelung and cites Prichard, but never seems to have fully grasped the concept that the languages of India, Persia, and Europe "sprang from some common source," as Jones had understood. Rather, he cites Horne Tooke with approval for the sentiment that a word is "always sufficiently original" in the language "where its meaning, which is the cause of its application, can be found." Etymology is not about finding the earliest source for a word or a root, but finding out its true, underlying sense. When that is found, "there appears no necessity to pursue the inquiry further."[62]

Richardson had not been educated at university, preferring to fulfill his vocation through teaching. He had only tangential connections with Trinity philologists and members of the Etymological Society. He had been master at the school on Clapham Common that Kemble had attended as a boy, recruiting him to assist in reading texts for the dictionary, and thereby, according to Kemble's sister Frances, instilling in him a love for philology.[63] Kemble was not a full member of the Etymological Society, however, and was out of Cambridge for much of the time between 1829 and 1831 and hardly likely to have impressed on Whewell, Hare, and the like the virtues of his former schoolteacher's work.[64] Two members of the Etymological Society, Hugh James Rose and his younger brother, Henry John Rose, were editors of the *Encyclopedia Metropolitana*, but they took over only after the death of Rev. Edward Smedley in 1836, and it is Smedley who Richardson thanks in the preface to the *New Dictionary* "who, for thirteen years, in character of editor of the Encyclopedia, has accompanied me page by page."[65]

The choice of Richardson's *New Dictionary* as the starting point for the Etymological Society enterprise is consequently unexpected, and it runs counter to the view that in the 1820s and early 1830s Whewell and his circle firmly sided with German comparative philology against an English tradition exemplified by Horne Tooke.[66] Preyer's contention that Whewell "frequently attacked Locke's notions on language and understanding and Horne Tooke's *Divisions of Purley*" is not true for the 1820s and 1830s, and his claim that his contributions to the *Philological Museum* "stressed the importance of the "Germano-Coleridgian" approach to historical and comparative philology, while also pouring contempt on Horne Tooke" is unsupported by the evidence.[67] There are no contributions from members of the Cambridge Etymological Society to the *Philological Museum* that deal with matters of comparative philology; the only substantial paper that attempts to trace etymological connections between different branches of the Indo-European language family (and incidentally showing familiarity with the work of the Danish comparativist, Rasmus Rask) was by an Oxford man, Edmund Head.[68]

Richardson's approach to etymology was not that of the Etymological Society according to Whewell's 1852 account, nor one that seems to fit in with their eagerness to categorize the words from the paths by which they entered the English language. The pages of Old English vocabulary included under the letter *F* show more independence. To illustrate, consider the words *fyr* and *fót* (modern English *fire* and *foot*), for which Richardson's etymologies are as follows

> FIRE A.S. *Fir, fyr*; Dut. *Vayr, vier*; Ger. *Feuer*; Gr. Πυρ, a Phrygian word, according to Plato. . . . It is difficult to suppose that our northern progenitors had no name for the element of *fire*, until they borrowed it from the Greeks; it is more probably that there was some common origin for both the Greek and Saxon in northern languages.[69]
>
> FOOT Dut. *Voet*; Ger. *Fusz*; A.S. *Fot*; Goth. *Foties*, which Junius derives from the Gr. Πους, Lat. *Pes* (ph?). It may be from the A.S. *Fettian*, to carry (sc.) that which carrieth, which beareth.[70]

The entry for *fyr* in the Etymological Society papers contains no speculation on its origins, but adds further comparisons, supplementing the Germanic forms with "New Guinea *faz*" and an illegible Arabic form.[71] For *fót*, the Germanic forms in Richardson are reproduced, and also the word in Romance languages (French *pied* etc.), followed by Greek, Persian (written in Arabic characters), Hebrew פצם, and concluding with "from Sans. *padas* A Foot; *pes*."[72] It is noteworthy that, even when improving on Richardson, there appears to be no recognition of the fact that the origin of these words could be explained by the supposition of a language family that included Sanskrit and Persian but not Hebrew or Arabic. Adelung and Schlegel had already shown that *daughter, eat, eight, father*, and *foot* all had close connections with the Sanskrit terms, more than two decades before the Etymological Society.

The defects of this approach to etymology in terms of the recent discoveries made by German scholars are apparent to us, and would have been clear to a scholar such as John Donaldson, who does not include any of the etymological speculations of Horne Tooke or Richardson in his account of the history and present state of philology in the *New Cratylus*. It is patent also that Whewell himself had a clear sense of the division between Indo-European and Semitic language families in his second paper in the *Philological Museum*, "On the Use of Definitions," since he uses the classification as an example of the necessity of categorization:

> If, for instance, we consider the languages of the earth, what a long and comprehensive labor of comparison was gone through before philologers had a clear view of the classes of languages that are now termed the Indo-European and the Semitic! And how little would it have contributed to the progress of philological knowledge, if, before this labor had been gone through, men had used the word *Semitic* and defined it to mean "the languages spoken by the

descendants of Shem," without knowing whether these languages resembled each other more than Arabic and Latin.[73]

It seems, however, that this basic principle of the categorization of language was not so clear to all the members of the Etymological Society, certainly not to whoever compiled their sheets for the letters D (perhaps Henry Malden), E (perhaps John Lodge), and F included in Whewell's papers. Aarsleff has ably shown how Horne Tooke's influence on philological work in England was remarkably persistent, not being seriously challenged until the 1830s when "the deplorable state of philology in England" became cause for "national reproach and shame."[74] We might have expected better from the Etymological Society, but in the context of the time it was understandable that they would rely on what they saw as the most up-to-date reference work on English, Richardson's *New Dictionary*.[75] Without knowing more about the circumstances, or indeed date, of the creation of these documents or their inclusion with the other Etymological Society material, there is little point in speculation, but one wonders whether the abandonment of the project was spurred by the realization that everything would have to be redone from scratch. Whewell's 1852 letter to the Philological Society can now be seen to have downplayed the extent of the Etymological Society's work, perhaps reflecting his realization that the enterprise was worthless in terms of the advances made in the history of languages.

Geology and Language

Whewell's vision of an etymological dictionary of English arranged so that words were grouped by their different origins has never been achieved. Although he may not have been fully aware of the latest developments in historical and comparative linguistics in the 1820s, his 1832 paper on adjectives is in some ways ahead of its time. His classificatory interests there could be read as looking forward to the turn toward descriptive and synchronic linguistics in the twentieth century. Whewell's most lasting contribution to the modern discipline of linguistics is probably to be found elsewhere: his success in framing the study of language change as what he called a "palaetiological" science, together with geology, ethnology, and other sciences, whereby the researcher could use the evidence of the present to "ascend to a past state of things . . . for the ascertained history of the present state of things offers the best means of throwing light upon the causes of *past* changes."[76] Whewell made known Charles Lyell's work on geology to a larger audience, coining the term *uniformitarianism* in his review of the second volume of Lyell's *Principles of Geology*. His description of language in geological terms caught on with other writers on language, notably the Trinity fellow and champion of the new philology, Donaldson, who wrote in his

New Cratylus, "The study of language . . . is indeed perfectly analogous to Geology; they both present us with a set of deposits in a present state of amalgamation which may however be easily discriminated, and we may by an allowable chain of reasoning in either case deduce from the *present* the *former* condition, and determine by what causes and in what manner the superposition or amalgamation has taken place."[77]

Although Whewell did not support the claim for uniformitarianism in linguistics, he was the midwife of the idea, which still remains fundamental to the description of language change.[78] The description of language through geological metaphors is also still frequent in the subject, and it is a suitable conclusion to this chapter to repeat one of the first, and perhaps most vividly expressed, examples, leaving Whewell with the last word: "The English language is a conglomerate of Latin words, bound together in a Saxon cement; the fragments of the Latin being partly portions introduced directly from the parent quarry, with all their sharp edges, and partly pebbles of the same material, obscured and shaped by long rolling in a Norman or some other channel."[79]

CHAPTER 7

Whewell and Tidology

MICHAEL S. REIDY

I have written to Lord Minto respecting to him that I had given great
time and pains to this subject [of the tides], that I believed the results
were considered as of use, that I expected to go on with the same kind of
work for some time, and that, besides any other claims I might have to
the Lowndean professorship, I ventured to apply for it, in order that this
appropriation of my time and thoughts might be a professional employment, recommended and awarded by a public office.

—William Whewell to Francis Beaufort, November 2, 1836

When William Whewell wrote to the hydrographer Francis Beaufort concerning his extensive and continued work on the tides, he was mobilizing
support to hold the next Lowndean Chair of Astronomy at Cambridge. He
felt he deserved it, he confided to his friend Richard Jones, "very much on
public grounds for Tidology."[1] This wasn't bravado. Whewell was writing
as the foremost tidologist in the world. He had recently presented to the
Royal Society his sixth series of researches on the tides—an account of his
multinational tide experiment of 1835—and was on the verge of receiving
the society's Royal Medal for his work. He had purposefully carved out the
tides as his own long-term research space.

Whewell has quite justifiably been viewed as a polymath. One need only
to look at the wide range of entries in this collection to appreciate his broad
pursuits. Yet his reputation as a "meta-scientist" is based in some respects
on his supposed lack of any sustained scientific research program. Richard
Yeo argued that Whewell was a "critic, adjudicator, and legislator of sci-

ence without being a major scientific practitioner or discoverer," suggesting that his relationship to scientific practice differed from other practitioners such as his close colleague John Herschel, "who could comment on issues of method or theory, often in relation to their own original investigation."[2] Menachem Fisch pushed this reasoning even further, arguing that "unlike Herschel, at no time did he ever fully devote himself to any one scientific project."[3] In summarizing this scholarship, Joan Richards concluded that "Whewell dabbled in it; he dropped in occasionally but on balance was more an observer than a participant in their enterprise."[4]

That historians have overlooked Whewell's tidal investigations makes sense. He worked from the Newtonian-Bernoulli equilibrium theory of the tides, a hydrostatic approach that has since been replaced by the modern harmonic method. His enduring contribution to tidal theory, however, is only part of what is important when studying his active participation in science. At the height of his studies, he was finishing his *History of the Inductive Sciences* (1837), formulating his philosophy, and positioning himself as a spokesperson for science. His twenty-year research project on the tides elucidates how he thought about the discovery process, about the practice of large-scale participation in science (both then and now), and how he envisioned the role of the newly invented *scientist*, a term he coined at the outset of his tidal research.

By analyzing Whewell's research on oceanic tides, several significant aspects about his life and work rise to the surface. First, we can chart how fundamental his tidal studies were in formulating his theory of scientific discovery, and how his study of history and philosophy in turn informed the manner in which he undertook his tidal studies. He made this clear in correspondence with his closest friends and collaborators. "I am meditating the returning forthwith and in earnest to my beloved Induction," he wrote to Jones in mid-November 1833, directly linking his *History* to his tidology. "I have been employed all the term hitherto upon a thumping paper on the Tides, which I intend to be a step of some consequence in the theory. I wish I could explain to you how useful my philosophy is in shewing me how to set about a matter like this, and how good a subject this one of the Tides is to exemplify it."[5]

Second, Whewell undertook his research as science played an increasingly active role in the political and economic ambitions of the British State.[6] His tidal work teaches us about the initial creation of the scientist and where such a devotee fit into to larger questions of State power and control. Tidology and the State formed a symbiotic relationship. Whewell's approach to the study of the tides would not have been possible if Britain had not been on its way to becoming the premier imperial power in the world. In turn, it was exactly a detailed understanding of the science of the sea, of its rhythms and laws, that states required to safely transfer goods, influence, and ultimately power between local ports and distant possessions. The close connection between physical astronomy and State power is on full display in Whewell's approach and methodology. Such an analysis reveals that the practice and process of empire specifically directed the kind of research Whewell undertook, including his choice of theories and his method of data analysis and discovery. Whewell's contributions to scientific methodology—including his concepts of "colligation" and "consilience," and his use and explanation of the graphical method—remain important for the philosophy of science, but in this reading, they also become powerful tools of the State as a means to control distant and far-flung geographies.[7]

Third, if we consider Whewell's tidology an example of early State-sponsored citizen science, and analyze the social and scientific backdrop in which he undertook it, we can better understand today's similar fascination with large-scale global participatory research and the hierarchical nature of such endeavors. Today's citizen science is often hailed as a democratizing process, a way to get more people involved in the practice and appreciation of science. Everyone can collate birds and butterflies, plant forests, and collectively help save the planet. By studying Whewell's large-scale participatory research, we gain an appreciation of the actual limits of such seemingly inclusive endeavors.

Whewell's history of induction and philosophy of scientific discovery outlined how each scientific discipline advanced through a "prelude," an "inductive epoch," and a "sequel." Similar to today's interests in conservation, in forests, birds, and butterflies, the early Victorian fascination with tides and currents was born from more immediate concerns of the culture of the day. For a prelude to Whewell's tidology, we must first turn to the changing rivers, estuaries, and dockyards of England.

Prelude to the Inductive Epoch

On the day of the full moon in February 1825, inhabitants living near the Thames River bore through the thatched roofs of their dwellings to climb on top.[8] They weren't lunatics. An extraordinarily high tide had raged up the Thames, breached its barriers, and inundated thousands of acres of land near the towns of Dartford, Crayford, and Erich (where the present-day M25 crosses the Thames). Similar devastation was reported in Amsterdam and as far away as Hamburg, Germany.[9] The entire North Sea seemed to erupt with exceptional fury. Flooding was not new to the Thames, but according to quayside inhabitants, the floods were getting both more numerous and more destructive. Devastating high tides returned a decade later, in January 1834, May 1835, and again in May 1836, causing severe flooding within the cities of London and Westminster. Boats maneuvered through narrow streets that now served as canals, entering buildings and houses to rescue inhabitants.

Those who lived within the environs of the tidal Thames were correct: the tides were indeed getting worse. But why, and what could possibly be done? While the cause of flooding appeared to be global, caused by some unknown convulsion located somewhere out at sea, it directly affected coastal shipping. Shifting shoals and sandbanks, wild eddies and currents, and unpredictable tides and tide streams made navigating rivers from sea to shore extremely precarious. With the introduction of steam engines and iron hulls, vessels were also getting heavier and lying lower in the water, making coastal shipping that much more difficult. Docking at most of the British ports, including many of the naval dockyards, depended entirely on the rise and fall of the sea, especially on the east coast.[10] This included entry into the Thames, the home of the largest port in the world.

Civil engineers turned the might of British industry toward the littoral, attempting to control the tides through large dock complexes, including embankments and seawalls, to standardize the depth of the river for Britain's coastal transportation system. These massive projects transformed the river's topography—and the rivers retaliated. With restricted areas and deeper causeways, the water flowed stronger and higher, leading to changing tidal and silting regimes. Constant dredging led only to the need for more dredging. The choice of the tides as a legitimate research topic in the early Victorian era was anthropogenic, caused by human environmental changes to the physical topography of the rivers and coastlines of Britain.[11]

The risks associated with stranded or sunken vessels were of immediate importance for those who manually labored in Britain's shipping trade, but they were also of financial interest to insurers and underwriters, dock trustees and ship owners, as well as bankers and merchants. When the Society for the Diffusion of Useful Knowledge was formed in 1826, it entered the almanac trade, producing an inexpensive and widely distributed *British Almanack* its first year. After being caught stealing their tide tables from other almanacs that first year,[12] and having botched all the results the following year,[13] the society enlisted the help of John William Lubbock, a Cambridge graduate and successful London banker, to infuse their tables with scientific principles. Lubbock, tellingly, was also a subscriber (financier) of the newly built St. Katherine Docks, opened in 1829, and thus financially invested in their safe use.

When Lubbock took up the study of the tides in 1829, he began with Newton's general theory, which had been advanced by Bernoulli in the eighteenth century, known in Lubbock's time as the Newtonian-Bernoulli equilibrium theory of the tides. It explained how the astronomical forces of the sun and moon influenced the tides, but it was no simple matter to transform that knowledge to actual tide predictions on specific coasts or ports. Lubbock was the first researcher to use the equilibrium theory to produce tide tables (at least the first to publish those methods). He also hired an expert calculator, Joseph Foss Dessiou, who was the first to then compare actual observations with the published methods (at least the first to publish). Lubbock eventually published a textbook on the tides, based on his and Dessiou's work, but perhaps his most enduring achievement in tidology was to introduce the subject to his former tutor at Trinity College, William Whewell.

Whewell had sought an active role as a researcher from his first entry into science in the early 1820s. "When I was admitted into the Royal Society," he wrote to Herschel in 1823, "I intended, if possible, to avoid belonging to the class of absolutely inactive members, and I have since been on the look out to find among the speculations in my way some one which might possible be worth presenting to it."[14] He began by writing textbooks, publishing *An Elementary Treatise on Mechanics* in 1819 and *A Treatise on Dynamics* in 1823, both of which he revised with new editions in the latter half of 1831 and republished in 1832.[15] These textbooks contained histories (material he subsequently incorporated into his *History*) that were proscriptive, guiding Whewell on how to approach a burgeoning field of research with the greatest prospects of advancement.

Whewell then darted in and out of several sciences, including mineralogy and meteorology, always studying the discipline's history, its nomenclature, where it was in the progress of science, and how, in comparison to other specialties, the field could be advanced. His textbooks and his science also helped him form a methodological approach to scientific discovery (subsequently incorporated in his *Philosophy of the Inductive Sciences* [1840]), but at this early stage he used his study of history and philosophy as a guide to his own research.

Whewell became intrigued by the study of the tides in 1829. He had reviewed Lubbock's work that fall, complaining to him that he should make his work "more general in its status," as tidology had masses of data but no laws explaining the phenomena.[16] By the spring of 1831 he was already giving Lubbock material that he could use. "With care and labour," Whewell confided, "I dare say something may be generalized out of them." This was one approach Whewell eventually took: generalizing from Lubbock's empirical work. He further outlined his possible research trajectory. "I think the best thing to do would be to make but such broken and conjectural lines of contemporaneous tide," he wrote, admitting that he "should really like to have the rummaging of the subject and the bringing together of the facts."[17] The "bringing together of the facts," a process he later described as "colligation," formed a central part of his philosophy of discovery. A year later in

April 1832, he fully revealed his intentions, asking Lubbock if he would object "to my mixing my labours with yours and supposing that I could satisfy myself so far as to write a paper on the subject?"[18] Lubbock was reticent to generalize his results beyond specific ports. Although Beaufort argued that the study was a "national object" and for the "urgent necessity of acquiring proper data for the construction of our Tide Tables," no systematic effort had ensued.[19] According to Whewell, the field of tidology was "still to be begun."[20] And he was set on beginning it.

The Inductive Epoch in Tidology

While the prelude to the study of the tides was confined to the coastlines of Britain, the inductive epoch entailed an expansion to all of the world's oceans. Whewell transformed the study from a temporal to a spatial science. For this he needed as much data as possible from as disparate sources as he could find. "I had rather set off doing any thing about cotidal lines till I can collect more materials and give more time to the subject than I can now," he wrote to Lubbock, "but if it appears that the only way to set people in distant parts to work upon the subject is to make some conjectural assertions, (not concealing that they are so) I should have no objection to take this method."[21] Whewell's first publication on the tides was a two-page circular, "Suggestions for Persons Who Have Opportunities to Make or Collect Observations of the Tides" (1832). He was aiming to "set people in distant parts to work." By early 1833, he was publishing "Memoranda and Directions for Tide Observations" in numerous installments of the *Nautical Magazine*.[22]

He also created a single sheet of instructions for distribution among possible observers living near the coast.[23] "Such a paper will be of great use," Beaufort assured Whewell, "and if you will supply me I will distribute it extensively, or, at any future time I will with your permission reprint it for circulation among our officers."[24] What to measure and how to measure it were often fortified in these types of instructions to travelers that codified a set of practices that had largely been created at the centers of calculation.

Whewell also reached beyond Beaufort and the British Admiralty. He brought his sister into his research. She had read in the journal of the Church Missionary Society about seemingly peculiar tides that ebbed and flowed only once per day on the coasts of Tahiti.[25] Whewell assured her that no such tidal rhythm could exist. "But if you do find in your missionary books, or any other books, any notices about the tides," Whewell responded, "let me know, for I believe I shall have to write something about them, and all information on the subject will be useful."[26] No discussions were too trivial; all interested him. He even struck a deal with her: he would subscribe to every society that helped him with his observations.

Along with the Admiralty and Missionary Societies, both of which had members scattered throughout the globe, Whewell also began a campaign closer to home. He read Lubbock's paper on the tides at the second British Association for the Advancement of Science (BAAS) meeting, held in Oxford in 1832, initiating a fruitful connection between the BAAS and tidology. During its third meeting, its first at Cambridge and the first with Whewell serving as vice president, the BAAS offered its first grant for scientific research: £200 for the reduction of tidal data. Other grants for tidal research followed: an additional £50 in 1834 and £250 in 1835. Whewell also gave presentations at the meetings. At the first meeting he attended in 1832, in a well-attended evening lecture, he focused on how local observers could contribute to the science. It was published as an appendix to the *Third Report of the Association* and was similar to the separate sheet he published for officers in the Admiralty.

That same year Whewell also presented his first paper on the tides to the Royal Society of London. As was becoming standard practice for Whewell, he outlined previous advances in the field (history), defined the correct terminology (nomenclature), and delineated his anticipated research that he believed would most successfully advance the field (philosophy of discovery). The last section of the paper, "Suggestions for Future Tide-Observations," outlined the gaps in observations while simultaneously foreshadowing his future research programs. The first observations that were required were "good and long-continued observations" at specific ports, to use those observations to create tide tables, and then to compare the tables to future observations. The process of comparing observation with tables, according to Whewell, could yield the laws that accounted for the phenomenon. Only then could researchers move to unearthing the theory of why such laws held. The history of physical astronomy suggested this approach, though tidology had yet to advance in that direction.[27] Beginning in 1833, four years before the publication of his *History* but during the time of its initial writing, Whewell embarked on this exact route.

There was a major problem with this approach, however. It required upward of two decades of observations at every port, and such observations simply did not exist except for a few select places. As Whewell explained: "Mathematicians have not yet learned to make accurate and trustworthy tables for any place, without having a long and careful series of observations *made at that very place*."[28] This could be overcome by the second set of observations that Whewell suggested were required, combined with a novel approach to analyzing the observations. "But in the meantime," Whewell wrote, "no one appears to have attempted to trace the nature of the connexion among the tides of different parts of the world."[29] This was his innovative leap in tidology.

Previous tidal studies had relied on the relation of *time*; Whewell's niche in tidology entailed focusing on the relation of *space*. As Whewell explained: "Continued observations at the same place are connected by relations of time; comparative observations at different places are connected by relations of space. The former relations have been made the subject of theory, however imperfectly; the latter have not."[30] In this way he transformed the study into a global, geographic science. He attempted to formulate a theory based on the progression of the tide throughout the world's oceans, enabling him to extrapolate from one port to the next and eventually to all ports in Europe and beyond. The upside was that he would no longer require long-term observations at each and every place. But there were also downsides. He now required short-term observations made at as many places as possible along the coastline of the world's oceans. If this wasn't difficult enough, there was an added hitch: the observations at every station had to be made simultaneously.

Shortly after this first publication, he wrote to Lubbock, hopeful that the tides would eventually advance to the level of other parts of astronomy "and have observers and calculators employed upon it in all parts of Europe." First, however, he limited his study to the coasts of Britain, starting out by demonstrating what could be done. "This is the way in which science generally begins, and we may as well make up our minds to it."[31] Whewell, it turned out, had reason to be hopeful. Richard Spence, a captain in the Royal Navy, had received Whewell's earlier circular that suggested comparative observations around Britain "may also be of great value." Spence informed Whewell that naval officers in the Coast Guard were stationed around the entire coast of Great Britain under the supervision of Captain William Bowles, the comptroller general. "Wouldn't they be perfect for helping gather the required observations?"[32]

This was exactly the type of mass participation Whewell required. Excited by the prospect, he shot a request off to Beaufort, who in turn contacted Bowles. "Mr Whewell, wants something more," Beaufort informed Bowles: "a consecutive line of observations along the coast of Great Britain made simultaneously on the same tidal wave and continued for a fortnight. Is there a possibility that you could accomplish such a grand operation for us at the whole series of your stations at one and the same time?"[33] Bowles, in turn, wrote to the deputy comptroller general, Samuel Sparshott, who responded that obviously not all the stations were needed, "many of them not being more than three or four miles apart."[34] Beaufort assured Sparshott otherwise, insisting that Bowles send the instructions to each officer in every district.[35]

In June 1834, following the instructions written by Whewell, the orders given by Beaufort, the organization of Bowles and Sparshott, and the participation of servicemen of the British Coast Guard, over five hundred stations in England, Scotland, Wales, and Ireland measured the changing surface of the sea every fifteen minutes, day and night, for two weeks. It was the first time such a grand and organized set of observations had ever been made. Thousands of observations inundated Whewell's lodgings in Trinity College, masses of tables, charts, and (as will be discussed in the next section) curves. "The returns of last June are more consistent and accurate than I could have anticipated," Whewell noted in his paper published in the *Philosophical Transactions* the following year, "made in many instances with ingenious and suitable contrivances."[36]

Whewell had cast his net widely. From missionary societies to the BAAS to the Admiralty and its network of observers around the coasts of Great Britain and Ireland, he had successfully marshaled hundreds of observers and tens of thousands of observations, along with ingenious and suitable contrivances, including models for tide gauges, possible methods of observing the tides far from the coast, and plans and suggestions for future measurements. Whewell worked hard to bring together the results, hoping to induce the Coast Guard to repeat the observations the following year. But Beaufort had other designs. He wrote to Whewell in February 1835, "Would it not be delightful appendage to the batch of Coast Guard Tides which are to be observed this year if we were to procure simultaneous observations along the shores of the Netherlands and France, Newfoundland, Nova Scotia and North America? If so, no time should be lost in determining on the periods at which the operations should take place. Suppose from the 9th to the 27 of June?"[37] A month after severe high tides had inundated the city streets of London and Westminster, requiring boats to maneuver through canalled streets to save its inhabitants, Beaufort and Whewell began planning "the great tide experiment," what Whewell boasted as including the most multiplied and extensive observations yet encountered in science.

Whewell's move from a temporal to a spatial approach transformed the study of the world's tides. He combined this novel approach with the new graphical method taking root in the physical sciences at that time. In a geographical study such as the tides—and ocean currents, atmospheric streams, and terrestrial magnetism—the graph served as the perfect means to find the relationship between complex variables, transforming massive amounts of data through space into a synoptic visual that could fit on a mariner's table. "We may in a very few years be able to draw a map of cotidal lines with certainty and accuracy," Whewell boasted, "and thus to give, upon a single sheet, a tide table for all ports of the earth."[38] The cotidal map served as a powerful tool used to consolidate power for both its creators and its users. They in turn supported the authority of the graphs, "bringing together" the motivations of the scientists and the State.

This intermingling of science and the State was everywhere apparent in the processes involved in the 1834 measurements around the coasts of Great Britain. Instructions had traveled back and forth across the chains of command, between Whewell and Beaufort, scientists and mariners, commanders in London and servicemen on the coast. The organization of the 1835 observations similarly flowed across hierarchical lines of command, but on a dizzyingly broader basis. Beaufort initially wrote to the Board of Admiralty, initiating the secretary of the board to write to the Foreign Office, leading the Duke of Wellington, the foreign secretary, to make the application to foreign governments through the British consul in each nation.[39] Whewell wrote the instructions for taking the observations, which Beaufort then personalized into detailed directions and printed forms for each station.[40]

Through these and similar efforts, the "great tide experiment" commenced. Nine countries—including England and Scotland (318 stations), Ireland (219), the United States (28), France (16), Spain (7), Portugal (7), Belgium (5), the Netherlands (18), Denmark (24), and Norway (24)—participated. For the first time in history, observers measured the tides at the same time, on the same tide wave, owing to the same astronomical forces, every fifteen minutes, day and night, for twenty consecutive days. The measurements covered both sides of the Atlantic, along with numerous islands under British and French possession, including the Isles of Scilly, the Ilse of Man, Mauritius, Malta, Ceylon, and three of the Channel Islands. They also extended into the Pacific, with ports in South Africa, New Zealand, and Australia also contributing data. In a letter to Forbes over twenty years after the observations were taken, Whewell referred to this undertaking as his "great achievement in Tidology."[41]

Taking observations, of course, constituted only the first step in a much larger process. As the thousands of data points began to amass in the crowded rooms of the Hydrographic Office, and eventually in Whewell's lodgings at Trinity, they became more and more unwieldy. Some were measured from low water, others from high water, some with primitive tide gauges, others with sophisticated self-registering machines. Some were incomplete, while others were obviously fabricated (either copied from the previous day, from published tide tables, or from seemingly thin air). All measurements—over forty thousand for high tide alone—needed to be calibrated to a constant zero point and correlated with wind and weather and their position on the globe. Verifying and organizing the data fell to the gaggle of professional computers that Beaufort had put on the job, including Joseph Dessiou, Daniel Ross, and H. Brody. Whewell had his own unique means of verification and organization, the most important job he assigned to the computers under his charge. For every station he had them transform the verified and tabulated data into curves, relating the height of the water with the passage of time. "The inspection of these curves," Whewell explained, "afforded me the means of judging of the best mode of combining them so as to get rid of local and casual anomalies."[42]

The process of "judging" and "combining"—the "bringing together" of facts through theory—fell on Whewell. It was his job to transform the mass of confusion into a stable product. "It became me," as Whewell put it, "to turn to the best advantage the large mass of materials thus collected."[43] Curves formed into isomaps were the perfect instruments to visually display the lawlike and seemingly uniform nature of the winds, tides, currents, magnetic variations, and barometric pressures over the world's oceans and landmasses.

Tides (and winds and weather and magnetic variation) all moved on their own accord, uncontrolled by the scientist or the mariner. Isomaps, however, were stable, enabling scientists to view dynamic nature as fixed and concrete. This is the purpose of charts and graphs, to convey what is complex and variable into an easily comprehended form, bringing chaos under control. The graphs' stability makes it easily transferable. No matter the variables used, the chart itself can be presented to the Royal Society or fit on the cabin wall of a schooner. A graph of the tides of the North Sea, moreover, can also be extended to the English Channel, and later be easily connected with a chart of the entire Atlantic. These in turn can be superimposed with similar charts, such as those of magnetic variations, covering similar large or small geographies. They combine to form whatever space the scientist or mariner desires to understand and control. Owing to their stability, mobility, reproducibility, and combinability, they created a flat representation of a dense assemblage of forces, a seemingly innocuous end product of purely scientific practice.

Graphs require faith, relying on the power of science to master the chaos of nature. As James Scott has explained, maps narrow the vision of its users. "This very simplification, in turn, makes the phenomenon at the center of the field of vision more legible and hence more susceptible to careful measurement and calculation. Combined with similar observations, an overall, aggregate, synoptic view of a selective reality is achieved, making possible a high degree of schematic knowledge, control, and manipulation."[44] How this selective reality is achieved, leading to control and manipulation, can be found not only in the steady hand of the professional calculator or the organizing ability of the scientist, but also in other (much more social) factors that lie hidden behind the two-dimensional end product.

Whewell's cotidal maps were built on tens of thousands of data points, placed into thousands of tables, and transformed into hundreds of graphs. These graphs themselves relied on a wide range of instruments, from simple tide gauges made of a tube and float, to sophisticated self-registering gauges consisting of pullies and gears. Beneath these observations, graphs, and instruments were other tangible forces giving rise to the authority of the cotidal maps. From the Admiralty's perspective, this included the "national object" of bringing order to the tides for the safety of mariners. It also spoke to the role of the Hydrographic Office within the Admiralty and that of the Coast Guard during peacetime. All this figured into the acceptance of Whewell's cotidal maps. For Whewell and the scientists similar motivations were at stake. Whewell was driven by his desire to contribute to science as an active researcher, to demonstrate the power of a spatial science and its graphical method of data analysis, and the proscriptive philosophy of discovery he was then attempting to create. The graphical method focused all of this on the stable coastlines and curves of cotidal maps rather than the turbulent ebb and flow of the ocean.

Whewell added one last process to give authority to his cotidal maps: that of the definition of science itself. His work recapitulated the process of scientific discovery as outlined in his *History* and *The Philosophy of the Inductive Sciences* (1847). In both of these publications he presented science as a progressive climb where each branch went through specific stages—a prelude, an inductive epoch, and a sequel—a gradual ascent moving from observations to phenomenological laws to causal laws. The first step in all the sciences was the important act of "colligation," where the researcher was confronted with a mass of facts that needed to be organized into mathematical laws. The bringing together of facts through the power of the intellect represented the most important step in the discovery process, setting the stage for the establishment of causal laws. Through this process, scientists formulated new "conceptions" that were required for each inductive epoch.

Whewell's model science was formal astronomy, his favorite example of which was planetary motion, and his exemplary scientist was Johannes Kepler. Kepler had used Tycho Brahe's observations of the path of Mars to determine the phenomenological laws of planetary motion through the "inventive fancy" of introducing the elliptical path. The inductive epoch of Kepler itself formed the prelude to "the last and most splendid period in the progress of Astronomy," the formulation of the cause of such motion.[45] It is an important aspect of Whewell's *History* that any further advance in the mechanics of the heavens would have to incorporate the phenomenological laws of Kepler. As Whewell explained, this proved to be the case with the introduction of Newton's theory of universal gravitation. With surprising foresight Whewell also believed that this would hold for Newton's theory as well. "It is no doubt conceivable," Whewell wrote, "that future discoveries may both extend and further explain Newton's doctrines;—may make gravitation a case of some wider law, and may disclose something of the mode in which it operates."[46]

Whewell noted that the discovery of phenomenological laws may often constitute the whole of a science for a very long time, and it was only through "great talents and great efforts" that advance to causal laws could be made. The most important step was to add a new element into the research. As Whewell explained, "The labour and the struggle is, not to analyze the phenomena according to any preconceived and already familiar ideas, but to form distinctly new conceptions."[47]

This then brings us to how scientists formed such distinctly new conceptions. Whewell stressed that no rules existed that would immediately give the general form of the relation by which astronomical phenomena were connected. But, he added, "there are certain methods by which, in a narrower field, our investigations may be materially promoted;—certain special methods of obtaining laws from observations."[48] And it is to this section, "Special Methods of Induction Applicable to Quantity," to which Whewell turned. Whewell listed them as the methods of curves, of means, of least squares, and of residues. He then proceeded to describe in detail the modern graphical method, using his own research on tides as way of explanation.

The graphical method did not simply represent the data; it actually added an additional element, giving the data new form, creating meaning from "obscurity and complexity."[49] Through these methods, a researcher could create data points that were otherwise impossible to measure, in the deep ocean for instance, far from the sight of land. Moreover, it could also demonstrate which observations were the most accurate and those that were obviously erroneous. By drawing a curve, "not through the points given by our observations, but *among* them," he could "obtain data which are *more true than the individual facts themselves*."[50] The method enabled a researcher to invent data that, according to Whewell, was more trustworthy, more real, more true, than the original observations.

In Whewell's *History* and *Philosophy*, tidology fit into the progress of the sciences in two ways. It represented the final stage of the sequel to Newton's inductive epoch, the last pillar to be erected in verifying the law of universal gravitation.[51] The study of the tides, however, also had its own prelude, inductive epoch, and sequel. It required its own colligation, its own conceptions, its own inventive fancy. He purposefully and repeatedly delineated the study of the tides as "the last great bastion of physical astronomy," which required a researcher to "colligate" massive amounts of observations though "new conceptions" in order to formulate phenomenological laws. His act of colligation for his study of the tides, linking observation to theory, exactly mimicked Kepler's process for explaining planetary motion. From masses and masses of observations, Whewell superimposed on the data a visual graph in the form of cotidal lines that brought order and regularity to the observations. Whewell viewed himself explicitly as the Kepler of modern tidology.

The further question was how one knew whether the phenomenological laws were correct. Here, too, Whewell's work drew from the philosophy of science he was formulating early in his tidal researches and that he would later publish in his *History* and *Philosophy*. After his initial paper on cotidal lines in 1833, his second paper, "On the Empirical Laws of the Tides at the Port of London," established the phenomenological laws of the inequalities associated with the London tides.[52] He then extended this to a second port. By doing the same thing for the port of Liverpool, Whewell had a means of "testing and improving the formulae to which I was led by the London observations.[53] His fourth publication, "On the Empirical Laws of Tides in the Port of Liverpool," agreed with the London tides "with a precision not far below that of other astronomical phenomena." Such a result, Whewell affirmed, suggested that it "is impossible to doubt, under these circumstances, that the theoretical formula truly represents the observed facts."[54] Whewell would use similar reasoning (and language) in his *Philosophy* when discussing the manner in which hypotheses could be tested, where the second step in the consilience process was to "foretel phenomena which have not yet been observed; at least all phenomena of the same kind as those which the hypothesis was invented to explain."[55] The process of comparing calculated tables with actual observations, the focus of Whewell's own research in tidology, was the important step that had been followed by all previous advances in formal and physical astronomy. It is what made the science "inductive."

The Sequel to the Inductive Epoch

Twenty years after Whewell had politicked, unsuccessfully, to hold the Lowdean Chair of Astronomy based on his work in tidology, he was still deeply enmeshed in tidal studies. He had been organizing support for what he termed the "Great Atlantic Tide Expedition" since at least 1838, the year he successfully received the professorship of moral philosophy at Cambridge. He continued to do so throughout the 1840s and early 1850s, culminating in an appointment with Lord Aberdeen, the prime minister, in May 1853. With George Biddell Airy and Edward Sabine in attendance, Whewell laid out his most convincing argument for the tide expedition. By this point he realized that a true causal theory of the tides would entail a study of the hydrodynamics of wave motion in large basins, not the equilibrium of fluids, and only through a tide expedition to hunt them down could he make further advances.

What would have formed the sequel to Whewell's tidal investigations never left port, and he never moved beyond the Newton-Bernoulli equilibrium theory of the tides. His work was based on a false theory, but one that worked to predict the tides.[56] Uneasily, this is what Whewell admitted all along. The rival theory, Pierre-Simon Laplace's hydrodynamic approach, required especially robust mathematics, and the integration of the partial differential equations proved utterly impossible in practice, even with Laplace's simplification of an ocean of constant depth covering a perfectly spherical earth. Furthermore, the irregularity of the ocean floor and the local conditions of the coast made the hydrodynamical solution to the tides even more problematic.

In practical terms then Whewell followed the equilibrium theory because mathematicians had yet to solve the differential equations needed for a hydrodynamic approach. A hydrostatic approach seemed equally daunting, requiring an incredibly ambitious research program of both long-term observations at specific ports and short-term observations made simultaneously around the globe. Britain's imperial expansion, however, did make this approach possible. Whewell was forced to make the tides global, a decision conditioned by Britain's imperial reach.

There are other ways, of course, that Whewell's tidology was linked to the power of the State. The British did not demand legal sovereignty over the ocean or always attempt to colonize distant estuaries; rather, they intellectually possessed these geographical regions by mastering their tides, streams, and waves, mapping their magnetic and atmospheric curves, and topographically setting the boundaries and contours of both land and sea through the power of the isomap. In this way, the British symbolically owned the ocean. For Whewell, scientists were the specialists who undertook this important work. They transformed the unknown into the known, graphed it onto easily transferable graphs, and helped produce an intellectual mastery that translated directly to physical mastery of the coastlines on the oceans' outer rims.

As Whewell visualized ocean space through his cotidal maps, he was also simultaneously creating an intellectual space for the modern *scientist*, a term he coined at the outset of his tidal investigations. His science was based on broad participation, with those he termed "subordinate labourers" helping to build the instruments, take the observations, and reduce the data. Indeed, military personnel associated with the coastal and trigonometrical surveys of the nineteenth century were essential for the study of the tides. "There is, probably, no class of society which has more frequent opportunities of adding to the general stock of scientific knowledge," noted one contributor to the first volume of *Nautical Magazine* in 1832, "than that composed of persons in the royal and mercantile navy."[57] Along with naval captains and surveyors, military and colonial administrators, dockyard officials and engineers, harbormasters, and tide table calculators and publishers, all combined to make a citizen-driven science possible.

But the scientist alone colligated from the center. Whewell was clear about the hierarchical nature involved in scientific discovery. His *History* and *Philosophy* helped define the nature of science, ensuring that a hierarchical nature was ingrained in the process.[58] Whewell defined scientists as people who had a larger historical vision of the slow and steady advances required to transform observations into phenomenological laws and eventually into causal laws. They were experts at colligation and inductive inferences.

Whewell always had his own study of the tides in mind as he contemplated history and formulated his philosophy. Both guided his own research, serving as a heuristic for the progress of science more generally. Though the graphical method is perhaps the most obvious place where Whewell used tidology to exemplify his philosophy of discovery, the tides formed the backdrop to all aspects of his philosophy of discovery, including the act of "colligation," the formulation of "conceptions," and his concept of "consilience." We know he was steeped in his tidal studies at the exact time he was formulating these concepts. "So much for philosophy," he wrote to Herschel in 1836, directly linking his work on induction with his tidal studies. "When I talk of giving the rest of my life to it, I always reserve to myself the tides, as a corner of physics which I shall go on and work at till all is done that I can do."[59] He made the connections explicit by repeatedly referencing his tidal work in his major publications. Whewell was no dabbler in tidology, never content to sit on the sidelines. Rather, his science was central to all of his other pursuits as a meta-scientist. His hierarchical definition of science and his positioning of the scientist as central to the State also continue to influence our own perceptions today. They too are based on Whewell's twenty-year research project in tidology.

CHAPTER 8

Whewell on Astronomy and Natural Theology

BERNARD LIGHTMAN

William Whewell delivered four sermons at the University of Cambridge in 1837, later published in book form as *On the Foundations of Morals* (1837). In the preface to the book he identified William Paley as the chief target of the work. "In the following Discourses," Whewell asserted, "disapprobation is expressed of a work now in use in the Examinations of the University of Cambridge—Paley's Moral Philosophy." Whewell recommended that Paley's book be dropped from the Cambridge curriculum immediately. He declared that "the evils which arise from the countenance thus afforded to the principles of Paley's system are so great, as to make it desirable for us to withdraw our sanctions from his doctrines without further delay."[1] Whewell's condemnation of Paley carried tremendous weight, especially within Cambridge itself. In 1838, just before his book was published, he had taken up the chair of moral philosophy at Cambridge. A few years later, in 1841, he was appointed as master of Trinity College.

Whewell's negative attitude to Paley may be, at first glance, rather puzzling. Whewell was well known as one of the authors of the Bridgewater Treatises, a series of works on natural theology published between 1833 and 1836. English clergyman, philosopher, and Christian apologist, William Paley was the author of one of the classic nineteenth-century statements of natural theology, *Natural Theology or Evidence of the Existence and Attributes of the Deity* (1802). Whewell's *On the Foundations of Morals* therefore raises the question: Did Whewell's criticism of Paley's moral philosophy also extend to Paley's natural theology? In his recently published *The Trinity Circle*, William J. Ashworth depicts Whewell as a leading member of a group of Anglicans, along with Julius Charles Hare and Connop Thirlwall, who began to be anxious about the growth of materialism and faith in the post–Napoleon war period of the 1820s and 1830s. They opposed the manifestations of these ungodly tendencies in Ricardian political economy, abstract French algebraic mathematics, Benthamite utilitarianism, and evolution. Ashworth argues that Whewell viewed Paley as a utilitarian, both in his moral philosophy and natural theology.[2] Here I want to build on Ashworth's approach to this aspect of Whewell's thought, but focus on how it plays out in his astronomical work.

Whewell may have been a polymath, but astronomy played a particularly significant role in how he developed his version of natural theology. He found the strongest empirical evidence for design in the natural laws revealed by the study of astronomy. Natural theology was a type of theology in which knowledge of God was sought by reason alone, independent of God's self-revelation in miracles, prophecies, and scriptures. From the second half of the seventeenth century to the middle of the nineteenth century, natural theology was at its peak in the English-speaking world.[3] As a result, concepts such as final cause, design, laws, miracle, and Providence were essential components of scientific discussion. Natural theology was so central to scientific thinking that its teleological assumptions, as Richard Yeo has remarked, "provided the intellectual framework in which scientific thinking occurred."[4] But there were two versions of natural theology. Both drew more often on evidence from the organic world. The most popular version emphasized the usefulness of the characteristics of plants and animals in adapting to their environments. This was a "utilitarian" form of natural theology in that each instance of adaptation provided additional evidence for divine wisdom, power, and goodness. The second version emphasized the pervasive existence of intelligible patterns in the organic world. In this idealist form of the argument the history of life was understood as the gradual material realization of a premeditated and integrated plan formulated by a benevolent and rational God. Some natural theologians used both of these approaches in their defense of theism. Others emphasized either the utilitarian or the idealist version.[5] Paley adhered to the utilitarian version and in his *Natural Theology* drew attention to the intricate contrivance to be found in the specific organs of living beings, such as the eye. By contrast, Whewell gravitated toward the idealist formulation, though he focused more on the inorganic than the organic world. Others who embraced the idealist version included the biologist Richard Owen and the liberal Anglican Frederick Temple. Although Whewell seemed to switch targets in the 1840s from Paley's utilitarianism to the evolutionary theory of the *Vestiges of the Natural History of Creation* (1844), the theme connecting his astronomical work over the course of his life was his commitment to the idealist version of natural theology.

Astronomy and General Physics

Whewell's *Astronomy and General Physics* (1833) was the first of the Bridgewater Treatises to be published. Jonathan Topham's recent study of the Bridgewater Treatises traces how they came to be seen as emblematic of the attempt to reconnect the sciences with Christianity just prior to the Victorian period. New developments in astronomy, geology, phrenology, and evolution had made the relation of the sciences to Christianity increasingly problematic. The treatises helped to define the public face of the sciences in the mid-nineteenth-century period by depicting them as no threat to Christianity. Like Whewell, several of the Bridgewater authors focused on how knowledge of the laws through which God governed the universe revealed his ongoing role in nature. Whewell believed that knowledge of the sciences offered new insights into God's relationship with the creation. As Topham points out, Whewell consciously chose to write a "reflective" work, rather than a theological treatise in the Paleyan tradition, intended to point to the wider religious significance of modern astronomy and general physics. This made his book far more accessible to a wide range of readers.[6] Whewell's was the most popular of all the Bridgewater Treatises, reaching a seventh edition in 1864.

In his introduction to *Astronomy and General Physics*, Whewell affirmed that "new discoveries" and "new generalizations" led us to "regard nature in a new light." This in turn could modify the "conception concerning the Deity, his mode of effecting his purposes, [and] the scheme of his government." In natural philosophy the "point of view" of its most successfully cultivated departments was that nature was "a collection of facts governed by *laws*: our knowledge of nature is our knowledge of laws." Whewell's aim was to examine how this "view of the universe" aligned with "our conception of the Divine Author." He promised his readers to show how the laws prevailing in nature were remarkably adapted to their purpose and therefore provided evidence of "selection, design, and goodness, in the power by which they were established." But the ascertainment of laws had been discovered in only "very few departments of research" in science, astronomy being one of them. The motions of the sun, moon, and stars had been reduced more completely to their causes and laws than in any other class of phenomena. Astronomy was a component of natural philosophy developed completely enough that it could be used to "make out the adaptations and aims which exists in the laws of nature; and thus to obtain some light on the tendency of this part of the legislation of the universe, and on the character and disposition of the Legislator."[7]

Astronomy and General Physics was divided into three books, "Terrestrial Adaptations," "Cosmical Arrangements," and "Religious Views." Book 1 dealt with the relations between the laws of the inorganic world (the general laws of astronomy and meteorology) and the laws that prevail in the organic world (the properties of plants and animals). Whewell's goal was to show that a wise and benevolent design had been exercised in producing an agreement between the laws of these two worlds. The length of the year and the day, the mass of the earth, the magnitude of the oceans and the atmosphere, the constancy and variety of climates, the laws of electricity and magnetism, the properties of light with regard to vegetation, the nature of sound, the atmosphere, light, and the ether, all pointed in the same direction: the structure of the world was the "work of choice" of a "most wise and benevolent Chooser," rather than the consequence of chance. The constitution of the world was "fitted for the support of vegetables and animals, and in a manner in which it could not have been, if the properties and quantities of the elements had been different from what they are."[8]

In book 2 Whewell moved on to what he called "cosmical arrangements," which mainly concerned astronomical topics. He began this section of the book by noting that in the past "writers on Natural Theology" had asserted that "arguments founded on those provisions and adaptations which more immediately affect the well-being of organized creatures" were to be preferred to arguments "drawn from cosmical considerations." Whereas "the structure of the solar system has far less analogy with such machinery as we can construct and comprehend," previous natural theologians had argued, this was not the case with "the structure of the bodies of animals." In addition, they maintained, the immediate bearing of cosmical arrangements on the support and comfort of sentient creatures was unclear. Whewell rejected this approach to natural theology. Considering the universe as "a collection of *laws*," astronomy provided some advantages over the version of natural theology that focused on design in organic beings. Whewell believed that our knowledge of the laws of the motions of the planets and satellites was more complete, and exact, than in any other department of natural philosophy.[9] This was Whewell's response to those who, like Paley, made isolated instances of organic mechanism the touchstone of design in nature.[10]

Whewell then went on to illustrate his point by examining the structure of the solar system, the designed circular orbits of the planets, the nature of the satellites, the stability of the oceans, the nebular hypothesis, the operation of mechanical laws, the law of gravitation, the laws of motion, and the existence of friction. The circular orbits of the planets, for example, were remarkably regular, which "excludes the notion of accident" in their arrangement. But not only was the structure regular, it was also the only one "which would answer the purpose of the earth, perhaps of the other planets, as the seat of animal and vegetable life." The circular form of the orbit appeared to be "chosen with *some* design" to secure the welfare of organic life. In book

2 Paley was joined by Pierre-Simon Laplace as a target of Whewell's criticism. Although Laplace understood that the stability of the solar system was not the result of chance, he failed to draw the proper conclusion that the arrangement of the solar system was the work of an intelligent and powerful being. Even if Laplace's nebular hypothesis were demonstrated to be true—which Whewell was not yet willing to concede—it did not prove that the solar system was formed without the intervention of intelligence and design. Whewell asked, Was there anything to prevent "our looking beyond the hypothesis, to a First Cause, an Intelligent Author, an origin proceeding from free volition, not from material necessity?"[11]

In book 3, "Religious Views," the final part of *Astronomy and General Physics*, Whewell discussed the broader implications of the findings of modern physical science that demonstrated the lawful nature of the universe. In the process Whewell enlarged the scope of natural theology by bringing in moral issues and making them central. Since Whewell saw a strong connection between natural theology and moral theory, it is clearer why his critique of Paley the moralist inevitably implied opposition to Paley the natural theologian. Although contemplation of the material universe "exhibits God to us as the author of the laws of material nature," Whewell maintained that a consideration of the moral world—"the results of our powers of thought and action"—was of far great significance. Here the reader was led to the God of revelation. Whewell's point was that there was a connection "between the evidences of creative power, and of moral government in the world." Though natural theology proper did not include moral philosophy, it was Whewell's belief that it led "naturally" toward it. For the creator of the universe was also the creator of the human conscience. However not everyone, including all scientists, were able to perceive this connection, since it was not direct, though the two lines of reason "converge to the same point."[12]

Whewell then launched into an examination of how advances in knowledge had tended to impress upon the great scientists of the past the reality of the divine government of the world. Galileo, Newton, Pascal, Boyle, Black, and Dalton were all believers in a wise maker and master of the universe. Those scientists who were unable to perceive divine design in nature or morality were, Whewell asserted, by definition, mere second-rate figures. The difference between first- and second-rate scientists boiled down to their habits of thought. Scientific, and moral, consequences followed from the adoption of either inductive or deductive habits of thought. The great scientific minds used the more difficult habit of inductive thinking when they made their original discoveries. Genuine inductive thinking led to the belief in the existence of a supreme intelligence and purpose. Second-rate minds relied on "deductive habits" that were, in essence, a form of derivative speculation, dependent on the general laws discovered by first-rate scientists, that added to our knowledge of effects but not to our knowledge of causes. Whewell made it clear that the majority of scientists used deduction, and included well-known religious skeptics like Jean le Ronde d'Alembert and Laplace in that category. "We may thus," Whewell insisted, "with the greatest propriety, deny to the mechanical philosophers and mathematicians of recent times any authority with regard to their views of the administration of the universe." Whereas Laplace predicted that final causes would disappear as our knowledge grew, Whewell affirmed that the discoveries of science would transfer the notion of design "from the region of facts to that of laws."[13] This section of Whewell's Bridgewater Treatise, in effect, as Richard Yeo has stated, shifted the emphasis from science and the design argument to the moral character of scientists, a novel and controversial move within natural theology.[14]

Whewell's final point was both a condemnation of Paley's version of natural theology and an admission of the limits of the design argument. Paley's emphasis on the analogy between human and divine contrivance, Whewell argued, was exceedingly problematic. For Whewell it was essential that scientists, theologians, and the general reader go beyond the analogy of human contrivance when conceiving of God. It was true that God constructed "the most refined and vast machinery." But more importantly, God established "those properties by which such machinery is possible: as giving to the materials of his structures the qualities by which the material is fitted to its use." This gave priority to the structure of nature—the laws that ordered the universe. But Whewell acknowledged that his approach to natural theology cut both ways. If God were conceived of as the author of the natural laws that "make matter what it is," no "analogy of human inventions, no knowledge of human powers, at all assists us to embody or understand" God's actions. Science disclosed "the mode of instrumentality employed by the Deity" while making it impossible to assimilate God's actions to human ones. In the end science represented God "as incomprehensible." This was Whewell's way of subordinating reason to revelation.[15] He ended his treatise on a note of humility. This point complements Michael Ledger-Lomas's assertion, in his chapter "In the Chapel of Trinity College" in this volume, that Whewell's sermons showed that he was careful to place limits on the value of natural theology.

In his detailed analysis of the Bridgewater Treatises, John M. Robson asserted that the authors "did not see themselves as merely reworking Paley's *Natural Theology*, or as performing the relatively insignificant task of making Paley's thought available thirty years on."[16] I have been arguing that Whewell saw himself as being in a more confrontational relationship with Paley. Instead of reworking Paley, or making his ideas accessible to a new generation of readers, Whewell's aim was to undermine the utilitarian form of natural theology while emphasizing an idealist version grounded on the notion of natural law. Near the end of the third book of his treatise, Whewell spelled out the central point of the entire treatise. He argued that despite the human inability to conceive how the material universe is the work of God, "we can at least go so far as this;—we can perceive that events are brought about, not by insulated interpositions of divine power exerted in each particular case, but by the establishment of general laws." This was the view of the universe proper to science and was also "the view which, throughout this work, we have endeavoured to keep present to the mind of the reader." In this, science combined "harmoniously with the doctrines of Natural Theology."[17]

Indications of the Creator

Whewell's *Indications of the Creator* (1845) was a peculiar book composed of excerpts from some of Whewell's previously published works, including *History of the Inductive Sciences* (1837) and his Bridgewater Treatise. In the preface Whewell wrote that the extracts were published for those who would were not likely to read the larger works from which they were taken.[18] These extracts all dealt with the topics of "Indications of Design in the Creator, and of a Supernatural Origin of the World; and, as connected with this latter point, the consistency of the Inductive with the Revealed History of the World." These themes, Whewell affirmed, "belong to Natural Theology," and they were the same from age to age, though they changed "their aspect with every advance or supposed advance in the Inductive Sciences." Whewell made it clear that he had tried to refute the baseless assertion that "as science advances, final causes recede before it, and disappear one after the other."[19] But in the preface Whewell never revealed the real reason why these previously published excerpts needed to be brought to the attention of the public once more in the year 1845.

Whewell must have realized that he had left this question unanswered in the preface as he wrote to his friends to explain himself. The book was written to combat the pernicious influence of Robert Chambers's anonymously published *Vestiges of the Natural History of Creation*, an evolutionary tract that had become a sensation, particularly among middle-class readers.[20] On July 18, 1845, Whewell claimed to the economist Richard Jones that he was amused by the success of the *Vestiges* and reflected on how "truths of any broad philosophical kind" would not easily find admirers in the current intellectual environment. The popularity of the *Vestiges* merely revealed its lack of philosophical sophistication. "No really philosophical book," Whewell declared, "could have had such success: and the very unphilosophical character of the thing made it excessively hard for a philosophical man to answer it, and still more to get a hearing if he did."[21] The peculiar nature of *Indications of the Creator* reflected how difficult it was for an intelligent man to respond to an unphilosophical work. Whewell's letter to John Herschel on March 12, 1845, is even more revealing: he feared that *Indications* might be viewed as insubstantial as the *Vestiges:* He began defensively: "I think I must make an effort to rescue my little book from your contempt." Herschel, Whewell believed, looked on *Indications* as pandering to a trendy "love of little books that pretend to contain the essence of large ones, and daintily dressed for dainty people." Whewell flatly denied it. "I do not know," he asked, "if you have seen or heard (from your letter I should think not) of a book called *Vestiges of Creation*. It is anonymous, has been circulated with great zeal and mystery, and is much read and talked of." Widely considered materialist in tendency, Whewell had been urged, by whom he did not say, to answer it. Whewell claimed that he had refused to do so, "except by extracting passages from my previous books" in which "all the arguments of the *Vestiges* are discussed and answered." Therefore *Indications* was not, as Herschel seemed to believe, "a wanton selection of elegant extracts, but a compulsory selection of theological abstracts" intended as a response to a dangerous evolutionary text that had challenged the hold of natural theology on the public mind in the 1840s.[22] In *Indications* Whewell would once again draw on astronomy as a resource for demonstrating that the goals of true science and Christianity were aligned.

Previous scholars have already discussed how *Indications* was intended as a rebuttal of the *Vestiges*. In 1988 Cymbre Raub described *Indications* as a response to the deism of Chambers in that Whewell continually drew attention to the power and continuing presence of the Creator. The very title of Whewell's book asserted the existence of a Deity, while Chambers left God out of the scope of his inquiry.[23] Six years before Raub's article appeared, Donald McNally interpreted *Indications* as a refutation of *Vestiges*, pointing out that in the second edition the preface referred directly to Chambers's book.[24] The preface to the first edition, however, also rejected evolution but without mention of the *Vestiges*. Here Whewell discussed a "morphological doctrine of modern times which has attracted much notice" as it seemed to offer a solution to "the great difficulty of the uniformitarian theory in geology, namely, the appearance of new species and classes of animals as we proceed from the earlier to the later formations." This morphological doctrine asserted that the types of animals might be arranged in a series ascending from lower to higher, with each animal of a higher kind passing through embryonic states that were the final condition of the lower kind. Whewell insisted that the facts agreed on by zoologists, geologists, and physiologists "do not agree with this doctrine" or any other "doctrine of the development of the kinds of animals from one kind to another by the influence of external conditions." Instead, Whewell recommended the morphological views of the comparative anatomist, paleontologist, and natural historian Richard Owen, whose new discoveries of instances of creative design were examples of the "teleological turn of the Inductive Mind."[25] Chambers, then, had joined Paley and Laplace on Whewell's list of the chief threats to an idealist natural theology based on the concept of law.

The table of contents of *Indications* reflected the structure of the *Vestiges*. Astronomy played an important role in both books. In the *Vestiges* Chambers started with astronomy and the nebular hypothesis, moved to the geological development of the earth, the appearance of life on earth, and the early history and evolution of human beings. Chambers unfolded an evolutionary epic that synthetically drew on each of the sciences: astronomy, geology, and natural history. Similarly, the five sections of Whewell's book covered astronomy, physiology, geology, the philosophy of biology, and palaetiology. Whewell purposely avoided a structure that ordered the sections in any way that offered even a hint of evolution. But he wanted to demonstrate that all of the sciences offered support for natural theology.

The astronomy section contained two extracts, one from the *History of the Inductive Sciences* (book 5, chapter 3, section 4), which dealt with the Copernican System and Galileo; and the other from book 2, chapter 3, of *Astronomy and General Physics* that focused on the nebular hypothesis. The extract on the triumph of the heliocentric theory provided two lessons for readers about one of the most famous encounters between science and Christianity. First, the meaning a generation put on the phrases of Scripture depended on the received philosophy of the time. Therefore, those who thought they were defending revelation were really supporting current philosophy, in this case scholasticism. Second, those who adhered too rigidly to the traditional mode of understanding scriptural references to natural processes were always condemned by succeeding generations. The prosecutors of Galileo, Whewell asserted, were still held up to the scorn of humanity. Here Whewell signaled his open-mindedness to new scientific theories. His opposition to the *Vestiges* was not due to dogmatism or enslavement to a literal reading of the Bible.[26] The extract from his Bridgewater Treatise on the nebular hypothesis served a different purpose. It was intended to reject the tendency of Laplace, and Chambers, to narrate the beginning of the universe in purely naturalistic terms. If the nebular hypothesis was granted, it "by no means proves that the solar system was formed without the intervention of intelligence and design." The nebular hypothesis, as put forward by Laplace and Chambers, explained through purely natural causes how an orderly solar system was produced. But Whewell insisted that behind natural causes lay a "prior purpose and intelligence" who created the material astronomers referred to as nebular, and who, more importantly, also designed the laws governing the process described by Laplace's hypothesis.[27] By reproducing a section of his Bridgewater Treatise, Whewell implied that his idealistic version of natural theology from the 1830s was as effective an antidote to Paley's utilitarianism as it was to the new threat in the 1840s, *Vestiges'* deistic evolutionism.

Of the Plurality of Worlds

In a letter to the eminent astronomer John Herschel dated January 3, 1854, Whewell asked his friend if he had seen a recently published book titled *Of the Plurality of Worlds* (1853). Whewell thought that the anonymous author's rejection of extraterrestrial life "will be deemed to some extent heterodox in science" since it was "so much at variance with opinions which you have countenanced." But Whewell trusted that Herschel would agree that discussion on such matters should not be suppressed, and, in Whewell's opinion, "the author seems to me to have discussed the question very fairly." Moreover, Whewell believed, if rejecting the existence of life on other planets allowed "a more satisfactory view of the government and prospects of us, the dwellers on this Earth, many of us would deem the loss a gain." Perhaps, in light of the advantages to natural theology, Herschel would not be upset "if the inhabitants of Jupiter, or of the systems revolving about double stars which you have so carefully provided for, should be eliminated out of the universe." Whewell ended the section of his letter dealing with *Of the Plurality of Worlds* with his hope that "you astronomers will let us speculate on the one side as well as the other; which is all that my friend asks." Strikingly, instead of revealing to Herschel that he had written the book, Whewell pretended that a mysterious friend was the author. If Herschel had already read the book and guessed that Whewell had "anything to do" with it, he asked Herschel "to not encourage any body in the same opinion."[28] Taking a cue from Robert Chambers's efforts to protect his identity as author of the *Vestiges,* Whewell must have believed that his book was so sensational that he couldn't risk telling his secret even to one of his closest friends.

Of the Plurality of Worlds was connected to the *Vestiges* by more than just its anonymity. It was an indication of how Whewell continued to be worried about the negative impact of the *Vestiges* and evolutionary theory on natural theology. Whereas in the 1840s providing excerpts from previously published material in *Indications of the Creator* was seen by Whewell as a sufficient reply to *Vestiges,* by the 1850s he believed that a fuller response in a completely new work was desirable. Previous scholars have already discussed how *Of the Plurality of Worlds* was, at least in part, a response to Chambers. In his thorough analysis of the debate between Whewell and David Brewster, John Brooke argued that Whewell had rethought his earlier acceptance of pluralism because Chambers had associated it with evolution and a version of the nebular hypothesis hostile to Christian doctrine.[29] Michael J. Crowe, in his *Extraterrestrial Life Debate* (1986), agreed with Brooke that Chambers had had a powerful effect on Whewell.[30] More recently, Laura Snyder similarly asserted that Whewell changed his mind about pluralism due to the *Vestiges*, since Chambers had linked the existence of extraterrestrial life to evolution.[31] Here I want to treat *Of the Plurality of Worlds* as the continuation of Whewell's development of an idealist natural theology in astronomy. Whewell seemed to have decided by this point that he could safely leave it to Owen to martial the evidence to be drawn from natural history to respond to *Vestiges*. Owen was also thinking through an idealist natural theology in this period as a counter to materialist- and atheist-leaning theories of evolution.[32]

In his Bridgewater Treatise Whewell had accepted the possibility of life on other planets since this "enlarges and elevates" our view of the kingdom of nature.[33] But now, twenty years later in *Of the Plurality of Worlds*, Whewell declared that there was no scientific proof that intelligent beings like us existed on other planets. Humans were unique. "The earth and its inhabitants are under the care of God in a special manner," he wrote, "and we are utterly destitute of any reason for believing that other planets and other systems are under the care of God in the same manner." The assertion of the existence of extraterrestrial life was "opposite to the real spirit of modern science, and astronomy in particular." Whewell then made the unusual move of turning to the science of geology to show that the existing scientific evidence pointed away from pluralism. Geology had demonstrated that the existence and history of humanity was different from other animals. All current geological knowledge confirmed that, compared to other living things, humanity had "been fitted to be, the object of the care and guardianship, of the favour and government, of the Master and Governor of All, in a manner entirely different from anything which it is possible to believe with regard to the countless generations of brute creatures which had gone before him." Whewell's strategy was to use geology, a science in no way inferior to astronomy, to constrain pluralists who relied on astronomical speculations to support their position.[34]

But what of the evidence to be found in astronomy itself? Although he had high regard for the accomplishments of astronomers, Whewell believed that the topic of extraterrestrial life took us to the boundaries of genuine science. He examined all the alleged astronomical evidence for pluralism, starting from the largest and most stupendous astronomical phenomena, such as nebulae, and then moving to the earth itself. Contrary to the nebular hypothesis, all nebulae were resolvable into stars, as Lord Rosse's telescope had shown. Strikingly, given how much Whewell focused on design in his natural theology, he pointed out that a spiral nebula, compared with a solar system, was nothing other than a kind of chaos that did not resemble the state preceding an orderly and stable system. Nebulae were therefore not inhabited. The evidence that fixed stars were inhabited was, at best, "indeed slight." Clustered and double stars "give us but little promise of inhabitants." Whewell then moved on to planetary systems with a fixed star at the center. Here again, he raised questions about the attempt to find life in planetary systems. Those who argued that there was life assumed, with absolutely no evidence, that the stars at the center of these systems were like the sun. New stars were born, others died, while still others underwent changes during their lifetime. Fixed stars were not generally in the same permanent condition of stability as our sun. Whewell then turned his attention to the planets in our own solar system. Perhaps they were abodes of life? Whewell examined each planet, one by one, to show that none of them could support life. The moon was a barren place with no water or air. Jupiter and Saturn were both water giants where the force of gravity was much greater than on earth. Uranus and Neptune were similar to Jupiter and Saturn, only they received less light and heat from the sun. Mars resembled the earth is some ways, but there was no evidence of life. Venus and Mercury, too close to the sun, were not suitable for life. Only the earth was placed in that region of the solar system "in which the planet-forming powers are most vigorous and potent," between "the hot and fiery haze on one side, the cold and watery vapour on the other." Only this region was fit to be a "domestic hearth, a seat of habitation," and here a planet had been placed "by a series of creative operations."[35] This observation naturally led to the chapter "The Argument from Design."

Whewell began the chapter with a strong statement on the importance of natural theology. "There is no more worthy or suitable employment of the human mind," Whewell proclaimed, "than to trace the evidences of Design and Purpose in the Creator, which are visible in many parts of the Creation. The conviction thus obtained, that man was formed by the wisdom, and is governed by the providence, of an intelligent and benevolent Being, is the basis of Natural Religion, and thus, of all Religion." Whewell was assuring his readers that a rejection of pluralism did not weaken a commitment to natural theology. In fact, Whewell hoped that "new lights" would be thrown on the "traces of Design which the Universe offers even in the work now before the reader," even though the views contained within were different "from those which have of late been generally entertained." Whewell reminded his audience that new views of the constitution of the universe brought to light by scientific researchers had in the past disturbed the thoughts of religious men but that after a season of controversy, "the old argument for design was capable of being translated into the language of the new theory, with no loss of force."[36] Natural theology was more resilient than some realized.

Whewell acknowledged that the proof of Design was seen most clearly in the structure of organized things, for example, in the structure of the human eye and in the human skeleton. These proofs were so strong that, even if evolution were proved to be true at some point in the future, "the evidences of design in the anatomy of man are not less striking than they were, when no such gradation were thought of." But these instances of design were all taken from the world of organic creation. Whewell wanted to draw his readers' minds to "the evidence of design in the inorganic world,—in the relation of earth, air, water, heat, and light," which, he admitted, were "less striking and impressive" to most persons than evidence from the organic creation. However, mechanisms operating in the inorganic realm, like those that provided the rain to water the earth, the mechanisms that Whewell had explored in detail in his Bridgewater Treatise, were at work throughout the entire solar system. This made them even more remarkable and impressive than the design to be found in earthly living beings. The design perceived in the organic world was "often affected by laws which are of a much wider sweep." Whewell's overarching aim in *Of the Plurality of Worlds* was to show that whatever the validity of pluralism or evolutionary theory, the crucial insight to learn from a study of nature was the centrality of law to the operation of nature. "Law implies a Lawgiver," Whewell declared, "even when we do not see the object of the Law; even as Design implies a Designer, when we do not see the object of the Design. The Laws of Nature are the indications of the operation of the Divine Mind; and are revealed to us, as such, by the operations of our minds, by which we come to discover them."[37] As Crowe has observed, Whewell proposed "fundamental reorientations" in natural theology, "such as greater stress on laws and overall patterns."[38]

Reflections on an Idealist Natural Theology

Shortly after the *On the Origin of Species* was published on November 24, 1859, Whewell received a copy from the author. He wrote to Darwin on January 2, 1860, to thank him for the gift. He told his correspondent that "it has interested me very much"; nevertheless Darwin would probably "not be surprised" that he could not "yet, at least, become a convert" to his doctrines. "But there is so much of thought and fact in what you have written that it is not to be contradicted without careful selection of the ground and manner of the dissent," Whewell asserted, "which I have not now time for."[39] Darwin was happy with the letter, thinking that, given Whewell's past rejection of the *Vestiges* and evolutionary theory, the letter could have been far more hostile. Writing to Charles Lyell two days later, Darwin remarked, "Possibly you might like to see enclosed note from Whewell, merely as showing that he is not horrified with us."[40] Whewell's opinion concerning the *Origin of Species* may have been more negative than he let on in his letter to Darwin. According to anecdote, Whewell refused to allow a copy of the book in the Trinity College library.[41]

There is no indication in Whewell's correspondence with Darwin about what he thought of the strategic placing of a quote from his Bridgewater Treatise alongside a quote from Francis Bacon opposite the title page of the first edition of the *Origin of Species*: "But with regards to the material world, we can at least go so far as this—we can perceive that events are brought about not by insulated interpositions of Divine power, exerted in each particular case, but by the establishment of general laws."[42] This quote, taken from the third section of *Astronomy and General Physics*, encapsulated not only the central argument of the entire book but also summed up Whewell's attempt from the 1830s on to shift natural theology away from Paley's utilitarian version toward an idealistic one based on the concept of natural law. It can be argued that the use of the Bridgewater Treatise quote was representative of how Darwin exploited Whewell's emphasis on natural law in order to reassure his readers that he was not putting forward a theory hostile to Christian doctrine.[43] In the final paragraph of the book Darwin compared the fixed law of gravity, discovered by the revered Newton, to the law of evolution, which produced beautiful and wonderful forms from simple beginnings. In the previous paragraph he argued that it "accords better with what we know of the laws impressed on matter by the Creator" if we think of the production and extinction of "the past and present inhabitants of the world" as being due to "secondary causes."[44]

After the publication of *Origin of Species*, Whewell still had more to say about natural theology and natural law. The final three chapters of *On the Philosophy of Discovery* (1860) dealt with the theological implications of his philosophy of discovery. The substance of these chapters came from material he had originally planned to include in *Of the Plurality of Worlds*.[45] Whewell was still thinking about the issues he had discussed in the early 1850s, believing that the discovery of scientific truth was a "progressive" process. Discovery of "the idealization of facts" went on "from age to age." But Whewell maintained that this account of the philosophy of science was incomplete without a reference to God. The "Fundamental Ideas" in nature that were progressively discovered by humans seeking knowledge, such as space, time, force, and matter, were in reality ideas in the "Divine Mind." This was the ultimate idealistic basis for Whewell's natural theology. The human mind was designed by God to have the ability to perceive the design in nature that had been put there by the divine mind. This was the only reasonable explanation for the human capability of uncovering necessary truths about nature. As Whewell put it: "The truths which exist or can be generated in man's mind agree with the laws of the universe, because He who had made and sustains man and the universe has caused them to agree." The correspondence between human ideas about nature discovered by the great inductive scientists in the past and the "facts of the world" could only be explained if "our Ideas are given us by the same power which made the world, and given so that these can and must agree with the world so made."[46] Scientific progress itself became another evidence for the existence of a deity, a meta-design argument.[47]

Whewell's attempt to move away from Paley's utilitarian emphasis on designed organs in living beings has been offered in the past as proof of the collapse of natural theology. Michael Ruse once argued that Whewell's difficulties integrating his religion, science, and philosophy were symptomatic of the breakdown of natural theology prior to the publication of the *Origin* and that this helps us understand why Darwin prevailed.[48] But historians of science later came to realize that Darwin did not deal a deathblow to natural theology. To the contrary, as Roberts declared, even a superficial examination of the history of Anglo-American Christian thought since 1859 "is sufficient to indicate that natural theology remained an ongoing, sometimes even thriving, enterprise." What Darwin actually did, Roberts maintained, was shape the future of natural theology by pushing theists to expand the scope of the design argument "from living things to the intelligibility of the natural world as a whole."[49] In other words, after Darwin, natural theologians began to emphasize that the regularity with which gravity and other natural laws operated provided the proof for divine design. But this was exactly what Whewell had been arguing since the 1830s, well before the publication of the *Origin of Species*. He had explored the merits of this more idealistic approach to natural theology through his astronomical works. Whether or not he anticipated that the life sciences, and the theory of evolution in particular, would come to provide the chief grounds for raising serious questions about natural theology is unclear. He was certainly aware of the danger that *Vestiges* and other theories of evolution posed for the future of the natural theology tradition, and his response was to try to outflank his opponents by resting his case for design on discoveries drawn from the physical sciences.

CHAPTER 9

Whewell, Bacon, and the History of Science

LUKAS M. VERBURGT

William Whewell is today still studied primarily as a philosopher of science. The relation between his two most influential books—*The History of the Inductive Sciences* (1837) and *The Philosophy of the Inductive Sciences* (1840)— has come to feature prominently in Whewell scholarship since the 1990s, and some scholars now identify Whewell as a founding father of the field of history and philosophy of science.[1] But while the historical underpinnings of Whewell's idealist philosophy of science are well recognized, comparatively little has been written on Whewell as a historian of science.[2] Despite the obvious benefits of holism in treating Whewell's polymathy, this is unfortunate for at least three reasons. First and most generally speaking, Whewell's work on history is important and interesting in its own right and not just as the basis of or testing ground for, or simply as an articulation of, his philosophy of science. Whewell himself suggested as much when, in the preface to the *History*, he insisted on the value of the book "in its independent character" as a "History of Science."[3] Second, the *History* contains philosophical and historiographical elements—on the level of methodology and narrative—that preceded, informed, and went beyond the *Philosophy*. To neglect this, for instance by dismissing the *History* as "premature," is to overlook much of "the sophistication of Whewell's historical writing."[4] Third, the study of Whewell as a historian of science can offer insights into his metascientific project as well as his philosophy of science that cannot readily be obtained from a study of the *Philosophy*, either alone or in connection with the *History*.

Perhaps the most striking example in this regard pertains to the long-standing issue of whether Whewell's views on science were Baconian or not. This has been endlessly (and often splendidly) debated by historians and philosophers alike who, struggling with Whewell's rejection of Bacon's work in the name of Baconianism, have either called Whewell "anti-" or "non-Baconian," or "ceremonially Baconian" at best, or abandoned the category of "Baconianism" altogether.[5] Whewell himself, however, thought of the *History* as standing to Bacon's *Advancement of Learning* (1605) as the *Philosophy* stood to the *Novum Organum* (1620).[6] From the early 1830s on he made it his life's task to "attempt to continue [Bacon's] Reform of the Methods and Philosophy of Science," firmly convinced that it might, "like [Bacon's], be most fitly preceded by, and founded upon, a comprehensive Survey of the existing state of human knowledge."[7] With John F. W. Herschel and John Stuart Mill, Whewell would come to struggle over the question of who was the legitimate heir to Bacon.[8] As part of this ongoing debate, which took place both publicly (with Mill) and mostly privately (with Herschel), Whewell presented himself as the Baconian that Bacon would and should have been had he been living in the nineteenth century. Hence, history was key to Whewell's philosophy of science not just for its actual development, as a guiding exemplification, but also for strategically presenting his own radical redefinition of *induction* as the realization of Bacon's spirit.

The aim of this chapter is to present Whewell as a historian of science. First, it asks why and how Whewell wrote the *History*. Here the focus will be primarily on the origins and early development of his views on the history and historiography of science. On this basis it then argues that Whewell was a Baconian, if the Baconianism with which he identified is understood as very much a nineteenth-century creation—one to which the *History* itself was directly instrumental.

Toward the *History of the Inductive Sciences*

Writing to Richard Jones in the summer of 1834, Whewell announced his plan for a series of three books, the first of which would deal with the "*History* of Inductive Science . . . historiographized in a new and philosophical manner," and the second, more "dry and hard," with the "*Philosophy* of Inductive Science."[9] Whewell had by that time been filling notebooks and writing drafts of works on induction for a number of years, starting in 1830–1831.[10] From December 1833 on he had begun to devote the bulk of his available time to writing a historical work on science.[11] Only a few years later the first stage of Whewell's project would reach fruition with publication of the three volumes of the *History*, comprising 1,447 pages in the third, and final, 1857 edition. The book would earn Whewell the title of "first modern historian of science,"[12] mainly for its historical detail and pioneering account of scientific progress in terms of a three-stage model (preludes, epochs, sequels). These and other aspects of the *History* have been studied in depth by other scholars, notably Geoffrey Cantor and Richard Yeo, and will here feature largely in the background.

This section describes the origins of Whewell's *History* by examining the background of his turn in the early 1830s to the history of science. The next section will discuss the development of his views on the subject, and more specifically on its historiography prior to 1837. What emerges is a picture of the *History* as much more than a test case for Whewell's early ideas on induction. Instead, it brought together several threads of his thinking from the 1820s to the 1830s that highlight the centrality of history to Whewell's metascientific project as a whole, of which the philosophy of science was one part.

Whewell's Metascientific Project

From the start of his illustrious career, "theorizing" was among Whewell's "favourite pursuits," and in his early letters to peers like John Herschel, Hugh James Rose, and especially Richard Jones, he made occasional mention of the "logic of science" and "inductive philosophy."[13] When he was in his late twenties Whewell was already privately developing ideas on science ("most true, philosophical, and inductive") and contemplating doing more sustained work on "the metaphysics of mathematics" and "the metaphysics of the history of mechanics."[14] Furthermore, Whewell pursued his scientific research on crystallography, mineralogy, and tidology for the sake not just of these sciences themselves but also that of the "higher philosophy . . . which legislates for [the] sciences." For instance, in October 1825 Whewell wrote of mineralogy as "one of the very best occasions to rectify and apply [the] general principles of reasoning."[15] These examples from the early 1820s show that Whewell was interested in the (interactions between) history and philosophy of science well before the 1830s. They also suggest that by that time Whewell's roles as scientist and meta-scientist were strongly overlapping.

It was only over the course of the 1820s–1830s that Whewell—not convinced of his scientific talents—eventually came to identify with his role as a "looker on" who considered science as his *object of study*.[16] From around 1830 to 1831 on, he deliberately fashioned himself as a *meta-scientist*, to use a term introduced by Richard Yeo; Whewell was thinking of founding a "school"—with Jones, Herschel, and himself as leading members—and started to put his views on induction on paper.[17] Whewell made it his mission not just to defend science against those who dismissed it as irreligious ("my theology"),[18] but to define and promote a particular religion-inspired view of inductive science ("my induction").[19] What connected these two sides of Whewell's "biblical-historical-philosophical-moral-scientific-theological"[20] project was the strange but powerful idea that, understood as opposing visions of science, deduction ("reasoning downwards from abstract principles") was inferior and potentially led to atheism, while induction ("reasoning upwards from observations") was superior and strengthened piety. Or, in Whewell's own words, induction made science "fall in with a contemplative devotion," whereas deduction rendered it "dogmatic" and "mad."[21] This made it possible, for instance, to introduce morality into natural theology and to give scientific methodology and mathematics teaching a deeply religious significance.

The idea itself already featured in Whewell's letters and occasional sermons from the second half of the 1820s but was first put into print in book 3 ("Religious Views") of his Bridgewater Treatise, *Astronomy and General Physics* (1833).[22] A prelude to the *History* and *Philosophy* that were yet forthcoming, this highly influential book presented the inductive science of Galileo, Isaac Newton, Blaise Pascal, and Robert Boyle—as opposed to the deductive science of Leonhard Euler, Pierre-Simon Laplace, Joseph-Louis Lagrange, David Ricardo, Richard Whately, and John Ramsay McCulloch—as *good* science, in both a moral and a methodological sense.[23] "I hope that . . . the contrast of induction and deduction will take hold," wrote Whewell in March 1833, "and then I do not care how soon people forget that [one day] they were blind to the difference."[24] At a time when "What is science?" was still very much an open question, Whewell made the spirit in which it was conducted—rather than, say, its products or utility—all important; and the right spirit was of course that of induction, first and famously expressed in Bacon's *Novum Organum*. This vision would echo throughout Whewell's entire metascientific oeuvre, not in the least in the *History*, where the focus would be on knowledge, theory, and science at the expense of application, practice, and art (or technology).[25]

Origins of the History

The precise origins of Whewell's *History* are difficult to trace, partly because his notebooks are not consistently dated. But it is clear that rather than "suddenly"[26] appearing around 1833–1834, the history of science emerged in his writings as early as in the early to mid-1820s. The first context was that of textbooks. Already in his *Elementary Treatise on Mechanics* (1819), Whewell drew attention to "the manner in which the first steps in Mechanics [were] made" in order to "illustrate the nature of scientific reasoning."[27] A few years later, in September 1822, Whewell told Jones that he considered "doing something" on the history of mechanics, following the example of Adam Smith's *History of Astronomy* (1795), "but with more historical facts."[28] Whewell here mentioned history as part of the "metaphysics" rather than the "logic" of science. Whereas the latter was ahistorical, the former was said to include the history of the development of the main "principles"—

for example, in the case of statics, that of the lever—of "pure" and "mixed" sciences such as geometry and mechanics.[29] In contrast to the late 1820s and early 1830s, Whewell de facto dismissed the idea of a historicized "logic" of sciences like chemistry or political economy. Instead, what mattered were histories showing the student that "pure," abstract sciences that appeared deductive, and whose laws seemed to be derivable a priori, were at root inductive. This strategy runs through all Whewell's textbooks, from the *Elements of Mechanics* to the *First Principles of Mechanics* (1832) and the *Mechanical Euclid* (1838). The *First Principles*, for instance, compiled speculative histories to show that central concepts of the discipline (force, motion, etc.) that appeared deductive had previously been established inductively.[30] And in the "Remarks" appended to the *Mechanical Euclid*, Whewell set out to show that his theory of induction allowed him to explain both mathematical and scientific reasoning.[31]

Whewell's central distinction between deductive science as bad and inductive science as good was at least partly strategic, for a closer look at his textbook oeuvre suggests that he did not dismiss deduction but sought to put it in its proper, historical place.[32] As Whewell remarked in October 1832 about the Bridgewater Treatise he was then drafting: "Each chapter (except one) consists of a historical inductive and a mathematical deductive portion, and I am persuaded that this is the way in which scientific books ought to be written."[33] This remark not only highlights that Whewell's early work on education, natural theology, and history flowed from the same metascientific source; it also points to the fact that one crucial reason for his eventual turn to history of science was to show *that* science is inductive. Whewell's "induction notebooks" from the early 1830s further confirm this. Following the model of his own textbooks, some of these sought to establish the inductive roots of a wide variety of scientific disciplines, from astronomy to music, by looking at their past. Importantly, all this happened at a time when Whewell was still very much in the process of finding out what the nature and scope of induction actually was. More specifically, by that time Whewell was gradually moving from the standard view of induction as generalization from particular observations to his own radically new view—one at odds with the letter but not, he argued, with the spirit of Bacon's writings on scientific methodology.

The second context in which Whewell mentioned the history of science in the 1820s was that of natural theology, more specifically the alleged harmony between science and religion. Writing to his friend, the Cambridge theologian Hugh James Rose, in the winter of 1826, Whewell remarked,

> I [do not] think that you quite fairly represent the nature of our progress in scientific knowledge when you talk of its consisting in the rejection of present

belief in favour of novelty. . . . I am persuaded that there is not in the nature of science anything unfavourable to religious feelings, and if I were not so persuaded I should be much puzzled to account for our being invested, as we so amply are, with the faculties that lead us to the discovery of scientific truth.

I hold that if inductive science be true it must harmonize with all the great truths of religion. . . . At least this appears in the highest degree probable—that our enjoyment of [truth] will be proportioned to the degree in which the previous advance of our intellectual cultivation has prepared us for it . . . [It] is by no means correct to consider advances in science as rejection of what was known for something new. The novelty, if the philosophy have been [*sic*] duly inductive, *includes* old truths and shews them from a new point of view.[34]

The wide-ranging debates on the nature of science in the early Victorian period took place within an intellectual context constituted by the dominant tradition of natural theology that viewed man's ability to know the world as an indication of his special place in nature.[35] It is not surprising, therefore, that Whewell's thinking on the history of science was deeply invested with theological and religious significance. Indeed, rather than philosophy, it was natural theology, alongside mathematics education, that provided him with a key rationale for turning to history.

Between the late 1820s and early 1830s Whewell's "theology" and "induction" projects developed as distinct, but mutually supportive parts of his nascent metascientific vision. When looking only at his published oeuvre from the 1830s to the 1850s, the theological line traced from *Astronomy and General Physics* to *Of the Plurality of Worlds* (1853), the inductive line was represented by the *History* and *Philosophy*, and they were joined, most prominently, in *Indications of the Creator* (1845), which contained "Extracts, Bearing upon Theology, from the *History* and *Philosophy* of the Inductive Sciences." There was a strategic division of intellectual labor between the two projects. The former was aimed at "friends of religion," showing that scientific progress harmonized with belief in God, whereas the latter was supposed to narrate the history of science without setting out from "a theological conclusion."[36] This meant that in the case of some claims about science made on behalf of religion, Whewell permitted himself to be programmatic. Perhaps the best example in this regard is exactly the opposition between induction and deduction. Whewell realized that the scientific "realities illustrating them" were yet to be provided.[37] More specifically, the historical conjecture that all genuine progress in science had been achieved inductively, and that all major scientific breakthroughs had been made by inductive discoveries, demanded a closer analysis of science's past. Here, showing *that* science is inductive was not enough; instead, a study of *how* induction drove science, that is, of "the process of thought by which laws of nature . . . had been discovered," was needed. This, of course, was "no easy matter, if it be possible," especially because the process had been "in so few instances successfully performed."[38] The next section will discuss some of the strategies Whewell seems to have had in mind for such research. Here, what matters is that Whewell's decision in 1833–1834 to write a historical work did not come out of thin air. For some time already, Whewell entertained ideas about the history of science, some of which were critical of earlier, "vulgar" histories of science.[39] Moreover, by the late 1820s he already had at least the intuition that history was a vital source for understanding how induction made scientific discoveries and progress possible.

Looking at Whewell's manuscripts of the period, it is impossible to neatly separate his study of the *that* and *how* of induction. The same holds for the historical and philosophical sides of this study, which developed largely in tandem. But when it comes to the origins of his historical-philosophical induction project, the historical *that* seems to have preceded the philosophical *how*: Whewell started out from the conviction that all sciences were at root inductive, in a roughly Baconian sense, and subsequently sought to find out how induction was actually applied in the sciences. In doing so, he gradually came to abandon some of his earlier views on inductive science to work toward his mature theory as put forward in the *Philosophy*.

There were other factors involved in Whewell's eventual turn to the history of science, in addition to the contexts of education and natural theology that formed the background to it. One factor may well have been "the rise of historical consciousness" in the 1820s–1830s, more specifically the increasing prominence of general surveys of science in Britain in this period.[40] Another, related but more tangible, factor was mentioned by Whewell himself in his "Address" at the third meeting of the British Association for the Advancement of Science (BAAS) in June 1833. Whewell praised the BAAS for its commissioning of several "reports on the recent progress and present condition" in various branches of science: "There is . . . no royal road to knowledge—no possibility of shortening the way. . . . We must all start from our actual position, and we cannot accelerate our advance by any method of giving to each man his mile of the march. Yet something we may do: we may take care that those who come ready and willing for the road, shall start from the proper point and in the proper direction."[41] The BAAS reports were Whewell's own brainchild, and in 1832 he himself had contributed the "Report on the Recent Progress and Present State of Mineralogy."[42] They provided another rationale for writing what would become the *History*. Much like the BAAS reports, one of the *History*'s main ambitions was to form "a platform on which we might stand and look forward into the future" as well as look back into the past "to remind us of what we have . . . and to increase our store."[43] Similarly, Whewell tellingly captured the aim of his 1833

"The Philosophy of the Progressive Sciences"—in terms similar to the BAAS reports—as that of a complete survey of "the present state and past history of many of the most complete branches of human knowledge."[44]

Pre-*History* History of Science

Because of other commitments, such as his work as tutor (1823–1839) and professor of mineralogy (1828–1832), it was only in late 1833 that Whewell was able to return "forthwith and in earnest" to his "beloved induction."[45] Only a few weeks later, in December of that year, Whewell made the decision to write a historical work on science.[46] By that time, he had already scribbled down hundreds of pages in his "induction notebooks," in which draft essays, some of book length, were composed and subsequently abandoned. At the heart of Whewell's thinking in these notebooks stood the problem of combining into a single notion of induction two elements of science—the ideal and the empirical, the pure and the physical, the subjective and the objective, or that is, the "Kantian" and the "Baconian." A lot has been written on how Whewell sought to bring these elements together, with scholars debating when he hit upon the solution (in 1831 or 1834?) and why, having found it, he decided not to publish his views on induction until 1837.[47] However important, such accounts tend to present Whewell's "induction" as a purely philosophical project and, ipso facto, focus on his philosophical views on methodology at the expense of and in isolation from his other pursuits of the period. But these pursuits, whether scientific research or history of science, were crucial to the development of Whewell's "induction." For by the early 1830s Whewell himself realized that he needed real-life scientific contexts, both past and present, to transform his abstract speculations on induction into a full-blown theory. As Whewell's philosophical views matured over the course of the 1830s, so did the historiographical ideas about the *what* and *how* of the history of inductive science that they came to inform.

History as Exemplification

From the official start of his "induction" project in 1830, Whewell turned to the sciences to look for occasions to define and promote what he, ever since the early 1820s, had called "inductive philosophy": the study of science that went beyond any particular discipline. During the 1820s student textbooks had been his preferred metascientific outlet, introducing histories to show that each branch of mechanics was partly "physical and inductive" and partly "pure and deductive."[48] Whewell also thought of his own work on crystallography, mineralogy, and tidology from this period in explicitly metascientific terms, and frequently expressed the hope that others belonging to his self-founded "school" would do the same for their disciplines—for example,

Jones for political economy and Herschel for astronomy. The two papers and three reviews Whewell published on political economy between 1829 and 1831 stand out as examples of scientific research that he pursued entirely for the sake of making a metascientific point: namely, that "the business of ['inductively'] reasoning up to principles and down [i.e., deducing] from them" should be kept apart.[49] Sometimes Whewell "got hold of a new science," such as meteorology—his "new pet" in July 1831—which he believed to be "admirable" for his work on induction because it was still in the process of becoming a science.[50] A few months later, in April 1832, he referred to his forthcoming BAAS report on mineralogy as being as "full of induction as anything in the world."[51]

Whewell's notebooks from these years listed many other sciences deemed worthy of metascientific study, including new coinages such as *photistics, thermotics, atmology, electrics,* and *galvanics.* Not all disciplines were considered metascientifically useful, and the focus was strongly on the physical sciences or, that is, on branches of "Natural Philosophy." Whewell dismissed the idea of writing on induction in the contexts of "moral, political and other notional sciences" as well as "intellectual philosophy," though he himself toyed with various other subjects (e.g., language, architecture, taste, beauty, law).[52] It was around this same period, early in 1831, that Whewell for the first time in print referred to history as a context for exemplification par excellence. As he wrote in his thirty-page review of Herschel's *Preliminary Discourse on the Study of Natural Philosophy* (1830):

> [Herschel's *Preliminary Discourse*] is much more than a mere survey of the state of [scientific] knowledge. . . . It is, we believe, one of the first considerable attempts to expound in any detail the rules and doctrines of that method of research to which modern science has owed its long-continued steady advance and present flourishing condition. It is certainly strange that this should be so; that while volumes upon volumes have been written upon the nature of human knowledge and the laws of human thought, this class of speculations should never have been constructed with a peculiar reference to those mental processes which have been *exemplified* in the progress of modern science. . . . We hope that the day will come when this shall be otherwise.[53]

All this metascientific searching for exemplification took place when Whewell had already written down many of his early ideas about induction in his notebooks. It suggests that what made him withhold from publishing his work, even after he had hit upon the new "antithetical" theory of induction that solved his problem, had to do with a profound worry he expressed in a letter from February 1831: "How can you expect to lay down rules and describe an extensive method with no examples to guide and substantiate your speculations? You may say a number of fine things and give rules that look wise and arguments that look pretty, but you will have no security that these devices are at all accurate or applicable."[54]

This had been the main source of the errors made by "the Master himself" (i.e., Bacon), and Whewell did not wish to repeat the mistake. Some of his induction notebooks from 1830–1832, however, contained exactly that: abstract topics in the history of philosophy of science, such as "General Propositions," "Abstract Terms," and "Terminology," and philosophical discussions of induction, for instance in "Of the Different Steps of the Inductive Method." Other notebooks from the period made more references to the sciences, but largely as illustrations of a preconceived and roughly Baconian theory of induction. For instance, in "Induction of Botany," drafted in 1830–1831, Whewell discussed whether "the Inductive Method, in the sense and in the point of view which we have hitherto assumed for it, does [here] receive a distinct and proper application."[55] The same holds for "Induction of Astronomy," "Induction of Chemistry," and similarly titled drafts of the period. Importantly, Whewell here used history in much the same way as in many of his 1820s textbooks: he started from an idea of what induction was and then looked at the history of science to show that these sciences were founded on induction.

It is impossible to pinpoint when and why exactly, but at some point in the early 1830s, Whewell must have realized that, rather than merely offering facts to test or illustrate philosophy, history of science was itself a source for philosophical thinking. Over a short period of time Whewell came to accept history not just as *the* study to "guide" and "substantiate" his views on induction but as containing "lessons [about] human knowledge"[56] for philosophy to unpack. It was from that moment on, that history was equally important as philosophy for his induction project at least. How else to explain that from 1833 to 1834 on, Whewell devoted himself to writing two books, one historical and the other philosophical, and that when he found himself, "in the course of my historical researches," becoming "metaphysical and transcendental," he turned to the second book?[57] And how else to explain the striking contrast between, on the one hand, the *First Principles of Mechanics* and his notebooks on induction and, on the other hand, the unpublished "Philosophy of the Progressive Sciences" (1833) and the *History?* Unlike the former, the later writings did not use history to show that science is inductive. Instead, they presented it as the best way to understand *what* induction is and *how* it made scientific progress and discoveries possible. This change in Whewell's general conception of history opened the door to other changes as well. From 1832 to 1833 on, his philosophical and historiographical views started to develop more and more in tandem, with the former turning to seeing induction as a *process* and the latter to approaching history in terms of a *general structure* over and beyond particular sciences.

Whewell's Pre-History Views on History of Science

Some commentators have argued that the *History* was premature, in that it was written when Whewell's philosophy of science was still incomplete. This may be true at the level of philosophical detail, but there are several reasons why it is doubtful that the *History* suffered from it. One is that, in terms of Whewell's role as public commentator on or "legislator" of science, the *History* likely did more than the *Philosophy*. Most importantly, the lack of fit between Whewell's two books once again highlights the need to take the *History*, including the process of thinking that went into it, on its own terms. This will be done here not by unearthing the major historiographical themes (e.g., progress, method, revolution) that Whewell critically engaged in the *History*. Instead, the focus will be on the development of Whewell's ideas about what the history of science should be about and how it should be written.

Whewell's process of reflecting on the *what* of history of science traced back as far as to the early 1820s. He then thought of it as part of the "metaphysics," rather than the "logic," of science and limited it to "pure" and "mixed" mathematics—for example, geometry and mechanics. During the 1820s Whewell adopted as his model the conjectural historiography of Adam Smith's *History of Astronomy*. It was, in fact, still used in the *First Principles of Mechanics*, where each chapter started with "speculations which led to the establishment" of a core principle of that science. The main historical insight the book offered was that, looking at the work of "the most intelligent speculators" (Aristotle, Stevin, Galileo), the history of science is one of trial and error: the history of the discovery of, for instance, the first law of motion is that of mistakes and errors in generalizing from experience.[58] Whewell's source material for his speculative history of science consisted of books of a popular genre that, as he wrote in an article from 1831, he considered "the best scientific history; the intellectual biography of great discoverers."[59] This genre would find an echo in the *History* through the book's concept of genius and its connection with the methodology of science and the morality of great discoverers.

Around that same time, Whewell started to shift his historical perspective from the individual to the structural, from the inductive origins of particular disciplines founded by particular scientists to the long-term developmental processes of all sciences. Importantly, in doing so Whewell came to critically engage with then-current historiographies of science, that is, with views on *how* the history of science should be written. The shift itself first became apparent in his early 1831 review of Herschel's *Preliminary Discourse*, a well-known but in this context often neglected source that contained many of the elements commonly seen as key to the *History*'s historiography.[60]

Whewell here not only put forward the notion that both facts provided by the world ("Observation") and ideas provided by the mind ("pure Reason") were required for induction to be successful and hence for science to develop.[61] He also introduced the basic narrative for a new history of science that would chart the progress ("inductive advance") of the sciences, arranged according to their place on an infancy-maturity scale of generality ("order of growth"). Whewell provided a rapid, seven-page sketch, starting with botany, "the least advanced of the sciences in the road of induction," as it was not more than an arbitrary nomenclature, and ending with astronomy, "the boast and glory of the inductive method," as it was based on one single, universal law.[62] Whewell's interest in historical periodization also became apparent when in 1832 he briefly caught the fever of "Saint-Simonianism," praising the work of the French social theorist Henri de Saint-Simon. What Whewell liked about the *Exposition de la doctrine de Saint-Simon* (1828), written by two Saint-Simonians, was not only, as others have noted, philosophical, but also historiographical. "I am entirely charmed," wrote Whewell in February 1832, "with the beauty and coherence of . . . his theory of organic and critical periods"—that is, periods of stability and reform.[63] This distinction could be said to have found its way into the *History*, as it pointed to the issue of the transition from one period to another. Whewell had already addressed this issue in 1830 when seeking to explain historical progress in architecture, another of his "pets" from the 1820s. *Architectural Notes on German Churches* (1830) did this through a new periodization of building styles; it determined the "epoch" of a particular style and the "transition" from, say, the Greek and Roman to, say, the Gothic. The central idea was that each epoch is defined by a "fundamental principle"—in the case of the Gothic style that of the pointed arch, which Whewell called a "revolution"[64]—and that later developments within an epoch should be understood as refinements of such an "*idea*, or internal principle of unity and harmony."[65] Before he began writing the history of science, Whewell was clearly sensitive to the need to appreciate the integrity of earlier periods, rather than judging it as deficient (or "barbarous") on the basis of present conventions. Moreover, he recognized that the study of the long-term development of intellectual phenomena asked for it to be distributed into separate periods. For instance, by 1830 he already spoke of "epochs," to which he added "preludes" and "sequels" in 1834.[66]

All this suggests that by the early 1830s Whewell had in mind a distinctive approach to the past, one that anticipated, in rough form, some of the elements that informed the new historiography of the *History*—for example, facts-ideas, generalization, periods.[67] Since these elements were developed in and across different contexts, what they highlight first and foremost is that history itself stood at the heart of Whewell's metascientific thinking at the time.

By 1833–1834, with his views on *how* the history of science should be written having crystallized, Whewell returned to the issue of *what* the historical book on science he resolved to write should be about. Importantly, in terms of scope the *History* differed significantly from the book he had envisioned in these years, let alone in the years before. "To some persons it may appear," wrote Whewell in 1837, "that I am not justified in calling *that* a History of *the* Inductive Sciences, which contains an account of the progress of the *physical* sciences only."[68] The origins of the limited scope of the *History* traced back to July 1834, when Whewell, having already drafted the one-hundred-page "Philosophy of the Progressive Sciences" (1833), decided to write *three* books on induction: *History, Philosophy,* and *Prospects* (this last one was never written). Prior to that, in fact still in late 1833, Whewell was drafting a work containing a chapter on the "Past Progress of the Pure Sciences." These sciences—geometry, arithmetic, algebra, (higher) analysis—would feature in the *Philosophy*, but they were omitted from the *History*, only to reappear as part of a historical narrative in *The History of Scientific Ideas* (1858).[69] The same goes for what Whewell dubbed the "hyperphysical" sciences, which included "morals, taste, politics, [and] language."[70] These, however, were already absent in the 1833 "Philosophy of the Progressive Sciences," as it provided a survey only of "the present state and past history of many of *the most complete* branches of human knowledge."[71] As Whewell added in the preface to the *History*, while the "hyperphysical" sciences "may properly be called Inductive," they were excluded because the process of induction was far less clearly exhibited in them than in the physical sciences.

It is undoubtedly true that the *History* and the *Philosophy* are very closely connected. But it is important to recognize that there are significant incongruities between them. Others have shown that it is debatable whether the *History* should be read as confirming the views found in the *Philosophy*, for instance because the notion of "inductive epochs" proved applicable in only a minority of sciences.[72] To this should be added that the *History* and *Philosophy* together did not form a completed whole and that the *History* did not offer enough history to carry the philosophical weight of the *Philosophy*. As Whewell himself would admit in 1858, the *History* contained only the history of science "so far as it is derived from *Observation*." As such, insofar as the book omitted the history of science "so far as it depends on *Ideas*," the suggestion that all of the *Philosophy* was based on, or drawn from, the *History* is simply false.[73]

Whewell, the "Imaginary Bacon"

There has long been much debate about whether Whewell was a Baconian or a non- or anti-Baconian. From a historical perspective, this issue is rather limited and artificial. It is limited because the focus again has been almost entirely on a comparison between Bacon and Whewell as philosophers of science—that is, between the *Novum Organum* and the *Philosophy*—to the exclusion of their respective views on the aims and scope of the history of science. The issue is largely artificial because it tends to neglect the fact that Whewell described himself as an "imaginary Bacon," and came to think of his own work, and his entire meta-scientific career even, as doing everything necessary to perfect Bacon's "Great Reform."[74] What Whewell considered necessary for this was to realize Bacon's philosophical ambition—to find the *"Organ"* of truth or, that is, the "Methods" of science—on the basis of historical evidence Bacon could only speculate about: namely, the "actual progress of science since his time."[75] Or, as Whewell wrote in the preface to the *Philosophy*: "The progress of [Physical] Science during the last three centuries has given us the means of inquiring, with advantages which [Bacon] did not possess, what that *Organ*, or intellectual method, is, by which solid truth is to be extracted from the observation of Nature."[76]

It suggests that Whewell's very strategy for renovating Bacon was itself Baconian. The *Philosophy* stood to the *History* as the *Novum Organum* had stood to the *Advancement of Learning* (1605): "Any attempt to continue and extend [Bacon's] Reform of the Methods and Philosophy of Science may, like his, be most fitly preceded by, and founded upon, a comprehensive Survey of the existing state of human knowledge."[77] This strategy in turn shows that, and in what sense, history of science was *the* fundamental element of Whewell's "imaginary" Baconianism. However widely it differed from Bacon's, Whewell took his philosophy to be that to which Bacon would have been led had he been a nineteenth-century contemporary. At the same time, the very fact that Bacon lived before the time of Kepler, Newton, and others meant that he could not be blamed for the fact that his *Novum Organum* was "now practically useless." Bacon's philosophical writings were passé, but the time for the realization of the Baconian spirit had only just arrived. In fact, by the 1830s–1840s, it was still to come: "A [Baconian] reform, when its Epoch shall arrive, will not be the work of any single writer, but the result of the intellectual tendencies of the age."[78] This points to an image of Whewell yet to be explored: that of a pioneer of the history of the philosophy of science.

CHAPTER 10

Whewell's Kant and Beyond

A Reassessment of Whewell's Philosophy of Science

CLAUDIA CRISTALLI

There is possibly no question about Whewell's philosophy of science so persistent as that of his debts to Kant's philosophy. This can be partially explained by taking into account the history of Whewell's reception. A towering figure in the philosophical landscape of the first half of the nineteenth century, after his death in 1866 Whewell quickly disappeared from mainstream philosophical discussions on science. This was not entirely unforeseen. John Stuart Mill's *A System of Logic* (1843)—a work that rivaled Whewell's own philosophy of science—had already been adopted as a textbook in Cambridge in Whewell's lifetime.[1] The early efforts of Isaac Todhunter and Janet Mary Douglas to make some of Whewell's writings and correspondence publicly available did not make him a popular thinker.[2] Although the following Cambridge generation was made aware of Whewell's erudition, they did not take up his philosophical problems.

In the twentieth century Whewell's thought sparked new interest among scholars from the United States and Europe, where Kant's and Hegel's tradition weighed heavier than Bacon's and Locke's. After Marion Stoll's work on Whewell's theory of induction,[3] Robert Blanché offered what was the first comprehensive assessment of Whewell's philosophy in the light of his "rationalism,"[4] and Silvestro Marcucci systematically engaged with what he perceived as Whewell's "scientific idealism."[5] Robert Butts offered a comprehensive account of Whewell's thought in the English-speaking world, with a focus on philosophy of science.[6]

Whewell's notions of "consilience of inductions" and "colligation of facts" were revived in modern debates on theory verification and theory formation, respectively, contributing to our understanding of evidence and of induction.[7] For philosophers interested in a historically informed philosophy of science, with a focus on scientific practices as well as the intellectual history of theories, Whewell offers an invaluable example; indeed, the rise in popularity of integrated history and philosophy of science may also explain some of the recent popularity of Whewell.[8]

Since the 1970s the scholarship on Whewell has grown steadily, helpfully branching out in many directions that take into consideration his numerous engagements with different sciences as well as his role in Victorian science and culture. The Whewell volume edited by Menachem Fisch and Simon Schaffer aimed at providing a "composite portrait" of his contributions to philosophy;[9] more recently, Richard Yeo, Laura Snyder, and Michael S. Reidy systematically engaged with Whewell's empiricism, his Victorian context, and his scientific practice as a "tidologist."[10] As Whewell's name started resurfacing in standard accounts of philosophy of science, the question of his philosophical roots became more pressing. Was Whewell to be understood as a representative of a watered-down version of Kant's philosophy, or as a properly British philosopher—that is, an empiricist? Was Whewell's philosophy "an imported hybrid," or "a plant native to the soil on which it grew"?[11]

This chapter argues for an interpretation of Whewell's philosophy of science as a sort of "native hybrid." On the one hand, Whewell conceived his philosophy as a tile in a broader mosaic of knowledge and cultural reform; together with some Romantic influences, this would inevitably root his philosophy to a specific "soil." On the other hand, Whewell's many interests called for a "hybrid" or eclectic approach to philosophy of science, which went from firsthand engagement with many sciences and their history to an eclectic incorporation of different philosophical traditions.

The empiricist interpretation has been championed by Snyder, who uncovered the Baconian elements of Whewell's philosophy and reconnected them to the Victorian discourse on the advancement of science and with Whewell's own scientific practice.[12] In contrast, Butts, Fisch, and, more recently, Steffen Ducheyne have focused on the ways in which Whewell's central notions can be clarified when compared with Kant's.[13] Even when adopting different solutions, Kant and Whewell were preoccupied by similar problems and their answers shared a core element: the notion of an active mind, endowed with the ability to form and organize our experience.[14]

The persistence of idealist and empiricist accounts of Whewell's philosophy testifies to the presence of elements from both traditions in Whewell's thought, and for the complex ways in which they are intertwined, as noticed by Margaret Morrison.[15] Morrison highlights the importance of the notion of a "sixth sense"—identified as the "muscular sense" or the awareness of the position of our body in space—for Whewell's account of the mind's activity; she thus reads Whewell's core thesis, the "On the Fundamental Antithesis of Philosophy" (1844) as a "distinctly Whewellian form of psychologism."[16]

After an overview of Whewell's philosophical works, I illustrate Whewell's "Fundamental Antithesis," a keystone of his philosophical project. I then examine Whewell's responses to Locke and Kant and his use of physiology in his argument for the idea of space. Finally, I return to the question of Whewell's legacy, arguing that Whewell had at least one follower in the pragmatist philosopher Charles S. Peirce. While Peirce's debt to Whewell has been pointed out before,[17] my claim is that Peirce's case presents a sympathetic commentary on Whewell's particular Kantianism as well as on his philosophy of science more broadly, and that his commentary should be part of our understanding of Whewell's legacy.

A Brief Sketch of Whewell's Philosophical Work

Whewell's first work in the philosophy of science appeared in 1840, *The Philosophy of the Inductive Sciences, Founded upon Their History*. History, in his eyes, was a precious source for the articulation of the method of science, while induction's meaning was enriched by an extensive historical inquiry into science's progress. In the preface to the second edition of the *Philosophy* (1847), after reminding his readers of his ambition to provide a modernized version of Bacon's *Novum Organum* (1620), he proudly affirmed, "Bacon only divined how sciences might be constructed; we can trace, *in their history*, how their construction has taken place."[18] The speculations of a philosopher, however insightful they might be, would never be able to match the wisdom of learning from the past.

Whewell did not aim to restrict the "inductive sciences" to the physical sciences only; however, he found it convenient in practice to limit his inquiry to "the Physical Sciences alone, in which the truths established are universally assented to, and regarded with comparative calmness." In this domain "we are better able to discuss the formal conditions and general processes of scientific discovery, than we could do if we entangled ourselves among subjects where the interest is keener and the truth more controverted."[19] If disciplines such as psychology and political economy do not appear in the *Philosophy*, this is more on account of the controversial nature of their conclusions than because of an assumed lack of inductive structure.[20] Nonetheless, Whewell was convinced that the physical sciences brought clarity to the notion of induction, clarity that would eventually trickle down to other fields, including morality and theology. His interest in a connection between these different domains of human thought lasted from the beginning to the end of his philosophical production.

The first texts where Whewell explicitly tackled philosophical questions concerning induction are his 1831 review of John Herschel's *Preliminary Dis-*

course on the Study of Natural Philosophy (1830) and in his Bridgewater Treatise, *Astronomy and General Physics* (1833). Herschel's *Discourse* quickly became "a minor classic" of his time,[21] and influenced thinkers such as Charles Darwin[22] and John Stuart Mill.[23] Whewell praised in one breath Herschel and Francis Bacon for their historical account of natural philosophy and for the adoption of induction as the key to scientific method.[24] However, the account of induction he gave was not Herschel's but his own, spelled out perhaps for the first time.[25] The differences are striking. For Herschel, induction was a matter of scrupulously classifying facts "under general and well-conceived heads or point of agreement (*for which there are none better adapted than the single phenomena themselves*)."[26] For Whewell, instead, induction *added* something to the phenomena that was not already present in them and that could not have been brought by the phenomena alone. Using the analogy of a string of pearls, Whewell broadened the powers of induction beyond the limits of Hershel's cautious empiricism: "If we consider the facts of external nature to lie before us like a heap of pearls of various forms and sizes, mere Observation takes up an indiscriminate handful of them; Induction seizes some thread on which a portion of the heap are strung, and binds such threads together."[27] The string binding facts together was not a "point of agreement" between them, emerging as it were from their classification, but it was the criterion for classification: it was something that could not be found in the facts themselves. The string was something that the mind added to the pearls: an idea. This element was crucial in Whewell's *Philosophy*, as he explicitly openly stated: "Perhaps one of the most prominent points of this work is the attempt to show the place which discussions concerning Ideas have had in the progress of science."[28] For Whewell, the lack of development of a particular branch of knowledge could depend roughly on two factors: a lack of (adequate) observations, or a lack of (sufficiently clear) ideas. Thus, conceptual or "metaphysical" debates were just as important for science's progress as the collection of careful and systematic observations. However, what justified the introduction of ideas, and where did they come from?

Whewell expanded his treatment of the connection between inductive reasoning and ideas in 1833, in the context of his Bridgewater Treatise, a work which aimed at showing that natural science was compatible with the Christian faith.[29] In the cosmological framework of the treatise, induction would point to the existence of something divine by contributing to the progressive clarification of laws of nature and ideas, such as time, space, and motion, that structure the natural world. In return, God was taken as the warrantor of the persistence, stability, and truth of the ideas. Although ideas (and natural laws) were contingent on God's design, they were also necessary because God created the world this way.[30] So, having abandoned an empiricist account of induction in 1831 made Whewell neither an idealist nor a

Kantian in 1833. In the Bridgewater Treatise, ideas were not seen as existing by themselves nor as conditions of possibility of human experience, but as something that came close to an insight into God's mind.

Only a year later, in his essay "On the Nature of the Truth of the Laws of Motion" (1834), Whewell expanded on the active role of ideas and changed the justification for it, moving it from God's mind to the very structure of our experience.[31] This essay would also be appended to the first edition of the *Philosophy*—the crowning achievement of Whewell's project of reforming natural philosophy. The "history" alluded to in the title had been the topic of Whewell's 1837 *History of the Inductive Sciences*, a project that Whewell conceived together with the *Philosophy*, as numerous notebooks and drafts show.[32]

This was also the year in which Whewell published the *Mechanical Euclid*, a work on the fundamentals of mechanics containing a final section with "Remarks on Mathematical Reasoning and the Logic of Induction." These remarks were later included as an appendix to the second edition of the *Philosophy*.[33] As it appears from this reconstruction, historical, philosophical, and educational expositions of science were deeply intertwined with Whewell's thought. He was not just a theorizer, and he consciously exhibited his credentials for writing such a broad account of science and its history:

> And if I may speak of my own grounds of trust and encouragement in venturing on such a task, I knew that my life had been principally spent in those studies which were most requisite to enable me to understand what had thus been done; and I had been in habits of intercourse with several of the most eminent men of science in our time, both in our own and in other countries. Having thus lived with some of the great intellects of the past and present, . . . I trusted, also, of understanding their discoveries and views, their hopes and aims.[34]

In his own view, to a solid academic preparation Whewell joined personal acquaintance with the works of science and the habits of scientific men. The writing of his *History*, conceived as the groundwork for his *Philosophy*, was allegedly based on firsthand experience, study of authors from the past, and constant dialogue with some of the most eminent scientists of his time.[35]

In successive editions, the contents of the *Philosophy* were further elaborated and expanded into three books with different titles: the *History of the Scientific Ideas* (1858), covering the first volume of the *Philosophy* of 1847; the *Novum Organon Renovatum* (1858), containing books 11–13 of the *Philosophy* and a section on aphorisms; and the *On the Philosophy of Discovery* (1860), containing book 12 of the *Philosophy* and additional material on the theological dimension of his philosophy of science. With this last work Whewell returned to the topics of theology and moral philosophy from which he had departed to build his philosophy of the inductive sciences in 1833.[36]

The many re-elaborations of the *Philosophy* suggest a thinker who, without ever losing sight of his systematic framework, was sensitive both to his critics' objections and to more sympathetic suggestions. Whewell often included essays published in the intervening years between one edition and the next as appendices;[37] some became centerpieces of his philosophical construction, such as his "On the Fundamental Antithesis of Philosophy."[38] A slightly re-elaborated version of this essay constitutes the first chapter of the second edition of the *Philosophy* and of the *History of Scientific Ideas*, as well as being added to the *Philosophy of Discovery* as Appendix E. The "Fundamental Antithesis" is among Whewell's most interesting contributions to the philosophy of science, together with his methodological reflections on the colligation of facts and the consilience of inductions. All these concepts speak of Whewell's interest not just in the classification and systematization of knowledge, but also and more importantly in its *growth*.[39]

The "Fundamental Antithesis"

The "Fundamental Antithesis" has sometimes been defined as the "simplified re-stating of the fundamental Kantian doctrine according to which knowledge is the result of the union of sensation and intellect."[40] How much he "simplified" or distorted Kant's theory and how much he transformed it into something else, namely his own theory of knowledge, was indeed open for debate since the first reception of Whewell's work. In the preface of the first edition of the *Philosophy*, Whewell avowed a Kantian influence—which is not to say full acceptance—in his analysis of the ideas of time and space: "Although I have adopted Kant's reasoning respecting the nature of Space and Time, it will be found by any one acquainted with the system of that acute metaphysician, that my views differ widely from his."[41] Kant's name doesn't appear in the first (1844) version of "Fundamental Antithesis" nor in its more systematic elaboration that constitutes the opening chapter of the 1847 edition of *Philosophy*. However, Whewell compared his theory with that of other important philosophers, who may be seen as embodying the two sides of his "Fundamental Antithesis": the empiricist Locke on the one side, and idealist philosophers such as Schelling, Goethe, and Hegel on the other.[42] In my discussion of the "Fundamental Antithesis," I focus on Whewell's reaction to Locke; this prepares the ground for an assessment of Kant's influence in Whewell's articulation of the idea of space in the next section.

Whewell used the expression *fundamental antithesis* to refer to the presence, in every type of knowledge, of two irreducible aspects. The articulation of these aspects can be philosophically loaded, such as "necessary and empirical truths," or appealing to common sense, as in the case of "thoughts and things." In his 1844 "Fundamental Antithesis," Whewell started with the first, but he moved it to second place in the *Philosophy*, where thoughts and things are defined as "the simplest and most idiomatic expression of the antithesis to which I refer."[43] Whewell explained: "In all cases, Knowledge implies a combination of Thoughts and Things. Without this combination, it would not be Knowledge. Without Thoughts, there could be no connexion; without Things, there could be no reality. Thoughts and Things are so intimately combined in our Knowledge, that we do not look upon them as distinct. One single act of the mind involves them both; and their contrast disappears in their union."[44] This thesis resurfaced in all the different illustrations of the fundamental antithesis. Next to the antithesis between thoughts and things was the antithesis between "Theory and Fact," which "implies the fundamental Antithesis of Thoughts and Things."[45]

Interestingly, every element of knowledge could count as a theory or as a fact, depending either on the degree of confidence of our knowledge of it or on the perspective from which it was regarded. As Whewell stated in the second to last section of the chapter "The Fundamental Antithesis Inseparable": "In a Fact [i.e., in something considered as a fact], the Ideas are applied so readily and familiarly, and incorporated with the sensations so entirely, that we do not see *them*, we see *through them*."[46] He provided examples to illustrate the role that expertise and habit would have in making us classify something as either "theory" or "fact": "A person to whom the grounds of believing the earth to revolve round its axis and round the sun, are as familiar as the grounds for believing the movements of the mail-coaches in this country, conceives the former events to be facts, just as steadily as the latter."[47] The possibility of assuming different perspectives on what is a fact or a theory was not, for Whewell, the gateway to "skepticism." As seen in the previous section, Whewell's *Philosophy* was deeply tied to his *History*, and it was about the development of scientific ideas in time. Thus "the absolute application of the antithesis in any particular case can never be a conclusive or immoveable principle."[48] Already in the first edition of the *Philosophy*, Whewell had stated that "fact and theory pass into each other by insensible degrees."[49] Being aware of the theoretical domain of even the most well-established fact would allow a critical perspective on it. After all, "did not the ancients assert it as a Fact, that the earth stood still, and the stars moved? And can any Fact have stronger apparent evidence to justify persons in asserting it emphatically than this had?"[50] Eventually, if the theoretical background supporting our understanding of a phenomenon as a certain "fact" no longer made sense, if the theory was a "false theory," then, Whewell declared, "the fact is no [longer a] fact."[51] The fundamental antitheses thus managed to reconcile the need for the truth-bearing element of scientific knowledge with the idea that it is always revisable, and that its growth may in fact consist precisely in the revision of what was previously assumed to be a "fact."

Whewell's Response to Locke and Kant

Whewell had ambitious plans for his writing. He aimed at *reforming philosophy*; this reform was to be based on "a criticism of the fallacies of the ultra-Lockian school,"[52] which could be summarized as the idea that all knowledge starts from sensation.

In *An Essay Concerning Human Understanding* (1689), John Locke famously described the intellect as an empty cabinet where images coming from the senses could be stored, filed, and kept for further reference.[53] Its functioning was compared to that of a "dark room" (camera obscura), with a small hole in one of its walls to allow images from the outside to be displayed on the opposite wall.[54] Locke however did allow some margin for the mind's activity: the mind would recognize its ideas by reflecting upon itself; furthermore, it would compose complex ideas by manipulating the materials coming from sensation in the same way in which we manipulate objects from the external world to make tools.[55]

While Whewell had an active view of the mind, he denied the possibility of a simple idea, directly deriving from sensation; for Whewell, every sensation was already *informed*—in the sense of "shaped"—by our ideas. Whewell recognized that he was using "Idea" in a very different sense from Locke's: "I do not say that my view is contrary to his [Locke's]: but it is altogether different from his." For Whewell, ideas were not objects in the mind, but rules for its operations: "We use the word Ideas . . . to express that element, supplied by the mind itself, which must be combined with Sensation in order to produce knowledge. For us, Ideas are not Objects of Thought, but rather Laws of Thought." Even if one admitted, with Locke, that ideas are nothing but transformed sensations, Whewell would immediately reply, "What is the nature, the principle, the law of this Transformation?" For Whewell ideas were *not* transformed sensations, but "laws of thought" and "acts of the mind" that added something to our sensory stimuli in order that sensation (as a whole) could emerge at all. Whewell believed that he could support this argument (1) philosophically, by appealing to a "Kantian" notion of space; (2) through a careful examination of the development of scientific ideas in history; and (3) by drawing from physiology and the most advanced theories of perception of his time. The latter were only alluded to in the "Fundamental Antithesis" chapter ("Who does not know how much we, by an act of the mind, add to that which our senses receive? Does any one fancy that he sees a solid cube?"),[56] but Whewell returned to the topic with more detail in his chapter "On the Perception of Space." In what follows, I first illustrate Whewell's notion of ideas in relation to Kant's philosophy, before circling back to the purpose of Whewell's reference to physiology in the next section.

As Wettersten notes, Whewell advanced four arguments in support of his notion of ideas: two of Kantian flavor, drawn from the necessity of laws and from the conditions of experience, respectively; one from contemporary physiology; and one grounded on methodological considerations.[57] Whewell's discussion of the idea of space perfectly illustrated all four; the first three, however, are the more relevant ones to shed light on the question of Kant's influence on Whewell. We will see that the necessity argument was reduced in the *Philosophy* to the argument from the conditions of experience, while the physiology argument stands on the side, providing independent corroboration for Whewell's thesis.

In the first chapter of the *Critique of Pure Reason* (1781), "Transcendental Aesthetics," Kant laid down the conditions of the possibility of sensibility as well as the lawfulness of propositions from geometry and mathematics; both depend on the "forms" of our intuition, either in its empirical use (for sensibility) or in its pure use (for geometry and mathematics).[58] According to Kant, the forms of intuition are space and time; space is a form of the outer sense, while time is of the inner sense.[59]

For Kant, space is both the condition of possibility of the representation of objects outside us and something that we can conceptualize without help from the world outside. Through the application of rules derived from concepts of the intellect, we derive a priori axioms and laws of geometry.[60] Whewell also argued "that our mode of representing space to ourselves is not derived from experience,"[61] although he admitted that some necessary elements of knowledge, named "axioms," are *progressively* attained in the development of knowledge: "There are scientific truths which are seen by intuition, but this intuition is progressive."[62] In fact, axioms and ideas are on a constant relationship of co-specification. The progressive intuition of a "scientific truth"—which can be, for example, a proposition in geometry, such as that the sum of all internal angles of a triangle equals 180°—may depend on considerable investment on our part (e.g., a certain number of hours studying geometry). At the same time, the idea of space, which constitutes the origin and condition of possibility of our axioms of geometry, becomes more and more clear as the geometrical investigation proceeds. This, however, should not reduce the axioms to empirical regularities. Propositions in geometry should remain necessarily true.

According to Ducheyne: "The Axioms are necessary because they give a partial description of the necessary conditions of perception."[63] They are not analytically true, as Snyder reads them, because ideas are not concepts but rather rules for the use of our faculties. Moreover, axioms cannot derive their necessity from a psychological analysis of the working of ideas because that would make such necessity nothing more than an empirical fact.[64] In dropping Kant's carefully drawn distinctions between forms and ideas, Whewell emphasized the active role of the mind in perception.

For Whewell, the structure of perception and its law-giving elements, the ideas, were responsible for the truth and necessity of axioms: "And thus it appeared that the source of geometrical truths was not definition alone; and we find in this result a confirmation of the doctrine which we are here urging, that this source of truth is to be found in the form or conditions of our perception; in the idea which we unavoidably combine with the impressions of sense; in the activity, and not in the passivity of the mind."[65] In Kant the "Forms of Intuition" *individuate* every phenomenon by assigning them coordinates in space and time. The "Categories of the Intellect" are derived a priori from the four main logical categories of judgment (according to Scholastic logic): quantity, quality, relation, and modality.[66] The result of this deduction is a table of twelve categories that, according to Kant, *regulate* the use of concepts of the intellect (such as causality) in the world of phenomena. Last, the "Ideas of Pure Reason" express ideals that, although incognizable (because they transcend experience), still *direct the use* of human reason itself. They are the "Ideas of Soul" (the unifying principle of all internal representations), "World" (the unifying principle of all external representations), and "God" (the ultimate ground of reality and hope of a foundation for moral behavior).

Whewell's "Ideas" covered both Kant's "forms of intuition" and his "concepts of the intellect"; they were not deduced from logical principles rather they were gradually discovered in the history of science. The terminological anarchy adopted by Whewell reflected a difference of emphasis in his analysis of knowledge, in which stressing the formative role of ideas was deemed more relevant than laying down their logical justification.

The perceived obscurity of Whewell's philosophy derived in great part from his attempt to maintain more than one apparently untenable position at the same time. He liberally integrated transcendental and empirical elements in his account of ideas and axioms: ideas should regulate our apprehension of the outer world, yet they are progressively clarified in experience. Likewise, axioms are intuited, but their intuition can be progressive. As if this was not enough, Whewell stated that ideas were rules for the mind's activity, but beside traditional philosophical arguments, he used empirical arguments from physiology to illustrate the activity of the mind. It is to this last controversial strategy that I now turn.

The Idea of Space and Physiology

Whewell already alluded to the possibility of a physiological explanation of the constructive role of ideas in cognition in the "Fundamental Antithesis" chapter, but he returned to the topic with more detail in chapter 7 of book 2 of his 1847 *Philosophy*, "On the Perception of Space." Here, Whewell emphasized that something else besides sense impression is necessary in order to perceive external objects as such: "A very little attention teaches us that there is *an act of judgment* as well as a mere impression of sense requisite, in order that we may see any solid object."[67] He exemplified this fact with cases from optical illusion and perspectival drawing, and he compared the activity of interpreting stimuli into solid figures to other acquired activities that seem immediate, such as the understanding of words of a spoken or written language.[68]

The idea that there is an active (usually inferential) component in perception was not new in philosophy,[69] and it had already been discussed in the context of vision by thinkers such as Ptolemy, Alhazen, English natural philosophers conducting observations with the microscope in the eighteenth century,[70] and George Berkeley, who was explicitly referred to as a source by Whewell.[71] Wettersten claims that Whewell's theory of perception is one of the points where the usual account of a general lack of influence of Whewell has to be readdressed, although it is far-fetched to claim that after Whewell "one could . . . no longer claim without further ado that perception was built up out of individual sensations."[72]

Whewell wanted to argue that even touch, which is usually considered a more immediate sense than vision, is in fact only able to give an account of the objects outside us through successive steps, such as threading our hands around them. This point was particularly important for Whewell because it enabled him to connect our ability to perceive with an active engagement with the external world, and to rule out the common notion of ideas as direct impressions of the outer objects on a passive mind. While this argument was based on physiology, it dovetailed nicely with both his philosophical criticism of Locke and his borrowings from Kant. Whewell's criticism of mind as a passive container of ideas was tied into a positive account of space perception, in which the very construction of geometrical objects is connected to our experience and situated in the movements of our body: "The apprehension of extension and figure is far from being a process in which we are inert and passive. We draw lines with our fingers; we construct surfaces by curving our hands; we generate spaces by the motion of our arms."[73] According to Whewell, all the evidence concurred to show that the images that we receive from perception are in fact the product of an activity of the mind; this activity, largely unconscious, is connected to an (even imperceptible) motion of the sense organs, and from there to an activity of the will: "Thus it appears that our consciousness of the relations of space is inseparably and fundamentally connected with our own actions in space. We perceive only while we act; our sensations require to be interpreted by our volitions."[74] Whewell introduced a sixth sense, called "muscular sense," as a link between our perception of objects in space and the bodily actions that enable such perception. For Whewell, "the muscular sense . . . is inseparably connected with an act originating in our own mind." This act is the application of an idea to the stimuli that are coming to our senses, while it is through our perceptual activity and the study thereof that we gradually conceive of the underlying idea: "The sensations of touch and sight are subordinated to an idea which is the basis of our speculative knowledge concerning space and its relations; and this same idea is disclosed to our consciousness by its practically regulating our intercourse with the external world." Whewell supported his analysis of perception with the results coming from physiologist Sir Charles Bell. According to Bell vision does not depend only on the projection of an object's shape on the retina, but also on the conscious feeling of muscular exertion of our eyes surveying the object. Since only a small portion of the retina allows a distinct perception of the object, our eyes naturally move around, focusing on the different parts of the object in succession. According to Whewell, Bell "has traced in detail the course of the nerves by which these [voluntary] muscles convey their information. The constant *searching* motion of the eye, as he [Bell] terms it, is the means by which we become aware of the position of objects about us."[75]

The argument from physiology was not new in the *Philosophy*, since Whewell had already used Bell's work on vision in his Bridgewater Treatise of 1833.[76] His recourse to physiological support for the thesis of the role of ideas in perception was a naturalist move, which clearly distinguishes his philosophy from (what would become the "orthodox" reading of) Kant's.[77] This latter point was highlighted by the logician Augustus De Morgan in his 1840 review of the first edition of Whewell's *Philosophy* in *The Athenaeum*: "Mr. Whewell states in his preface, that he has adopted Kant's reasoning respecting the nature of space and time, but that nevertheless he dissents widely from the views of that philosopher. It appears to us that he has borrowed the essential doctrines of the Transcendental philosophy, but that he has explained them in a manner incompatible with the clear conceptions of Emmanuel [sic] Kant."[78] Whewell replied by disjoining his notion of "act of the mind" from a psychological and physiological analysis of the process: "These special physiological analyses of the process of perception, whether true or false, do not disturb the general doctrine that all perceptions of external objects involves an act of the mind, and that in the conditions of this act lie the foundations of universal truths."[79] As in the case of his rejection of the terminological (and corresponding conceptual) distinction between the Kantian "Forms of Perception," "Categories of the Intellect," and "Ideas of Reason," Whewell appeared anxious to press the point of the mind's activity in cognition over and beyond the way in which this activity was to be interpreted. Results from physiology confirming Whewell's thesis of the active role of ideas were gladly incorporated in the system, and Whewell saw them as ways to corroborate his philosophical account. Nonetheless, it was not physiology, but the active role of the human mind that constituted the foundation of Whewell's epistemology and the source of his disagreement with more empirically minded authors, such as Herschel and Mill. If physiology stood in the way of a reader's acceptance of Whewell's theory of ideas, it could be dropped.

Charles S. Peirce, an American Follower

Whewell died in 1866 from the consequences of a fall from his horse. In spite of his determination to reform philosophy and of the universally recognized vigor of his arguments, most reactions to his *Philosophy* were lukewarm.[80] Nonetheless, there was at least one instance in which Whewell's work was reviewed both explicitly and enthusiastically, although the author of such praise would remain outside the mainstream philosophy of science for a long time: the American pragmatist Charles S. Peirce.[81] Unfortunately, contemporary Whewell scholars typically downplay Peirce's autonomy in his reception of Whewell's thought, and focus almost exclusively on Peirce's notion of "abduction," in which they see nothing but a plain restatement of Whewell's notion of "consilience of hypotheses." Interpreters such as Menachem Fisch and Henry Cowles argue that Peirce's debt to Whewell's philosophy is just as deep as it is unacknowledged, particularly in regard to Whewell's theory of induction, which Peirce might have appropriated under a new label, "abduction," but eventually didn't. Thus Fisch writes, "Peirce chose to ignore his mentor, to coin a new term to denote Whewellian induction, and to pose thereafter as if he had thought it out for himself."[82] Wettersten states that "nowhere in Peirce's writings do we find a serious, critical review of Whewell's philosophy."[83]

This narrative has recently been challenged by scholars who have looked more closely at the multifaced aspects of Whewell's influence on Peirce, without necessarily concluding that the latter just gave a superficial impression of the former. For instance, Tullio Viola acknowledges the formative influence of Whewell but goes on to explore at length Peirce's own uses of history and its relevance for his epistemology.[84] Drawing from Harro Maas and Mary Morgan's study of nineteenth-century graphical analysis,[85] Chiara Ambrosio shows the importance the "method of curves" in both Whewell and Peirce. Since Whewell's method of curves also illustrates of his notion of induction, Ambrosio explains, a study of Peirce's relation to this method indirectly contributes to the vexing question of Peirce's debt to Whewell's theory of induction.[86] Ultimately, the question of Peirce's relation to Whewell is interesting not just because of its bearings on the developments of logic and of a theory of induction and hypothesis making, but also because it allows an assessment of the role of a notion of "active mind" in the development of a practice-based philosophy of science. In the following I limit myself to a few observations connected with Whewell's selective borrowings from Kant, which further highlight the influence of Whewell's philosophy of science on Peirce.

Peirce's reception of Whewell is characterized by an appreciation of the latter's historical erudition, by acknowledgment of his contributions to science and by a friendly reading of Whewell's eclectic relationship to Kant's philosophy. In 1865 Peirce inaugurated his philosophical career with a series of lectures delivered at Harvard University, "On the Logic of Science." An entire lecture—not numbered—was dedicated to the "Theories of Whewell, Mill, and Compte [sic]," presented as the three major approaches to the logic of science. Mill was the champion of empiricism and nominalism, Auguste Comte of positivism, and Whewell—Peirce's clear favorite—was acknowledged as "the most profound writer upon our subject [i.e., the logic of science]."[87] As noticed by Snyder, Peirce opened the lecture with a statement on the controversial topic of Whewell's Kantianism, judging that Whewell's realism produced a deep divide between the two thinkers.[88] Nonetheless, he went on to describe Whewell's "fundamental antithesis in philosophy," which Snyder recognizes as the core of his philosophy, as "very Kantian."[89]

Peirce's treatment of Whewell was initially constrained by the topic of his lectures. Focusing on logic, he highlighted Whewell's notion of colligation as "the possible germ of a strictly logical theory of induction"[90] and criticized Whewell's lack of formal resources to analyze the validity of inferences. In his *Novum Organon Renovatum*, Whewell presented "inductive tables" that should exhibit "the relation of the successive Steps of Induction" in two sciences, optics and astronomy. The tables present the "several Facts" of a given science on the top of the page. Brackets unify a number of them under one historic figure (e.g. Euclid, Ptolemy, Snell, Newton) and their theories; those theories are eventually unified again with other brackets, "so as to form a genealogical Table of each Induction, from the lowest to the highest."[91] Those tables—Peirce observed—do not possess a formal structure from which one could assess the inferences' validity: "Is there any other conception besides that adopted which would have colligated those facts? How am I to know?"[92] It is impossible to gauge from Whewell's tables whether the inductive processes that they represent are formally right or wrong.

Peirce had more space to show his appreciation of Whewell and of his role in the philosophy of science in his 1869 Harvard Lectures "On the British Logicians." When it came to Whewell, Peirce showed a deep acquaintance with all his works, from the scientific to the historical to the philosophical. According to Peirce, Whewell's credentials as a "scientific specialist," gained through his work on tides, should reassure any reader curious about his philosophy: "Dr. Whewell's qualifications for treating of science could hardly have been better than they were, for he was not only a scientific specialist but an eminent scientific investigator, his works upon the tides containing research of no ordinary importance. Indeed they will never be forgotten."[93] Peirce returned also to the debate surrounding Whewell's Kantianism, again recognizing the limitations of the extent to which Whewell can be considered Kantian—"Whewell's theory does not involve the whole of Kantism"[94]—and proposing that the worth of his philosophy should not be measured in terms of its proximity to Kant's. Peirce argued that, precisely because of their differences, the existing points of contact between the two philosophers reinforced the plausibility of their main thesis, namely "the general proposition that cognition consists of two elements one of which is idealistic and the other empirical."[95] Peirce recognized the importance of Whewell's own formulation of this principle, articulated in the "Fundamental Antithesis" as the coexistence of "facts" and "ideas": "There is probably no one maxim of logic the ignorance of which by ordinary people produces such deplorable results as this that all Facts involve ideas."[96] Peirce embraced different aspects of Whewell's philosophy early on, including the core statement that knowledge entails an irreducible duality between an active mind and the outer world.[97] For Peirce, perception always involved interpretation, and even apparently immediate sensations were the result of (unconscious) inferential processes. In arguing for this, he followed in Whewell's footsteps, referring both to Berkeley and to the physiological research of his time to support his views.[98] Peirce's naturalism would eventually go much further than Whewell's, as he came to argue for an evolutionary account of Mind itself.[99] Nonetheless, Peirce maintained a lifelong methodological interest in the logic of discovery and in the possibility of formally analyzing inferences; in the role of history—and history of science especially—in teaching "lessons in logic," thereby illustrating the progressive refinement of our ideas; and in the moral, social, and communal value of science. As this brief review of appreciative remarks shows, if Whewell did not have many followers, he had a good one in Peirce. Yet, a full history of Whewell's influence on American pragmatism remains to be written.

This chapter has presented a concise account of Whewell's philosophy of science, focusing in particular on the problem of its relation to Kant's. A prominent theme of Whewell's vast philosophical production was his rejection of traditional empiricist accounts of induction and of the formation of ideas. Whewell wanted to change the meaning of "Idea" by making it an expression of a rule for the organization of our experience; at the same time, he would accept the naturalization of this activity—for instance, through physiology—to the point of proposing such arguments as a possible corroboration of his philosophical outlook. While the "Fundamental Antithesis" remains at the core of Whewell's original contribution to philosophy, it is at the level of ideas that the similarities and differences with Kant vividly emerge.

Whewell was neither an "English Kant" nor a traditional empiricist. His philosophy of science, grounded in the toils of a practicing scientist and nourished by a deep engagement with science's history, tried to tread a middle way between transcendental philosophy, idealism, and empiricism. Whewell invested considerable energy in clarifying and defending his position, but with little immediate success. While he had no direct heirs in England, his work had a profound influence on the American pragmatist Charles S. Peirce. Interestingly, also in Peirce's case the question of a Kantian influence remains a controversial issue.[100] Both Kant and Whewell attempted to provide a secure basis for the advancement of science. Yet, their common aim should not make us forget the different ways in which they chose to articulate it.

For Whewell, science is neither grounded in facts nor in a transcendental scheme. In fact, one of the salient aspects of his "Fundamental Antithesis" is that the line dividing facts and ideas is porous and can be revised, a case in point being the ancient belief that the earth stood still while the sun revolved around it. In virtue of this continuity between the activity of the mind and the outer "facts," past discoveries enable future ones not only by enlarging the basis of facts on which we are standing, but also by allowing the progressive clarification of ideas.

CHAPTER 11

Whewell, Gender, and Science

HEATHER ELLIS

When William Whewell has been discussed in relation to gender and science in nineteenth-century Britain, he has sometimes been seen as an advocate of women's involvement in science.[1] He is often remembered (erroneously) as having coined the term *scientist* as a gender-neutral term for a cultivator of science in honor of the famous female mathematician and science writer Mary Somerville.[2] In a review of Somerville's *On the Connexion of the Physical Sciences* in the *Quarterly Review* in 1834, Whewell does indeed offer fulsome praise for her book, referring to "Mrs Somerville's able and masterly . . . exposition of the present state of leading branches of the physical sciences."[3] On closer inspection, however, it becomes clear that Whewell was not a far-sighted visionary in terms of women's participation of science, but rather much more in line with the attitudes of contemporary men of science. If we look more closely at his praise for Somerville, both in the *Quarterly Review* and elsewhere, we can see that it is instrumentalized by Whewell to argue for a particular form of scientific masculinity, namely the model of gentlemanly science that the British Association for the Advancement of Science (BAAS) had been promoting since its foundation in 1831.[4] As a prominent Cambridge mathematician, Whewell was intimately involved with the establishment of the BAAS and is counted among the "gentlemen of science" who initiated the new association.[5] Referring to Somerville's achievement in the preface to the 1836 edition of his *On the Free Motion of Points, and on Universal Gravitation*, Whewell declares, "Our willingness to adopt a more extended study of the mechanism of the heavens into our academic system must needs increase, when these severer studies, thus shewn to be reconcilable with all the gentler train of feminine graces and accomplishments, can no longer, with any shew of reason, be represented as inconsistent with a polished taste and a familiar acquaintance with ancient and modern literature."[6] Here he is specifically calling for the natural sciences to be made a more integral part of university education, a topic on which he spent so much of his time, first as a fellow and tutor, and later as master of Trinity College, Cambridge.

Whewell, Women, and a Gentlemanly Mode of Science

The first element to explore here is Whewell's alignment of Somerville's "feminine graces and accomplishments" with what he refers to as "a polished taste and a familiar acquaintance with ancient and modern literature." There is an assumption operating here that a gentlemanly persona, characterized by "a polished taste and a familiar acquaintance with ancient and modern literature," partakes of and exhibits something of these very "feminine graces and accomplishments." Any model of masculinity that positions itself as celebrating in any way explicitly feminine virtues is at considerable risk of attack from those seeking to accuse its adherents of effeminacy and unmanliness. In this one statement Whewell encapsulates much of the fragile and unstable position in which British science found itself in the early nineteenth century with respect to its reputation as a masculine practice.[7]

A brief discussion of the position of science as a gendered practice in this period is necessary here to understand what was at stake when Whewell made these claims. When it was founded in 1831, the BAAS was seeking in part to revivify science and reform it as a socially acceptable practice for gentlemen. It was reacting to a debate that was raging in the periodicals at the time about the "decline of science" in Britain.[8] Science was still frequently aligned with the activities (and social disadvantages) of the pedantic university scholar, despite (as Whewell makes clear above) the relative weakness of the natural sciences (in comparison with classical studies and mathematics) at Oxford and Cambridge.[9] Scientists (like university pedants) were seen as unmanly, socially awkward, and disconnected from the cut and thrust of conversation and business. One of the aims of the BAAS was to transform the investigation of the natural world into a practice fit for gentlemen. It approached this task in a number of ways, including the relentless recruitment of England's gentry and aristocracy to host, support, and attend its peripatetic annual meetings. Ironically, this sustained attempt at courting aristocratic favor resulted in the new association, within a few years, acquiring a reputation for another form of effeminacy—this time one associated with the pampered, lazy, and excessive lifestyles of the aristocracy.[10] Such noble sponsorship had clearly not been traditional for scientific gatherings outside the Royal Society, underlining the very different social world with which scientists had previously been associated. At the 1837 BAAS meeting in

Liverpool, visiting men of science were treated to "mountains of venison and oceans of turtle," with the Cambridge geologist and close friend of Whewell, Adam Sedgwick, asking, "Were ever philosophers so fed before? . . . Twenty hundred-weight of turtle were sent to fructify in the hungry stomachs of the sons of science!"[11] Emulating the lifestyle of the aristocracy also opened up the BAAS to charges of effeminacy because one of its defining characteristics was a mixed-sex sociability in which women played a significant role, supposedly civilizing the manners and behavior of the men taking part and elevating the tone of meetings.

Whewell seems to have been personally invested in this attempt to transform the public image of science into a socially acceptable, gentlemanly practice. Unlike many of his fellow dons, Whewell had himself come from lowly origins and despite his best efforts had acquired a reputation in Cambridge for his (relatively) coarse expressions and social behavior. As Richard Yeo writes, "For a time Whewell was not only very much alone, but also seen as unusual. His manners and speech were considered rude or rustic. There is a report of Whewell's comment upon a herd of pigs being driven past the college gate soon after his arrival: 'They're a hard thing to drive—very—when there's many of them—is a pig.'"[12] From his early days as an undergraduate, Whewell was keen to get out of Cambridge and see London society. As Yeo records: "Having visited London for the first time in 1815 Whewell admitted to his sisters that he had only seen the city from 'the outside' because, not knowing anyone there, he could not 'see anything of its society.'"[13] It seems likely that Whewell would have felt doubly excluded from dominant models of gentlemanly masculinity: first on the grounds of his (relatively) poor, rustic origins, as Richard Yeo has highlighted,[14] and second as a university scholar. The social distance apparent between the aristocracy, courted so assiduously by the BAAS in its early years, and many practicing men of science confirms the survival well into the nineteenth century of the substantial social gap between the scholar and the gentleman that Steven Shapin observed in the seventeenth and eighteenth centuries.[15] As late as September 1840, when the geologist Charles Lyell wrote to Whewell informing him of his nomination for the BAAS presidency in the following year,[16] Whewell was extremely reluctant to be put forward for the role, on the grounds that he was not a man of sufficiently high social rank and influence. "It could only produce failure and ridicule," he wrote to Roderick Impey Murchison, "to have me put in a place which should be occupied by some person of great local position, influence and popularity," in short, a "person coming nearer to the usual conditions, and likely to give the business its usual attractions."[17]

Thus, when we look more closely, we see that Whewell's endorsement of Mary Somerville and of women's participation in the world of science generally, was more about recommending a particular gentlemanly mode of science, characterized by patterns of mixed-sex sociability, fashionable among the aristocracy, designed to counter still powerful critiques of men of science as unworldly pedants, than it was about explicitly promoting women's involvement in scientific research. If we place Whewell's comments within the context of contemporary debates about the participation of women in the activities of the BAAS (debates in which Whewell himself took an active part) this distinction becomes clearer. Mixed-sex sociability was an important part of aristocratic culture and central to the civilizing role of knowledge during the Enlightenment. Jack Morrell and Arnold Thackray specifically identify the involvement of women as a "major factor in the change from natural knowledge as a remote and cloistered virtue to science as a public resource."[18] The involvement of women in the early years of the BAAS was also key to the transformation of the public reputation of men of science from retiring, effete scholars to active, socially engaged gentleman scientists. The first ever conversazione, which emulated elite mixed-sex social gatherings on the continent, was held in 1831 at the first BAAS meeting at York. The *Yorkshire Gazette* remarked on the presence of "elegant females" and "fashionable ladies" and how "the charms of beauty and the varied stores of philosophy seemed united."[19]

A powerful case could be made for admitting women to sectional discussions at BAAS meetings in terms of boosting the masculine reputations of the male scientists presenting there. While masculinity could certainly be validated by peers within an all-male audience, women also had an important role to play in terms of confirming the masculine reputations of male speakers. The obvious admiration of his largely female audiences had been a significant factor in establishing Humphry Davy's reputation as a Romantic hero of science. As Jan Golinski writes, "His deportment as a lecturer at the Royal Institution made use of conventions of masculine display before an audience that was, to a significant degree, female. The command of his audience that Davy achieved was a significant resource in making his reputation as a discoverer."[20] Some leading BAAS members in the early years of the association's history commented similarly about the role of women at meetings, that they stimulated the assembled men of science to fresh exertions. During the 1832 meeting at Oxford, Adam Sedgwick described the ladies' gallery as "that blazing crescent which had decorated the meetings" and spurred the philosophers on to new efforts. Whewell and Sedgwick pushed hard for the increased presence of the wives and daughters of scientists at meetings as they were convinced it encouraged a gentlemanly atmosphere.[21]

This view of the potentially positive role of female audiences in helping to construct the public masculine reputations of male scientists is confirmed in recent research carried out by Charles Withers and Rebekah Higgitt. Considering female audiences for Section E (geography) of the BAAS, they write that "women provided a successful foil to the heroic, manly explorers they flocked to hear." In this way they helped to reinforce the gendered dichotomy central to the BAAS's self-understanding in its early years between "male expert/female audience."[22] Withers and Higgitt argue that, in general, women were content to adopt a passive, admiring role when watching and listening to male scientists at BAAS meetings. "Seeing and describing the scientific lions took a prominent place in women's accounts of BAAS meetings."[23] Reflecting on the masculine qualities of the various men of science they encountered was a favorite activity according to a study of women's diaries. They would try to discern the mental character of particular scientists or traces of the hardships they had endured by scrutinizing their faces and deportment. Some of the thoughts recorded by women attending BAAS meetings at the time confirm this impression. Sara Jane Clarke, for example, wrote that she was "truly impressed by the manner and presence" of scientists like Thomas Romney Robinson and David Brewster.[24] Harriet Martineau thought that women chiefly attended BAAS gatherings "to sketch the savans."[25] *The Times* likewise reported of the 1836 meeting in Bristol that the "softer portion" of the audience were "on the full gaze, to see what kind of creature a philosopher was."[26]

This admiration on the part of female audiences, moreover, seems to have been directly encouraged by the men of science themselves. Their perceived attractiveness to women became part of their masculine image, and something they worked hard to secure. Caroline Fox, for example, records Adam Sedgwick, as "saying many soft things to the soft sex" at the 1852 meeting in Belfast.[27] We gain a little more detail from a letter written by John Herschel to his wife in 1838, relating an earlier example of Sedgwick's flattery: "Sedgwick said, in his talk on Saturday, that the ladies present were so numerous and so beautiful that it seemed to him as if every sunbeam that had entered the windows in the roof (it is all windows), had deposited there an angel."[28] When writing to Charles Daubeny about the arrangements for the 1832 Oxford meeting, Charles Babbage stressed the "importance" of "enlist[ing] the ladies in our cause." The participation of ladies guaranteed a gentlemanly atmosphere, he maintained, ensuring that "scientific men mix more in general society, and that the more intelligent amongst the upper classes . . . get a little imbued with love for science." He positively extolled the value of female admiration. "Remember the dark eyes and fair faces you saw at York," he urges Daubeny, "and pray remember that we absent philosophers sigh over the eloquent descriptions we have heard of their enchanting smiles."[29]

It is important to remember, however, that this fondness for the presence of ladies did not generally extend to their active participation in scientific research. The geologist William Buckland, as BAAS president in 1832, made clear his view that women ought not to attend the scientific part of meetings; he confessed, however, that they were vital to the public image of the association: "Their presence at private parties is quite another thing," he declares, "and at these I think the more ladies there are, the better."[30] It was, I would suggest, in this way that Whewell thought Mary Somerville's scientific accomplishments were most useful, not in the first instance for their own scientific merit, but because they seemed to make the practice of science more acceptable in gentlemanly circles, more consistent with "a polished taste and a familiar acquaintance with ancient and modern literature." In other words, Somerville made it easier (and more socially acceptable) for a university scholar like Whewell to take part in scientific research.

"A Sex in Minds": Whewell and Mary Somerville

Despite praising Somerville's *Connexion of the Physical Sciences* as "able" and "masterly," Whewell also denies it any claim to be considered as containing original discoveries. "Mrs Somerville's work," he writes, "is, and is obviously intended to be, a popular view of the present state of science."[31] As evidence of this he quotes at length Somerville's own dedication in which she claims to aim at nothing more than "to make the laws by which the material world is governed, more familiar to my countrywomen."[32] Whewell goes to quite some length to reassure his readers (and one suspects himself) that Somerville's work is no threat to men of science. Indeed, he fully endorses her expressed hope that women will learn much from her work, although not without raising the question *whether* the women of England have yet progressed far enough in their general knowledge and understanding of science to receive the full benefit of Somerville's instruction. More than this, when asking this question, Whewell highlights precisely those tropes and stereotypes about the female character that cast them as beautiful endorsers and inspirers of male scientific activity but as potentially incapable of the firmness of mind needed to be original inquirers themselves. Indeed, he adopts a kind of flirtatious and chivalric (but ultimately condescending) attitude toward them here that reminds us strongly of the attitude toward women at early BAAS meetings: "And if her countrywomen have already become tolerably familiar with the technical terms which the history of the progress of human speculations necessarily contains; if they have learned . . . to look with dry eyes upon oxygen and hydrogen, to hear with tranquil minds of perturbations and eccentricities, to think with toleration that the light of their eyes may be sometimes polarized, and the crimson of their cheeks capable of being resolved into *complementary colours*;—if they have advanced so far in philosophy, they will certainly receive with gratitude Mrs. Somerville's able and *masterly . . . exposition*."[33]

Whewell continues to "praise" Somerville's work in this style by going so far as to say that even men of science—and note the phraseology cho-

sen—"individuals of that gender which plumes itself upon the exclusive possession of exact science . . . may learn much that is both, novel and curious
in the recent progress of physics from this little volume."[34] This statement
needs some unpacking. On the surface, Whewell is praising Somerville's
book, saying that its usefulness and interest go far beyond the limited audience of "countrywomen" she addresses it to; indeed, that there are "few"
men who will not benefit from reading it. However, two things are worthy
of note. First, Whewell does not use the word *men* but rather "individuals
of that gender which plumes itself upon the exclusive possession of exact
science," reinforcing not only that "exact science" is normally considered the
"exclusive possession" of men, but also that this is something that is important to their identity as men (they "plume" themselves upon it). Hence, Mary
Somerville's intervention into the field of "exact science" is not in truth to
be welcomed, but rather represents a threat to the status quo. This is not
explicitly said by Whewell of course (as this would not be gentlemanly), but
it is apparent in his somewhat condescending reference to Somerville's "little
volume."

More evidence of Whewell's underlying discomfort at Somerville's
achievement is provided by the extensive discussion he offers on the fact that
the *Connexion of the Physical Sciences* has been authored by a woman. It is
so extensive that his biographer Isaac Todhunter, writing a few short years
after Whewell's death, feels the need to suggest to his readers that "perhaps
too much stress is laid on the fact, which is brought prominently forward,
that such a work had been written by a woman."[35] A substantial part of this
discussion is taken up with the exceptional nature of Mary Somerville as
a "person of real science" (he clearly does acknowledge her ability at some
level).[36] He calls upon his (male) readers to validate his views about Somerville's exceptional status: "Our readers cannot have accompanied us so far
without repeatedly feeling some admiration rising in their minds, that the
work of which we have thus to speak is that of a woman." "There are various prevalent opinions concerning the grace and fitness of the usual female
attempts at proficiency in learning and science," Whewell writes, implying
it is generally not considered "graceful" or "fitting" for women to engage in
scientific research.[37] Somerville, however, proves the rare exception to this
rule, not principally because of her "real and thorough acquaintance with
these branches of human learning, acquired with comparative ease," but
rather because these abilities are "possessed with unobtrusive simplicity"
that, Whewell suggests, befits the female sex. She does not, like men, "plume"
herself on her achievements. In this "remarkable circumstance," Whewell
writes, "all our prejudices against such female acquirements vanish,"[38] but
only because Somerville takes care to act within the prescribed boundaries of what (male-dominated) society deems as acceptable female behavior.

Addressing her work only to her "countrywomen," she takes care not to put herself in direct competition with men of science, to avoid being seen as a threat.[39] In feminist literary criticism, this has become known as the "modesty topos."[40]

Somerville's modesty is praised by Whewell in a number of other places. A reviewer of his *History of the Inductive Sciences* (1837) expressed a wish ("foolishly" Whewell's biographer Todhunter remarks)[41] that Mary Somerville should have been included in the work. Whewell's reply is clear: he does not see Somerville as an original discoverer and besides this, he believes her proper feminine modesty would never allow her to permit her name to be included—even if she deserved it: "With regard to the excellent and accomplished lady whose name the critic has thought proper to introduce into his pages . . . I will only say, that if I had employed my office of historian for the purpose of complimenting her with a place among discoverers in astronomy . . . I am persuaded that her clear sense and genuine modesty would have disapproved of the introduction of such a passage into my work."[42] His review of Somerville's *On the Connexion of the Physical Sciences* concludes with two pseudo-chivalric sonnets praising Somerville, presumably composed by Whewell himself.[43] The form of the sonnets places Somerville firmly in the traditional female position as the object of male adoration while also underscoring Whewell's own credentials as a literary gentleman. The content of the sonnets is as interesting as their form, as it is Somerville's modesty that is once more the focus. In the second of the two poems, while she is described as "learned" she is also "popular"; though she "instructs the world," she remains modest and "unobtrusive," so that she retains her femininity intact and is "dubbed by none a Blue [stocking]."[44] Whewell's sonnet thus contains a veiled threat of what might happen to Somerville if she were not modest: she would be subject to gendered insults ("dubbed a Blue"), just as BAAS members were called unmanly as a result of their elaborate feasting. Her achievement would not be acceptable but for her modesty. In a similar way, the first of the two sonnets seeks to lessen the threat that Somerville appears to represent. While she is praised for having precisely that clarity of thought that men, according to Whewell at least, lack ("Full of clear thought; free from the ill and vain / That cloud our inward light"),[45] she appears—almost godlike—on a pedestal, the romantic subject of the chivalric "lays" of an "earlier time," removed from any direct comparison or competition with Whewell himself.

In his 1834 review of Somerville, Whewell pronounces: "Notwithstanding all the dreams of theorists, there is a sex in minds. He [that is, man] learns to talk of matters of speculation without clear notions; to combine one phrase with another at a venture; to deal in generalities; to guess at relations and bearings; to try to steer himself by antitheses and assumed maxims. Women never do this: what they understand, they understand clearly: what they see at all, they see in sunshine."[46] As far as it goes, this seems to equate to the "equal but different" school of thought regarding men's and women's respective intellectual powers; yet Whewell goes on to state that although women could, on occasion, exhibit a similar "power of understanding" to men (albeit of a different "kind" and "mode"), "it may be, that in many or most cases, this brightness belongs to a narrow Goshen; that the heart is stronger than the head."[47] In other words, for the vast majority of women, although they may in some respects possess greater clarity of thought than men, this is focused only in a narrow area, and their feelings predominate over their reason. With his final sentence Whewell makes his position clear: "It certainly is to be hoped that it is so."[48] A few exceptional women, such as Mary Somerville, there may be, and if they remain modest and adhere to the social expectations of femininity, then so be it. However, Whewell is clear that this should not be the case for the majority of women. These remarks reflect Whewell's wider thoughts on women's education. He certainly supported their education and did not feel that it should be confined to traditional female accomplishments. This is clear from his endorsement of the educational value of Mary Somerville's work for other women, as well as his support for initiatives such as the Queen's College for General Female Education located in Harley Street, London. His sister-in-law, Lady Monteagle, who was one of the "Lady Visitors" of that College, invited him to lecture there on Plato, which he agreed to do.[49] Elsewhere, he expresses the view that "ladies" as well as men could benefit from reading Euclid.[50] However, when writing to his friend, Mrs. Austin, about his plan to lecture to ladies about Plato on May 13, 1857, Whewell declares that while he did not believe there to "be any difference of power of understanding in men and women . . . of kind and mode of understanding there may be and is."[51]

Further doubt about Whewell's enthusiasm for women of science arises when we consider the stories he tells about the few exceptional women he knows who have risen to the heights of mathematical and scientific knowledge. Although presented merely as historical accounts, read through a critical feminist lens, their grizzly ends send a warning to Whewell's female contemporaries of the fate awaiting women who do not keep within their boundaries. Most prominent here is the story of Hypatia, which Whewell describes as "unhappily as melancholy as it is well known," as though underscoring the point that the women of his day really should know the dangers of stepping outside the bounds of femininity. "She was the daughter of Theon," he continues,

> the celebrated Platonist and mathematician of Alexandria, and lived at the time when the struggle between Christianity and Paganism was at its height

in that city. Hypatia was educated in the doctrines of the heathen philosophy, and in the more abstruse sciences; and made a progress of which contemporary historians speak with admiration and enthusiasm. Synesius, bishop of Ptolemais, sends most fervent salutations "to her, the philosopher, and that happy society which enjoys the blessings of her divine voice." She succeeded her father in the government of the Platonic school, where she had a crowded and delighted audience. She was admired and consulted by Orestes, the governor of the city and this distinction unhappily led to her destruction. In a popular tumult she was attacked, on a rumour that she was the only obstacle to the reconciliation of the governor and of Cyril the archbishop. "On a fatal day," says Gibbon, "in the holy season of Lent, Hypatia was torn from her chariot, stripped naked, dragged to the church, and inhumanly butchered by the hands of Peter the reader and a troop of savage and merciless fanatics: her flesh was scraped from her bones with oyster shells, and her quivering limbs were delivered to the flames."[52]

The most telling line in this story is perhaps when Whewell explains to his readers (who are likely to have included women interested in a review of Mary Somerville's work that was especially addressed to them) that it was precisely Hypatia's "distinction" (talent, that which made her distinct and different from other women) that "led to her destruction."

Until now, Whewell assures us, Mary Somerville has maintained that modesty required of women; yet, for all her modesty, she is proclaimed to be as "rare" as Hypatia, one of only a tiny number of women who have excelled at mathematics in the same way as men; she is likened directly to Hypatia; and in many ways Somerville's actions—writing and publishing books under her own name, even delivering scientific papers at meetings of male scientists (albeit through her husband)[53]—do fly in the face of conventional expectations (including, as he admitted, Whewell's own) of female behavior and ability. His inclusion of the story of Hypatia and the graphic and horrific detail of her torture and death (which serve no clear purpose in the review) can reasonably be interpreted as evidence that Whewell felt personally threatened by Somerville, particularly as she was encroaching on what he viewed as very much his own territory, the communication and explanation of the history of science to a popular audience. We remember the quotation mentioned earlier, when Whewell, in a reply to a reviewer asking why Somerville has no place in his history, highlights the elevated status of "my role as historian" and his right *not* to recognize her achievement.

Whewell and the "Great Man" Theory of the History of Science

From the discussion above, where we see that Whewell did indeed feel threatened by Mary Somerville, we can move on to focus on his relationship with other male scientists. One of the reasons that Whewell appears to have felt threatened by Mary Somerville is because she adopted a role (and made a name for herself) in an area close to Whewell's own—explaining the development of scientific knowledge to a popular audience. When we consider the wider scientific enterprise in England at the time, this type of role seems to sit rather awkwardly in terms of its perceived value, importance, and manliness. In the scientific community, there was a strong sense in the early nineteenth century that the great men of science were those who made great discoveries, not those who merely communicate knowledge of those discoveries to others. We see this clearly with regard to Whewell himself when Francis Galton describes why he is not included in the "Men of Science" section of his work *Hereditary Genius* (1869), just three years after Whewell's death. One criterion was particularly important for inclusion, according to Galton in the introduction to the section: "The fact of a person's name being associated with some one striking scientific discovery."[54] Galton admits that Whewell's "intellectual energy was prodigious, his writing unceasing, and his conversational powers extraordinary." Moreover, "his influence on the progress of Science during the earlier years of his life" was, we are told, "considerable"; yet for all this he was not selected by Galton because "it is impossible to specify the particulars of that influence, or so to justify our opinion that posterity will be likely to pay regard to it. Biographers will seek in vain for important discoveries in Science with which Dr Whewell's name may hereafter be identified."[55] In other words, his chosen role as "historian" of science, an explainer—the same role Mary Somerville was identified with—was not sufficient to admit him to the pantheon of elite scientific men. We remember Whewell's own reply to a reviewer of his *History*, that he had not given Mary Somerville "a place among the discoverers of astronomy" because she was none such and to do so would be mere flattery.

That Whewell was sensitive during his life to the view that Galton expressed about him after his death, while not certain, is strongly suggested when we examine his response to his brother-in-law Frederic Myers's *Six Lectures on Great Men*.[56] While praising his project in general terms (particularly "that it may help to correct the tendency of the present times to moral cowardice"[57]), Whewell criticizes Myers in a letter of March 1848 for focusing on "great men," those he identifies as "bold and vehement" as opposed to "good men,"[58] who, though less remarkable for their individual achievements, nonetheless contribute to noble projects that benefit humanity as part of a much larger collective effort: "The difference between us I have sometimes expressed . . . by contrasting worship of heroes with reverence for ideas. It appears to me that a reverence for the ideas of truth, justice, humanity, and for the forms in which they have been embodied—law, institutions, books, national habits, including, of course, religious light and heat—that these are more truly deserving of reverence than any man's character."[59] Whewell tries to argue that "the progress of mankind . . . consists in the progress of these things" and "not in the energy with which at intervals this man or that labours to promote their progress." "Only a slow progress is granted to man," he continues, adopting an elevated religious tone, "and only a slight share to any one man and the men to whom the greatest share is due are not, I think, those whom you call great men."[60] In other words, whatever the world might think of individual "great men," it was, Whewell was suggesting, rather those who labored humbly and inconspicuously for the greater good who were really more deserving of credit and renown.

And yet, in his own *Philosophy of the Inductive Sciences* (1840), Whewell does indeed label certain individual discoverers as "great men" just in the way that Galton does in *Hereditary Genius*. He describes Bacon, for example, as the "Hero of the revolution in scientific method," standing "far above the herd of loose and visionary speculators who, before and about his time, spoke of the establishment of new philosophies."[61] Drawing on the language of classical myth, Bacon is for Whewell "not only one of the Founders, but the supreme legislator of the modern Republic of Science, not only the Hercules who slew the monsters that obstructed the earlier traveller, but the Solon who established a constitution fitted for all future time."[62] Indeed, Whewell describes his own approach to the history of science in strikingly similar terms to Myers's lecture series—as an evaluation of the contribution of "the great men of the past" and their "discoveries."[63] When covering the ancient world, he endorses the speech of Pliny praising Hipparchus and others in a similar vein: "Great Men! elevated among the common standard of human nature, by discovering the laws which celestial occurrences obey, and by freeing the wretched mind of man from the fears which eclipses inspired."[64] Thus, in both his *History* and *Philosophy*, Whewell plays his part in constructing that image of the history of science as, in Dena Goodman's words, "a mythical history of masculine reason."[65]

In the end, we are left with the impression of a man who was somewhat ill at ease with himself and his own masculinity (like many of his contemporary men of science); he sought to be recognized as a gentleman rather than a scholar, although he doubted his credentials even when he became master of Trinity. He did not see himself sufficiently aristocratic to be elected as president of the BAAS. He idealizes great men such as Bacon and Newton, "discoverers" in the history of science, but struggled to live up to this ideal himself. He tried at times (particularly in his response to Frederic Myers's lectures on great men) to construct an alternative vision of good, humble men, team players rather than heroes, who work together collaboratively for the collective good of science and humanity. But he did not seem to believe fully in this alternative ideal, although it is the one he was aligned with by Galton after his death. Ultimately, as we have seen in his comments about Mary Somerville, modesty and humility were, for Whewell (and his contemporaries), fundamentally feminine virtues—and this fact was hard to reconcile with a vision of himself as an influential, independent man of science.

CHAPTER 12

William Whewell and the Palaetiological Sciences

MAX DRESOW

In 1837 the forty-two-year-old William Whewell became the fifteenth president of the Geological Society of London, succeeding Charles Lyell. It was, at first blush, a curious selection. Although Whewell had accompanied Adam Sedgwick on geological field trips, and was later appointed chair of mineralogy at Trinity College, he was no one's idea of a practicing geologist.[1] True, he had twice descended the Dolcoath mine in an attempt to determine the average density of the earth, but these experiments failed, in part because Whewell and his collaborator kept dropping their instruments.[2] He redeemed this failure with a masterful essay on mineralogical classification and later carried to completion a large-scale project measuring the ebb and flow of tides across Europe.[3] However, neither of these accomplishments belonged to geology as Whewell was soon to define it: as a "palætiological science."[4] Perhaps for this reason, Whewell felt compelled to protest his election in an 1836 letter to Sedgwick: "I am not the proper person for your generalissimo."[5]

The mystery surrounding Whewell's election dissipates when we learn that it was largely a political machination designed to deny the presidency to older geologists like William Buckland. (Buckland was just ten years Whewell's senior, but his well-known eccentricities had become something of an embarrassment to younger geologists like Lyell and Roderick Impey Murchison.) Still, a residuum of mystery remains, for if all that was needed was a "young[er] man," what was it that recommended Whewell in particular? Here it pays to remember that British geology in the 1830s was dominated by "gentlemanly specialists"—basically, talented aristocrats who operated outside the structures that came to define geology in later decades.[6] During this interval, access to the inner circle of the geological elite was mediated as much by social relations and matters of patronage as by institutional factors. This created an environment in which a talented and well-connected man like Whewell could rise, even to the level of the presidency, with few distinct accomplishments to his name.[7] But Whewell had another thing going for him. As a member of the Royal Society and the author of a string of textbooks, Whewell's scientific opinions were charged with a credibility that most geological amateurs could scarcely pretend to. Factor in his growing reputation as a "critic and reviewer, adjudicator and legislator of science," and Whewell was a force to be reckoned with, even near the periphery of his omni-competence.[8]

So much for why the Geological Society wanted Whewell. Now why was Whewell interested in the Geological Society? What reason did the mathematically inclined Whewell have for attending to developments in geology? And why did Whewell single out geology as the best example of a palaetiological science? These are the questions this chapter will confront. And happily, answers are not far to seek. In fact, the problem is that there are *so many* answers, all jumbled up, that need to be examined in relation to different aspects of Whewell's metascientific vocation. For example, in his capacity as a historian and methodologist of science, Whewell was interested in geology as an exemplar of palaetiology: a relatively new form of science whose foundations remained to be philosophically clarified. This complemented his interest as a critic and legislator of science, which was to define the scope of geology, including, most importantly, limits to its speculative competence. Whewell's interest in the limits of geology was in turn related to his role as a cleric and theologian and gave substance to his view that geology provides an indirect support for natural theology.[9] Finally, Whewell the nationalist and university man was interested in geology as a means of heading off narratives about the decline of British science.[10] This list does not exhaust Whewell's motivations (he was also interested in a number of internal geological debates), but it does provide a framework for understanding his interest in geology in the 1830s, when most of his writings on the subject were composed.

The remainder of this chapter uses this framework to explore Whewell's writings on the palaetiological sciences. It begins with a consideration of Whewell's analysis of palaetiological science, paying special attention to the status and position of geology among the inductive sciences. Then it considers Whewell's views on the scope and limits of palaetiological science by examining his treatment of the limitations of geology. This is followed by a brief section that considers Whewell's role as a defender of British science in the wake of Charles Babbage's much-discussed *Reflections on the Decline of Science in England* (1830). Finally, it considers Whewell's intervention in a methodological debate in geology about the proper way of determining the causes of past events and its relevance for geological theory.

Palaetiological Science

The palaetiological sciences are those sciences "in which the object is to ascend from the present state of things to a more ancient condition, from which the present is derived by intelligible causes."[11] Included among their ranks are geology, palaeontology, glossology, and comparative archaeology. But Whewell's discussions were mostly limited to geology, so in keeping with the themes of this chapter, mine will be as well. Geology around 1830 was a youthful science experiencing "its first and greatest boom in conceptual innovation, empirical expansion, and public approval and interest."[12] It had come a long way since its consolidation in the second half of the eighteenth century. But as everyone recognized, it still had a long way to go before it could be regarded on par with a fully mathematized science like Newtonian mechanics. This set a challenge to those like Whewell who wished to locate geology within the span of the inductive sciences. First, geology needed to be characterized and delimited. Second, it needed to be shown that geology deserved its place in the inductive sciences despite its relative youth and lack of mathematical development.

As Jon Hodge observes, when Whewell sat down to write about geology in his *History of the Inductive Sciences* (1837), there "was no consensus concerning geology as a site on the map of knowledge or as a division in the classification of the sciences."[13] Nearly everyone agreed that geology was a science in some suitably permissive sense of that term. But whether geology was yet a properly inductive science was open to question, an ambiguity that Whewell confronts without totally resolving in the *History*. For Whewell, every inductive science has one or more "Fundamental Idea" particular to it.[14] So, the idea of cause is particular to the science of mechanics, and the idea of space particular to geometry. However, geology has no fundamental idea of its own, only a variant of the idea that Whewell assigns to mechanics. This is the idea of a *"historical cause,"* meaning a cause that "occupies some certain portion of time [but not another]." Nonhistorical causes, by contrast, operate "at all times and under all circumstances." They are truly permanent or *"mechanical causes"* that, one gathers, embody the Fundamental Idea of causation in its purest form, free of the contingencies of historical causation.[15]

The relevance of this distinction for geology's status as an inductive science is made clear by some observations that Whewell recorded in his notebook in 1831. Here Whewell distinguishes between what he terms "permanent" and "progressive" causes: causes of things continuing as they are, and causes of things beginning to be. Permanent causes include the force of gravitation and the luminiferous ether. They are the mechanical causes of the *History*. Progressive causes, by contrast, include the true account of the rising of the Alps (whatever that is) and Pierre-Simon Laplace's nebular theory (if correct). These are historical causes. Perhaps surprisingly, Whewell can think of no reason to regard permanent causes as better candidates for scientific investigation than progressive ones. Still, it remains the case that sciences of progressive causation, like geology, "[differ] somewhat from the sciences so called."[16] This is because the main business of science is to determine "general laws" that are "universally and constantly true";[17] but geology seeks mainly to explain particular things that happened at particular times in the past. If geology is a science, then, it is a science whose proximate aim is not to arrive at universally and constantly true laws, but instead to arrive at a true account of the etiology of some part of the earth.

By the time his *History* appeared, Whewell had apparently overcome his reservations about including geology among the inductive sciences. He nonetheless placed it at the end of his survey, in part to indicate that geology "differs somewhat from the sciences so called," and in part to illustrate that geology provides a kind of portal connecting science and natural theology (more on this later). Geology belongs among the inductive sciences because it is based on an idea of causation, and so conforms to a picture of induction that requires ideas to connect the facts of experience (in this case, traces of the past). Yet it differs from the main run of the inductive sciences in that it is based on an idea of causation that orients it to particular chains of events as opposed to universal relations.[18] The term *palaetiological science* serves to codify this difference, and to highlight the status of geology as a separate-but-perhaps-not-quite-equal partner among the inductive sciences.

While the palaetiological sciences differ from the traditional inductive sciences in lacking a fundamental idea, they resemble them in other respects. For example, in both cases there is a descriptive or phenomenological portion (think Kepler), as well as a dynamical or causal portion (Newton).[19] The phenomenological portion involves the classification of phenomena according to "the true principles of natural classes," as well as the formulation of laws concerning such things as the "general form of mountain chains; the relations of the direction and inclination of different chains to each other; the general features of mineral veins, faults, and fissures; [and] the prevalent characters of slaty cleavage."[20] This sets the stage for etiological research, whose aim is to formulate the laws responsible for the "facts" recorded in the phenomenological portion. In the science Whewell terms "Geological Dynamics," researchers treat "not only [the] subterraneous forces by which parts of the earth's crust are shaken, elevated, or ruptured, but also [the] causes which may change the climate of a portion of the earth's surface . . .

[and those] which modify the forms and habits of animals and vegetables."[21] But geological dynamics is not an end in itself—instead, it is a kind of intermediary between phenomenal geology and physical geology, or true geological theory. Unsurprisingly, given that geology in the 1830s remained wet behind the ears, Whewell did not think that any "such theory [yet] exists on any [geological] subject."[22]

I will return to the subject of geological theory in the final section. Before coming to this, however, it is necessary to treat what was perhaps Whewell's central interest in relation to the palaetiological sciences: determining their limits.

The Limits of Geology

In his astute essay on Whewell's geology, Jon Hodge observes that Whewell's most persistent preoccupation as a writer on geology was not the adjudication of internal geological disputes, nor the articulation of a sound geological methodology.[23] Instead, it was to place limits on geology's rightful sphere of operations. Since for Whewell geology included not only stratigraphy and geological dynamics, but also biogeography and speculations about the origin of species (including man), it was liable to come into conflict with religious thinking on the same matters. This created the need for some way of resolving (or better, avoiding) territorial disputes. Who has the right of way in these cases? Or, to put a different spin on things: How can we identify the limits of geology and theology, respectively, such that confrontations on matters of shared concern can be decisively headed off?

There was no shortage of literature on geology and religion in Great Britain, much of it concerned with assessing the correspondence of recent geological findings and the Genesis narrative, variously interpreted. What made Whewell's project noteworthy was its approach. He sought what Hodge terms a "delimitational adjudication" by means of "epistemological explication": in other words, a way of adjudicating competing claims on shared questions by means of philosophical reflection.[24] If Whewell could show that philosophical reflection implied a limit to geological speculation, then the door would be open to a conciliatory position that granted geology a large scope of operations while at the same time releasing it from the hopeless task of establishing a correspondence between rocks and Scripture. It would also guarantee that the progress of geological science would not exclude the operation of a creative first cause: something that would presumably make it easier to abandon older, literal interpretations of Scripture for more geologically plausible ones. Far from encouraging unbelief, such concessions would be nods to the advance of an unthreatening palaetiological science, in keeping with the Thomistic maxim that truth cannot conflict with truth.

So what did Whewell take the limits of geological investigation to be? Despite his reputation as a conservative thinker, Whewell granted that geology can "go very far back" in its inquiry into past causes: it can "determine many of the remote circumstances of the past sequence of events;—ascend to the point that, from our position, at least, seems to be near the origin;—and exclude many suppositions respecting the origin itself."[25] Still, it remains "no irrational opinion" to say "that in all those sciences which look back and seek a beginning of things, we may be unable to arrive at a consistent and definite belief, without having recourse to other grounds of truth."[26] The reason is that, while the palaetiological sciences urge us

> by an inevitable consequence . . . to look for the beginning of the state of things which we thus contemplate . . . in none of these [areas] have men been able, by the aid of science, to arrive at a beginning which is homogeneous with the known course of events. The first origin of language, of civilization, of law and government, cannot be clearly made out by reasoning and research; and just as little, we may expect, will a knowledge of the origin of the existing and extinct species of plants and animals, be the result of physiological and geological investigation.[27]

The same holds true for all reasonings about the present geological order, from whose beginnings we are "separated by a state of things, and an order of events, of a kind altogether different from those which come under our experience."[28] It follows that the palaetiological sciences can have nothing to say about beginnings. As we approach the beginning of the present causal order, the "thread of induction respecting the natural course of the world snaps in our fingers."[29] The most geology can say is that there *is* a beginning; it cannot say what caused it.[30]

In a sense, this is a curious argument. If the palaetiological sciences can carry their investigations "very far back," then why are they incapable of carrying them all the way to the beginning? Why must they run into a wall where the chains of filiation they are engaged in tracing have their origin? In short, why do things like the origin of language and species fall outside the bailiwick of palaetiological science? Whewell seems to suggest that because the palaetiological sciences have so far failed to account for genuine originations, they are incapable of doing so. But this seems unduly pessimistic. Whewell does not say that geologists are incapable of achieving a physical geology because no palaetiological science has yet framed a complete and true theory. So why should we take the failure of the palaetiological sciences to penetrate to the beginning of things as a mark of incapacity?

Presumably the answer has something to do with the nature of those causes responsible for originations—first causes or truly creative causes. These are a kind of historical cause in that their operation is confined to a particular interval of time. But unlike ordinary historical causes, which cause events to occur in keeping with extant physical laws, creative causes bring new regimes of order into being, and thus involve the advent of new causal laws. True beginnings involve changes that transcend, and indeed alter, the prevailing causal order. This is why inquiry into them causes the string of induction to snap in our fingers.[31] When a new species is born, or a language created, something basic changes, and it is beyond the capacity of inductive methods to account for this change. At most the palaetiological sciences can *point to* the operation of a creative cause; they cannot tell us what the cause *is*. And since science "can teach us nothing positive respecting the beginning of things, she can neither contradict nor confirm what is taught by Scripture on that subject."[32] Properly circumscribed, neither geology nor theology can step on each other's toes, "and it is unworthy timidity to fear contradiction," just as it is "ungrounded presumption to look for confirmation in such cases."[33]

It is important to observe that Whewell's delimitation cuts both ways. It is not only that he limits geology to make room for God, as Kant found it necessary to deny knowledge to make room for faith. In addition, Whewell thinks that "religious . . . commentator[s]" must sometimes alter their interpretations of Scripture to accommodate geological findings.[34] He cites a rule on this subject, "propounded by some of the most enlightened dignitaries of the Roman Catholic church, on the occasion of the great Copernican controversy begun by Galileo," to regulate the interpretation of Scripture.[35] The rule states that "when a *demonstration* shall be found to establish the earth's motion, it will be proper to interpret the sacred Scriptures otherwise than they have hitherto been interpreted in those passages where mention is made of the stability of the earth and movement of the heavens."[36] Likewise, when a geological proposition can be established by adequate proofs, the interpretation of Scripture must be modified to accommodate this finding. Whewell warns that this should not be done hastily, "as [when] the supposed scientific discovery is doubtful"; "but when a scientific theory, irreconcilable with [an] ancient interpretation, is clearly proved, we must give up the interpretation, and seek some new mode of understanding the passage in question, by means of which it may be consistent with what we know."[37] The scope of conflict between geology and Scripture is thus limited to haggling over what constitutes a clear proof of a geological thesis, of the sort that compels a traditional interpretation of Scripture to be given up.

There is a risk that my treatment in this section will make Whewell seem more modern and ecumenical than he actually was. In fact, Whewell makes clear throughout his writings on geology that only the "distinct manifestation of creative power, transcending the known laws of nature" can account for true origins.[38] Geology can in no case demonstrate that God did this or that, but it *can* show that certain occurrences "are quite inexplicable by the aid of any natural causes with which we are [presently] acquainted."[39] In these cases, as Whewell famously said of the species problem, geology "says nothing, but she points upwards."[40] By showing the impossibility of discovering "a beginning which is homogenous with the known course of events," the palaetiological sciences demonstrate the necessity of seeking a cause that is inhomogeneous with the known course of events, which is to say, a supernatural one.[41]

Despite the fact that Whewell's strategy for delimiting geology cannot be regarded as wholly reactionary, many will regard it as conservative in drift, as imposing unwarranted limitations on scientific inquiry for the sake of preserving a maximally wide role for the first cause. In other contexts, however, Whewell acted as a stalwart defender of science, including that most fashionable of early nineteenth-century sciences, geology.

Whewell and the Declinists

In 1830 a remarkable book appeared under the title *Reflections on the Decline of Science in England*. Written by Charles Babbage, then Lucasian Professor of Mathematics at Cambridge, the book was a frontal attack on British science, bristling with allegations of slipping standards and misplaced priorities.[42] It was particularly hard on the Royal Society, which Babbage saw as little more than a social club whose members showed little or no interest in conducting original research. Yet the problems were not confined to this once venerable institution. Equally troubling was the absence of government support for the sciences, which contrasted with the far superior situation in France where university professors were paid for teaching *and* research. In Britain, by contrast, overworked professors often had to privately tutor students in order to supplement their relatively meager salaries. This left little time for original work, a theme David Brewster amplifies in his review of *Reflections on the Decline of Science*:

> Mr. Babbage has asserted that "the great inventions of the age are not, with us at least, always produced in universities"; but we go much farther and maintain, that the great inventions and discoveries which have been made in England during the last century have been made without the precincts of our universities ... nor need we have any hesitation in adding, that within the last fifteen years not a single discovery or invention of prominent interest, has been made in our colleges, and that there is not one man in all the eight universities in Great Britain who is at present known to be engaged in any train of original research.[43]

It comes as no surprise that Whewell and his friends at Cambridge were hostile to Babbage's book, and no less to Brewster's fulminations against the English universities.[44] Writing in 1831, Whewell scolds "a perverse generation of critics" of British science for their "unprovoked and unfounded attacks" that, at any rate, misconstrued the primary function of the university as one of discovery as opposed to education.[45] Yet Whewell does not omit to detail, at great length, the discoveries of those university professors that Brewster had so "unjustly accused of being backward and inert."[46] And, significantly, in building his case he begins with "one of the most modern and most progressive of sciences, Geology."[47] Professor Buckland at Oxford University, for example, "has for a long course of years been engaged with a vigour never interrupted and never exceeded, in some of the most original paths of research to which the study has given rise."[48] Ten years had passed since Buckland had reconstructed the Kirkland Cave fossils: work that earned him the prestigious Copley Medal and, in Humphrey Davy's words, established a "distinct epoch . . . in the history of the revolutions of our globe."[49] But the accomplishments of British geology were not Buckland's alone. No less important were the contributions of William Conybeare, "one of the creators of the science [of geology] . . . whose fame has received the valuable stamp of foreign recognition, by his appointment as corresponding member of the French Institute."[50] Turning from Oxford to Cambridge, Whewell praises his friend Adam Sedgwick, whose "energetic geological operations in the field are interrupted only by the official obligation of delivering the most animated and instructive courses of lectures."[51] Drawing all this together, Whewell concludes that "it [is] strange, that the Universities, which thus unquestionably have been and are the first and most zealous nurses of this youngest of the sciences, should be exposed to the trite and unmeaning charge of cherishing antiquated dogmas and neglecting the advances of modern discovery."[52] Geology disproved any generalization about the decline of science in Britain and the backwardness of universities. Indeed, it placed British universities at the vanguard of scientific discovery and innovation.

There is no need to belabor the point. Whewell regarded geology as a feather in the cap of recent British science. While other nations may have overtaken Britain in their cultivation of the "exact sciences" (the main focus of Babbage's polemic), the geological sciences had been cultivated on the British Isles "with a sagacity, perseverance and success, which must make it hereafter appear one of the most remarkable passages in the history of science."[53] Surely this supplied an additional motivation, if one was needed, to find a place for geology in the span of the inductive sciences. But it was chiefly useful in 1831 as a means of heading off narratives about the decline of British science and the backwardness of English universities.

Whewell and Lyell on Geological Theory

So far I have concentrated on Whewell's work as a philosopher and advocate. But he also played an important role in several internal geological debates, including the debate sparked by Charles Lyell's *Principles of Geology* (1830–1833). Famously, it was Whewell who gave the antagonists in this debate their monikers: the "Uniformitarians" and the "Catastrophists."[54] But historians have shown that these labels are often more misleading than useful, with one going so far as to claim that "there was no Uniformitarian–Catastrophist debate."[55] I will thus mostly avoid them in this section. Instead, I will focus on Whewell's criticisms of Lyell's "speculative geology" and the methodological ideas that underpin it.[56]

Whewell and Lyell were not enemies. On the contrary, Whewell was always quick to praise Lyell for placing the science of geological dynamics "in its proper prominent position."[57] Lyell was not the first cultivator of geological dynamics; a notable predecessor was the German geologist Karl Ernst von Hoff, who in several large volumes had "collected from ancient and modern writers a very large body of curious information."[58] Yet it was only with the publication of Lyell's *Principles* that "the full effect of such [information]" was made manifest, since this work "attempted to present such assemblages of special facts, as examples of general laws."[59] For taking this auspicious step, Lyell was to be praised, even though Whewell suspected that "several generations must elapse" before geological dynamics could reach the status of an exact science.[60]

Where Lyell was to be faulted was for a certain hastiness in generalization: for a penchant to vault from laws, imperfectly established, to highly speculative theories. An example of the latter is the claim "that all the phenomena of the earth's strata may be considered as produced by a continuous series of events, of which changes now occurring are fair examples."[61] This is uniformitarianism, as Whewell understood it, and it is guilty of a variety of sins. A particularly egregious one concerns Lyell's claim that the "doctrine of uniformity" has "a great degree of *à priori* probability" that can be established in advance of any of its applications:

> It is highly unphilosophical, it has been urged, to assume that the causes of the geological events of former times were of a different kind from causes now in action, if causes of this latter kind can in any way be made to explain the facts. The analogy of all other sciences compels us, it was said, to explain phenomena by known, not by unknown, causes. And on these grounds the geological teacher recommended "an earnest and patient endeavour to reconcile the indications of former change with the evidence of gradual mutations now in progress."[62]

In responding to this line of thinking, Whewell grants that geologists should eschew gratuitous hypotheses about past geological changes. Whether past changes to the earth's crust have been gradual or catastrophic "must be collected . . . from the facts of the case," and we "must suppose the causes which have produced geological phenomena, to have been as similar to existing causes, and as dissimilar, as the effects teach us."[63] But this is not what Lyell recommends. For Lyell, it is a *principle of reasoning* that geological agencies have never acted with different degrees of intensity than those they presently exert. He states as much in a letter to Murchison, in which he claims that his book "will endeavour to establish the principle[s] of reasoning in . . . [geology] . . . which [are] that no causes whatever have from the earliest time to which we can look back, to the present, ever acted, but those now acting; and that they never acted with different degrees of energy from that which they now exert."[64] But this, Whewell thinks, is an "altogether arbitrary and groundless" assumption.[65] "We must learn from an examination of all the facts, and not from any assumption of our own, whether the course of nature be uniform."[66] Nothing is gained by yoking oneself to an a priori view about the rate and intensity of geological processes, especially when this is at variance with the evidence from the rocks. He drives the point home with a characteristic flourish: "*Time* . . . can undoubtedly do much for the theorist in geology; but *Force* . . . is also a power never to be slighted: and to call in one to protect us from the other, is equally presumptuous, to whichever of the two our superstition leans. To invoke Time, with ten thousand earthquakes, to overturn and set on edge a mountain-chain, should the phenomena indicate the change to have been sudden and not successive, would be ill excused by pleading the obligation of first appealing to known causes."[67]

How had Lyell made such a blunder? Whewell's answer, strangely enough, is that he had been *too* Newtonian, at least in the sense of sticking too close to what Newton said in the "General Scholium" (1726). Here, Newton presented his rules for reasoning, including the famous first rule of natural philosophy: "No more causes of natural things should be admitted than are both true and sufficient to explain their phenomena."[68] In Thomas Reid's influential explication, this means (1) that the causes "ought to be true, to have a real existence, and not to be barely conjectured to exist without proof"; and (2) that they "ought to be sufficient to produce the effect."[69]

Now, (2) is uncontroversial. It just states that causes should be capable of producing the effects they are adduced to explain. So, if a cause cannot produce the relevant effect or phenomenon, then it is not a real cause of that phenomenon. By contrast, (1) is more ticklish. It is often interpreted as a prescription stating that we must assume no other causes than those we know "*from other considerations* . . . to exist."[70] Put differently: causes must be known from evidence other than the evidence they are called upon to explain. But such an injunction is "an injurious limitation of the field of induction," Whewell thinks, for "it forbids us to look for a cause, except among the causes with which we are already familiar."[71] And this seems perverse, especially in an area of science like geology whose task is to resurrect lost worlds. If the upturned strata of the Isle of Wight seem to indicate a period of convulsive elevation, why should we limit ourselves to explaining them by gradual upward nudges, ignoring all signs of violence? In short, "why should we not endeavor to learn the cause from the effects, even if it be not already known to us?"[72]

Whewell is not entirely out on Newton's first rule; he just favors a different interpretation: "When the explanation of two kinds of phenomena, distinct and not apparently connected, leads us to the same cause, such a coincidence does give a reality to the cause, which it has not while it merely accounts for those appearances which suggested the supposition."[73] This is the criterion of *consilience*, so dear to Whewell's heart, and so seemingly promising in its application to the palaetiological sciences.[74] In light of this promise, it might be surprising that the most celebrated example of a consilient palaetiological argument—Darwin's argument in the *Origin of Species* (1859)—did not meet with Whewell's approval.[75] But in part this is because the *Origin* transgressed the limits of the palaetiological sciences that Whewell had labored so hard to construct. For Whewell, an inductive science of origins was almost a contradiction in terms. The palaetiological sciences could resurrect a whole series of lost worlds, but they could have nothing to say about beginnings, about the *transitions* between those worlds. In this respect Whewell remained a geological thinker more at home in the early decades of the nineteenth century than in its frenzied middle years.

Outstanding Questions

William Whewell was a minor figure in the history of nineteenth-century geology. Still, he saw further than many of his contemporaries, at least when it came to the more general features of geological reasoning. His characterization of geology as a palaetiological science epitomized the novelty of geological inquiry as concisely as any expression before or since, and might even have caught on, were it not so difficult to say. His treatment of geology in relation to theology was reasonable and even-handed, even if it was somewhat marred by his mystery making about beginnings. And his meticulous dissection of Lyell's *Principles* was a model of philosophical acumen that ought to have left no one in doubt about the scientific bona fides of "catastrophist" geology. In all this Whewell wore a variety of hats, but none more prominently than those of the systematizer and the critic. That Whewell could play these roles in a science whose empirical details he was only imperfectly acquainted with is both a testament to his talents and a reminder that geology in the 1830s was still in a period of relative adolescence.

Several important questions about Whewell and the palaetiological sciences remain unanswered. The first concerns Whewell's engagement with geology in the 1820s. Just what was he doing on those excursions with Sedgwick, and how did this field experience shape his writings about the palaetiological sciences? We know that Whewell, for all his bookishness, was not averse to outdoor recreation. But we know next to nothing about William Whewell as a field geologist. What sites did he visit? How did he "pick geological rubbish" out of Sedgwick's eyes?[76] And who else did he venture into the field with, hammers in tow?

There is also the matter of his engagement with geology after 1840. Whewell became the Master of Trinity College in 1841, not long after the end of his stint as the president of the Geological Society. His last work to treat geological topics at any length was his *Philosophy of the Inductive Sciences* (1840, reaching its third edition in 1860). But this did not mark an end to his interest in geology, even if it marked an end to the opportunity to freely exercise that interest. So in what ways and to what extent did Whewell remain connected to the geological community during his time as Master of Trinity College? Did his views on important geological topics undergo any substantive changes after the appearance of his *Philosophy*?[77] Much ink has been spilled on Whewell's reaction to Darwinian theory, but Whewell was also present for early discussions of the glacial theory at the Geological Society and subsequently maintained an interest in the physics of glacial motion.[78] It is hard to believe that Whewell paid no attention to a theory that played so crucial a role in the conceptual development of the earth sciences. This could form a promising locus for future historical research into Whewell and the palaetiological sciences.

CHAPTER 13

Whewell and Scientific Classification

ALETA QUINN

Whewell's philosophy of classification followed on the enormous progress made in the classificatory sciences of zoology and, especially, botany in the nineteenth century, as well as efforts toward a natural organization of chemical elements and minerals. Recent scholarship has addressed a wide-ranging shift in the eighteenth and nineteenth centuries toward viewing classification as a methodology for discovering natural properties and arrangements of the classificatory objects.[1] Yet, the exact meaning of *natural system* is both historiographically and philosophically challenging.[2] This chapter analyzes Whewell's views about natural classification and methodology in the classificatory sciences. To introduce the importance of the topic, it is helpful to briefly note ongoing discussions.

The meaning of modern-day natural classifications of plants may seem comparatively clear, give or take some complications: botanists make and defend claims about the branching order of lineages in the tree (or network) of life. This conceptual framework was effected largely following the mid-twentieth-century works of Willi Hennig and Lars Brundin. Hennig offers several examples to motivate the claim that there is such a thing as "natural classification." While the waterways of Europe might be classified on the basis of their navigability or water management, classification based on drainage is a fundamentally different type of project. An archeologist might classify shards on the basis of color, material, or decorations, but the task of reconstructing the original vessels of which the shards are fragments is fundamentally different—it is reconstructing a natural system.[3] These examples suggest that the "natural classification" reflects important truths about the nature of the objects in question—why they are the things that they are—including information about their origins.

The examples do not involve "essences" in the sense of core properties that determine observable properties, in the manner of the Quinean tradition of "natural kinds" (e.g., as the molecular structure H_2O determines the properties of water).[4] Nor do the examples point to a particular epistemic function of classifications, that they should allow us to make as many general propositions as possible, which has been claimed to be the ultimate philosophical interpretation of "natural" classification.[5]

The view that natural classifications are built around identification of natural kinds in the Quinean sense is not present in Whewell's work.[6] The view that classification should render general propositions possible is present in his work (and that of his interlocutor John Stuart Mill),[7] but for Whewell it is not at the heart of understanding classificatory science. He would not have accepted the idea that organisms should be classified on the basis of their evolutionary histories, given that he did not accept the view that species result from descent with modification.

What, then, was natural classification to Whewell? In this essay I will explicate Whewell's ideas about "the natural system" via an exploration of his philosophy of classification, in turn via analysis of his multiple contacts with classificatory science: mineralogy, as a stage in the palaetiological sciences (in particular architecture), and biological systematics (botany and zoology). Whewell's approach to classification hinged on his view that *natural affinity* is both relationship and method. The tight connection between methodology and classificatory schema is also demonstrated in Whewell's classification of the sciences themselves.

I begin as Whewell did, with his assigned task as chair of mineralogy (1828–1832) to impose order on the species and systems of minerals.

Mineralogy

Whewell published *An Essay on Mineralogical Classification and Nomenclature* (1828) in the same year he was appointed chair. The work built on his travels in Germany and Austria (which also proved formative on his philosophy of historical sciences—see "Classification in the Palaetiological Sciences" below), in particular his encounters with Friedrich Mohs.[8] In a letter to Hugh James Rose,[9] Whewell wrote that his travel plans "will depend upon the motions of Professor Mohs." Whewell met Mohs in Freiberg in July 1825, and subsequently moved with him to Vienna through September: "Some of my principal employments at Vienna will be seeing mineralogical collections and sitting at the feet of Professor Mohs while he discusses them. . . . I do not know whether I shall be able to give as good an account of the German schools of Mineralogy, as you have given of their Theology, but there is much to be said thereupon. They have got a Natural History system of my science, as they have got a Naturalist scheme of yours: and a very dangerous heresy it is considered to be."[10] Whewell viewed mineralogy as much confused (compared to the biological classificatory sciences) and placed the blame on an overreliance on chemical analysis as the basis for classification. In this he followed Mohs, who devised a classification based on physical rather than chemical properties, taking a stand on one of the then-key controversies in mineralogy.[11] Understanding the elemental makeup of minerals would surely be useful to some extent, because the characteristics of minerals must depend in some sense on this material composition.[12] However, the *arrangement* of those materials must play a crucial role, as evidenced by the fact that (apparently) natural sequences can be identified in which minerals appear to remain "the same" despite substitutions of the compositional materials (isomorphy). Classification on the basis of material composition is mainly useful because it constitutes an alternate line of reasoning, distinct from the method Whewell emphasized: analysis of crystalline properties in terms of the concepts of *symmetry* and *polarity*.

Laura Snyder analyzed the convergence of distinct lines of inductive reasoning as a verification criterion (consilience), whereas Sandoz interpreted Whewell's remark about chemical tests as crucial to the process of discovery: "If, having formed such a classification independently of Chemistry, and compared it with the results of chemical analysis, we do find any general chemical properties which prevail in our mineralogical classes . . . it is manifest that these laws then contain important knowledge."[13] Indeed, as I will discuss below ("Botany and Natural Affinity"), Whewell's analysis of successful classificatory methodology placed great emphasis on the convergence of distinct lines of reasoning.

In Whewell's view attention to crystallographic properties was crucial because of the nature of the classificatory objects (minerals) and the reason that a natural system obtains. Minerals are what they are because of their crystalline structures. Whewell compared crystals to biological organisms (and grappled with the concept of individuality) by noting a kind of coordination of parts. Just as changes in one part of an organism are linked to changes in the organism considered as a unified whole, changes in one part of a crystal are linked to changes in other parts. An increase in one angle between adjacent planes, for example, is necessarily linked to changes in other angles or else to other geometric changes (such as addition of facets). But only a subset of the total geometrically possible changes are observed in practice; there is some set of unknown rules governing the reasons there are the crystals that there are in the world. This set of unknown reasons constitutes a natural system that can (and should) be used to classify the observed entities.

Knowledge of the natural system hypothesized via crystallographic study would be powerfully confirmed in the event that it matched reasoning on the basis of elemental composition. This agreement of independent lines of evidence would go beyond efforts to resolve the apparent problem of isomorphy (that substances with the same atomic structures may have different chemical compositions).

Methodologically Whewell advocated study of minerals in terms of planes of symmetry. Other crystallographic properties, such as cleavage and optometric properties, give evidence about planes of symmetry, but the symmetry relations are themselves crucial to the natural system of crystal formation.

Whewell's advocacy included the development of a system of notation and methods for measuring and calculating these crucial properties. In a series of papers published between 1821 and 1832, and particularly a paper from 1825, Whewell "laid the foundations of mathematical crystallography."[14] Herbert Deas argued that Whewell succeeded in his aim to produce a system that would be uniform across different kinds of crystals, universal in application, and mathematically simple. He also argued that Whewell's work incorporated a concept of *lattice*, which was unfortunately dropped by his immediate successors but would later (with the help of more advanced mathematics) become central to the science of crystallography. George Rapp noted the importance of Whewell's students in developments in mineralogy after Whewell stepped down as chair in 1832. Rapp also observed that Whewell's definition of mineral species—"having the same elements combined in the same proportions with the same fundamental form"—is similar to the modern definition: "that a given mineral has a defined composition and a unique crystal structure."[15]

With respect to mineralogy, Whewell developed systems of notation, identifying and measuring characters, and terminology. Whewell viewed such methodological advances as core components of classificatory science.[16] It seems then that his philosophy of classification was developed in the context of his actual practice of science. Some recent scholars, however, have argued that Whewell was more an observer than a participant in science—a "meta-scientist" rather than a scientist.[17] Others have objected, arguing that Whewell's contributions to tidology constituted significant scientific progress, while conceding that Whewell did not establish a theory that could explain tidal phenomena.[18] In some cases, such as that of historical architecture, to which I will now turn, Whewell did develop a scientific theory.

Classification in the Palaetiological Sciences

Whewell seems to have been a pioneer in the idea of classifying together the sciences that reconstruct history (in Whewell's terms, *palaetiological sciences*;

more recently, historical sciences).[19] Whewell construed the study of architecture as a historical science that traces the development of architectural styles as embodiments of more or less fundamental concepts. In particular he argued that Gothic architecture represented the progressive development of the concept of *vertical space*, from which all the characteristic features of Gothic architecture could be traced following a temporally dependent sequence.[20]

Whewell developed his theory of Gothic architecture in connection with his experience visiting Gothic churches. His objective was not only the actual theory of the development of Gothic architecture but the methodology needed to construct such a theory. Thus his *Architectural Notes on German Churches* (1830) included extensive guidance about what features to observe and what terms to use in describing those features. Such guidance formed the core of what Whewell, in his *Philosophy of the Inductive Sciences* (1840), would identify as a first stage in classificatory science: developing a *terminology*.

Though the terminology must be formed first in order to organize the inquiry, the terminology must also be open to reassessment and reform in successive stages, because "in fixing the meaning of the terms, at least of the descriptive terms, we necessarily fix, at the same time, the perceptions and notions which the terms are to convey; and thus the terminology of a classificatory science exhibits the elements of its substance as well as of its language."[21]

Whewell recognized that fixation of terminology affects decisions about what characters are important to consider in forming a classification. Entities to be classified may be scored as having a resemblance versus a difference with respect to a character, depending on how the character is described. Progressive knowledge of the natural classification—the next stage, forming the "plan of the system"[22]—may entail adjustment of the characteristics to be considered and the terms with which they are described. Yet, classificatory science cannot begin without an initial terminology, a key scientific achievement that requires careful analysis, above and beyond the recognition that some characteristics are likely to be trivial.

Although Whewell implemented his philosophy of classification in the case of architecture, he did not consider architecture to be (essentially) a classificatory science. Rather, (comparative) architecture was to be organized under the concept of *historical causation* and thus the methodology proper to historical (palaetiological) science.[23] A preliminary classification of churches as Gothic (versus Romanesque) was a necessary prerequisite to the central task, the construction of a historical account of why Gothic churches came to be as they are: "So long as men were blind to this character of unity and connexion in the members of this form of art, it was impossible that they could speculate with any distinctness or success on the conditions and causes of its rise and progress; so long as they did not perceive clearly *what* it was, they could not discern *how* it came to be; so long as they did not understand the language of Gothic Architecture, they could not trace its phrases to their roots."[24]

Whewell's discussion of geology as a historical science also indicates that the classificatory stage is a prerequisite to the main activities of historical science: "The Phenomenal portions of each science imply Classification, for no description of a large and varied mass of phenomena can be useful or intelligible without classification."[25] The use of classificatory ideas and methodology follows Whewell's comment that "in our classification [of the sciences], each Science may involve, not only the Ideas or Conceptions which are placed opposite to it in the list, but also all which *precede* it."[26] The list (p. 281) places the palaetiological sciences (e.g., geology, distribution of plants and animals, glossology, and ethnography) opposite the concept of historical causation. The classificatory sciences (systematic botany, systematic zoology, and comparative astronomy) precede the palaetiological sciences, and are organized via the concepts of degree of likeness and natural affinity.

The palaetiological sciences are also preceded by psychology and biology, in keeping with Whewell's analysis of the nature of Gothic churches. What it is to be a Gothic church—the nature of the classified object—is a result from a particular historical lineage. That historical lineage in turn represents the progressive unfolding of an *idea* (vertical space) and is thus partly a psychological narrative. With respect to architecture, then, the nature of the natural system reflects human psychology—"the tendencies, faculties, principles, which direct man to architecture."[27]

Additional messiness might be introduced to the concept of a natural system with respect to architecture because of the nature of the classificatory entities. An individual church might belong to multiple different styles of architecture. Indeed, this is often the case for cathedrals, whose construction required dozens or even hundreds of years. The system for classifying churches may thus be less "natural" than in the classificatory sciences proper: the causal factors underlying church classification are not expected to form a single, fully coherent system.

For its part the naturalness of the system of geological strata would depend on the coherence of the causal factors in geological history. As Whewell's views about geology are discussed elsewhere in this volume,[28] here I will mention only that formation of a geological system requires the idea that strata occurring in very different locations and conditions are meaningfully the same (Cambrian, Ordovician, etc.). Whewell considered this a step that required "genius and good fortune," having been the task "only of the most eminent persons."[29] The formation of a preliminary classificatory terminology is a considerable scientific achievement.

Whewell discussed terminology—a fundamental element of his philosophy of classification—in connection with architecture and geology. However, the fullest development of Whewell's philosophy of classification is provided in his discussion of the classificatory sciences proper, to which I now turn.

Systematics and the Method of Types

Whewell's account of "Types" must be understood in the context of the development of rules for biological nomenclature, in particular the contemporary development of the Strickland Code.[30] Whewell had been asked to serve on the British Association for the Advancement of Science (BAAS) committee that produced recommendations to standardize the use of names, including rules for assigning reference in the face of taxonomic change.[31] Though he declined, the committee was familiar with his work (citing his *Philosophy* in their final report) and he with theirs.[32] The Strickland Code articulated the idea of the *type method* later adopted by the international committees, a version of which is still in use today:[33]

> When a genus is subdivided into other genera, the original name should be retained for that portion of it which exhibits in the greatest degree its essential characters as at first defined. Authors frequently indicate this by selecting some one species as a fixed point of reference, which they term the "type of the genus." When they omit doing so, it may still in many cases be correctly inferred that the first species mentioned on their list, if found accurately to agree with their definition, was regarded by them as the type. A specific name or its synonyms will also often serve to point out the particular species which by implication must be regarded as the original type of a genus. In such cases we are justified in restoring the name of the old genus to its typical signification, even when later authors have done otherwise. We submit therefore that
>
> § 4. The generic name should always be retained for that portion of the original genus which was considered typical by the author.[34]

Ambiguity in what the original author regarded as the type could be resolved by standardizing the practice of designating a type when generating names: "§ G. It is recommended that in defining new genera the etymology of the name should be always stated, and that one species should be invariably selected as a type or standard of reference."[35] This recommendation was adopted, and the principle was later applied to species names as well (so that the definition of a species is fixed by reference to a type specimen). Ideally the type is selected because it demonstrates the character states thought to be typical of the taxon, but what matters for fixing reference is the act of designation. The organisms included in the meaning of a name are the type and whatever are the organisms that belong with the type.

This system thus does not specify the meaning of taxon names in terms of possession of characteristics (i.e., definite descriptors).[36] As Whewell put it: "Though in a Natural group of objects a definition can no longer be of any use as a regulative principle, classes are not therefore left quite loose, without any certain standard or guide. The class is steadily fixed, though not precisely limited; it is given, though not circumscribed; it is determined, not by a boundary line without, but by a central point within; not by what it strictly excludes, but by what it eminently includes; by an example, not by a precept; in short, instead of a Definition we have a *Type* for our director."[37] The Strickland committee report concluded by crediting "Linnæus, Smith, Decandolle, and other botanists," and remarking that "the language of botany has attained a more perfect and stable condition than that of zoology; and if this attempt at reformation may have the effect of advancing zoological nomenclature beyond its present backward and abnormal state, the wishes of its promoters will be fully attained."[38] Whewell also considered botany more advanced than zoology, and so his most developed philosophy of classification followed on his analysis of the history of botany.

Botany and Natural Affinity

Whewell analyzed botany as progressing via the application and refinement of concepts of *likeness* and *natural affinity*.[39] His philosophical account of botany followed his historical analysis, which relied heavily on Augustus P. de Candolle's *Théorie élémentaire de la botanique*.[40]

To Whewell, botany was a subdivision of natural history, which he distinguished from "biology."[41] The subjects were clearly related, and Whewell expected the relationship to be clarified over time, in keeping with his commitment to the unity of science.[42] But Whewell's "Philosophy of Biology" was primarily concerned with what we now call "physiology,"[43] and operated under its own independent "Fundamental Idea" that gave rise to concepts:

> Thus we have seen how the first confused mechanical conceptions (Force and the like) were from time to time growing clearer down to the epoch of Newton;—how true conceptions of Genera and of wider classes, gradually unfolded themselves among the botanists of the sixteenth and seventeenth centuries;—how the idea of Substance became steady enough to govern the theories of chemists only at the epoch of Lavoisier;—how the Idea of Polarity, although often used by physicists and chemists, is even now somewhat vague and indistinct in the minds of the greater part of speculators. In like manner we may expect to find that the Idea of Life, if indeed that be the governing Idea of the science which treats of living things, will be found to have been gradually approaching towards a distinct and definite form among the physiologists of all ages up to the present day.[44]

Whewell's focus on teleo-functional approaches to biology must thus be interpreted with care in the context of his analysis of botany and the other classificatory sciences. In particular, analysis of the adaptive function of parts with respect to the organism's external needs may be distinct from analysis of the connection of functions within the organism. Physiological interdependence may be the *justification* for the method of natural affinity—functional analysis is not itself the primary principle of classification.

To put the point another way, one might think that the best way to classify organisms is to start with the needs and functions that are most important to the organism's survival and functioning in the external environment. Instead, Whewell started from analysis of what makes an organism the kind of thing that it is, and cited Kant's definition—in the *Critique of Judgment*—of an organism as "that in which all the parts are mutually ends and means."[45] Whewell conceived of living beings as systems of mutually dependent functional systems: nervous system, respiratory system, and so on. In his view, the dependency relation often entails that a change in one part of the organism necessitates a change in another part of the organism, just as a change to one part of a crystal may require (by symmetry, by whatever are the causal laws governing crystalline form) change to other part(s) of the crystal.[46] Such dependence could refer to development, which was De Candolle's approach, or to Georges Cuvier's "conditions of existence." In Whewell's view, whatever the causes of change in form, the fact that such changes occur in organized, interdependent ways indicates that groups are "related by Affinity."[47] Thus it is possible to identify individual bones of the hand as "the same" across vertebrates and to identify changes in the shape of the bones when considering different groups of vertebrates. In an unpublished chapter drafted for the *Philosophy*, Whewell described the sameness of bones in the hands of humans, apes, bears, pigs, cows, and horses.[48]

A further question is the reason for the existence of the dependency schema. Whewell considered the existence of the dependencies, as well as particular adaptative "contrivances," evidence of divine design. The direction of Whewell's reasoning (as in his Bridgewater Treatise of 1833) was from the results of science to claims about a Creator, in keeping with the natural theology of his day.[49] Discovery of the natural classification, to the extent that this was possible, would rely on the regular process of science; claims about design would not contribute directly to the project. In Whewell's view natural systems and natural relationships exist because of a creator with a design, but human investigation of biological ends is distinct from the project of natural theology.

The key to the project of classification was *natural affinity*. This term was often used to mean one (or more) particular kind of relationship among organisms, but to Whewell the term also referred to a *method* for discovering those relationships.[50] According to Whewell—in his *History of the Inductive*

Sciences (1837), where he closely followed De Candolle—the history of classificatory sciences constitutes the gradual elucidation of the idea of natural affinity. In its clearest form the method relies on the concordance of classifications constructed on the basis of independent lines of reasoning. In the case of biological classification, because organisms are integrated functional systems, the method of natural affinity constructs classifications on the basis of distinct functional systems. When such classifications agree, despite having begun from different starting points, the agreement is evidence that the systematist has correctly identified (part of) the natural system of organisms.

To illustrate, we can consider a core example from De Candolle's *Théorie Élémentaire*. Nutrition is a key function not only for survival but in terms of what makes an organism the thing that it is. De Candolle pointed out that nutrition is one of the functions that distinguishes living from nonliving beings; the way in which nutrition is carried out is then crucial to being the kind of living being that one is. The idea here might be supported by thinking about the conversion of nutrients into organic matter. Distinct plants take in the same soil, sunlight, and air, and convert these ingredients into tissues that are specific to the kind of individual plant.[51] With respect to nutrition, De Candolle claimed that the most important division of plants is on the presence or absence of *vaisseaux*—that is, the tissue types and arrangements characteristic of vascular plants (*végétaux vasculaires*) which have roots (*racines*) distinct from stems (*tiges*). Candolle's *végétaux cellulaire* (mosses, liverworts, lichens, fungi, algae) lacked these distinct structures.[52]

For Whewell reproduction is also a key function in terms of what makes an organism the kind of organism that it is, following the principle that species breed true. De Candolle had argued that the most important distinction in terms of reproduction is the presence or absence of cotyledons. Further, the resulting division turns out to match the primary division on the basis of nutrition: the vascular plants are the *végétaux cotyledonés*, and the nonvascular plants and the *végétaux acotyledonés*. De Candolle claimed that therefore, the division is natural, and proceeded to identify subclasses of each of the main groups following the same procedure: classify on the basis of (presence/absence, number, arrangement, etc. of) key parts for key functions, and then compare the classifications made on the basis of distinct functions.[53]

Just as in his history of botany, Whewell's philosophy of classification closely followed De Candolle's. Classifications are built on the fundamental idea of *likeness*, but systematists are successful when they recognize and assign importance to likeness on the basis of natural affinity. Systematists' appeal to natural affinity was latent or obscure at first,[54] but has become progressively more refined along with the study of functional systems. Whewell expressed some doubt about the complete concordance of classifications made on the basis of all functions, but endorsed this method of classification as the basis of the natural system:

> Such an assertion is perhaps more than we are entitled to make with confidence; but it shows very well what is meant by Affinity. The disposition to believe such a general identity of all partial natural classifications, shows how readily we fix upon the notion of Affinity, as a general result of the causes which determine the forms of living things. When these causes or principles, of whatever nature they are conceived to be, vary so as to modify one part of the organization of the being, they also modify another: and thus the groups which exhibit this variation of the fundamental principles of form, are the same, whether the manifestation of the change be sought in one part or in another of the organized structure. The groups thus formed are related by Affinity; and in proportion as we find the evidence of more functions and more organs to the propriety of our groups, we are more and more satisfied that they are Natural Classes. It appears, then, that our Idea of Affinity involves the conviction of the *coincide of natural arrangements formed on different functions*; and this, rather than the principle of the subordination of some characters to others, is the true ground of the natural method of Classification.[55]

Classifying the Sciences

In much of the above I have emphasized the relationship between natural classification and the nature of the classificatory objects, but Whewell's philosophy of classification ultimately centered on natural affinity as a *method*. My account thus agrees with Sandoz's argument that Whewell classified the sciences based on methodology, rather than the objects of scientific study (an Aristotelian approach) or the faculties of mind (a Baconian approach).

It should be noted, however, that in the instances of classificatory science explored above, the nature of the classificatory entities (minerals; churches; organisms) both informs and justifies the classificatory methodology (symmetry; historical causation; likeness and natural affinity). In other words, the meaning of *natural system* partially depends on the nature of the classificatory objects. The reason that a natural system obtains differs in the cases of minerals (whose properties are caused by chemical/physical/developmental integration), churches (whose properties are caused by the operation of historical causes), and organisms (whose properties are caused by functional integration).

It has been pointed out that Whewell's intent to classify on the basis of methodology was also reflected in successive organizations of the British Association for the Advancement of Science. Natural history was one of its five committees, to which a sixth (on statistics) was added at the 1833 meeting.[56] At least two of the other sections—Section 2, "Chemistry, Mineralogy,

&tc.," and Section 3, "Geology and Geography"—would also involve a central role for natural classification.[57] The importance of classificatory science is also demonstrated by Whewell's efforts to build up natural history collections, courses, and research.[58]

The philosophy of classificatory science, then, would have been viewed as a critical component of philosophy of science. In Whewell's own words: "To classify is to divide and to name; and the value of the divisions which we thus make, and of the names which we give them, is this;—that they render exact knowledge and general propositions possible."[59] Whewell's contemporary and critic, John Stuart Mill, similarly emphasized the process of classification as central to the formation of general propositions; book 1 of the *System of Logic* (1843) develops Mill's philosophy of science via his philosophy of language and classification. Snyder, among others, argued compellingly that the debate with Mill was crucial to the development of Whewell's philosophy of science.[60] Snyder also urged that scholarship of these nineteenth-century philosophers must consider their views in context of the philosophers' own concerns and interests. Further work on Whewell's philosophy of classification thus has the potential to provide new understanding of his philosophy in general.

It was and is not only Whewell who viewed classification as central to the practice of science. Whether he was himself primarily a practitioner or an observer or looker-on of science, his views on classification reflected careful study of the development and practice of classificatory sciences. Further work on Whewell thus also has the potential to inform historical and philosophical scholarship on classification, such as how the concepts of *species* and *natural system* relate to biological theories,[61] and on Whewell as an early proponent of cluster concepts of natural kinds.[62]

CHAPTER 14

Whewell and Moral Philosophy

DAVID PHILLIPS

Though his work is almost never read by moral philosophers today, William Whewell was the most important philosophical opponent of utilitarianism in the early Victorian era.

As John Rawls observes at the start of his 1981 foreword to Henry Sidgwick's *The Methods of Ethics* (1874), it seems fair to say that "since the middle of the eighteenth century the dominant systematic moral doctrine in the English-speaking tradition of moral philosophy has been some form of utilitarianism."[1] As Rawls goes on to say, one reason for this is "the long line of truly brilliant writers who belonged to this tradition."[2] Another is the adaptability of utilitarianism. It was combined over that period with a striking range of metaethical views: the naturalism of Jeremy Bentham and (at least as he is often read) of John Stuart Mill; the nonnaturalism of Sidgwick; and the noncognitivism of R. M. Hare.[3] Different aspects of the utilitarian normative doctrine have at different periods seemed crucial. In his notorious "proof,"[4] Mill argues in favor of hedonism but just assumes the truth of consequentialism. By contrast writers today are apt to claim that the core of utilitarianism is consequentialism and that hedonism is an optional extra.[5] Some utilitarians were moral reformers, others provided utilitarian arguments for conventional morality.

Utilitarianism's protean character means that the debate between utilitarians and their critics also changed over time. Those acquainted primarily with more recent phases of this debate will find a mixture of familiar and less familiar themes in the exchanges between Whewell and his utilitarian interlocutors. William Paley was the predecessor against whom his program in moral philosophy was initially directed. Whewell's principal aim in writing his main ethical works was for them to displace the theological utilitarianism of Paley's *Principles of Moral and Political Philosophy* (1785) from the Cambridge curriculum (in which he succeeded).[6] John Stuart Mill was Whewell's most important contemporary philosophical opponent (as Whewell was Mill's). Mill wrote a major review criticizing Whewell's moral philosophy and defending Bentham against him.[7] Whewell responded in later editions of *The Elements of Morality* (1854).[8]

By the mid-Victorian period philosophers as different (and antagonistic) as Sidgwick and F. H. Bradley could share a picture of the received options in current moral philosophy: the inductive utilitarianism of Mill or the intuitive anti-utilitarianism of Whewell. The opposition between them encompassed both questions in normative ethics of what you ought to do and questions in metaethics of how you know what you ought to do. Utilitarians like Mill thought that what you ought to do is whatever will bring about the greatest amount of happiness for everybody, and they thought that you know this through experience. Intuitionists like Whewell thought that what you ought to do is to always follow absolute, nonutilitarian moral rules and that you know what these rules are because they are self-evident in the same way that mathematical truths like 2 + 2 = 4 are self-evident. One of Sidgwick's most important philosophical contributions was to reject this constrained menu of options, by combining utilitarianism with intuitionism. Sidgwick devoted a substantial portion of *The Methods of Ethics* to criticizing an absolutist deontological view he called alternately "common-sense morality" or "dogmatic intuitionism." Though Sidgwick did not identify this or any of the other methods of ethics he discussed with a single proponent, it is tempting to take Whewell to be the model for the dogmatic intuitionist. Whewell's fall into obscurity in moral philosophy is largely explained by the perceived success of Sidgwick's critique of dogmatic intuitionism.

I will explore Whewell's contributions to moral philosophy by focusing in turn on his relations with his main utilitarian opponents.

Whewell versus Paley

The three core features of Paley's view are reflected in his definition of *virtue*: "the doing good to mankind in obedience to the will of God, and for the sake of everlasting happiness."[9] Utilitarianism is the criterion of rightness, God's will is what makes acts right, and the obligation and motivation to behave rightly are supplied by divine rewards and punishments that make right action always in each person's individual interest.

When he first took up the position of professor of moral philosophy at Cambridge, before writing *The Elements of Morality*, Whewell urged that Joseph Butler be added to the curriculum as an antidote to Paley. To this end he produced an edition of Butler's sermons.[10] In the preface he contrasts them, to Butler's clear advantage:

> The points of opposition between Butler and Paley are obvious enough. Paley declares his intention . . . to omit the "usual declamation" on the dignity and capacity of our nature; the superiority of the soul to the body, of the rational to the animal part of our constitution; upon the worthiness, refinement and delicacy of some satisfaction, or the meanness, grossness, and sensuality of others. Butler, on the contrary, teaches that there *is* a difference of *kind* among our principles of action, which is quite distinct from their difference of strength . . . that reason was intended to control animal appetite, and that the law of man's nature is violated when the contrary takes place. Paley teaches us to judge the merit of actions by the advantages to which they lead; Butler . . . teaches that good-desert and ill-desert are something else than mere tendencies to the advantage and disadvantage of society. Paley makes virtue depend upon the consequences of our actions: Butler makes it depend upon the due operation of our moral constitution. Paley is the moralist of utility; Butler, of conscience.[11]

We can distinguish two elements in this contrast. There is, first, the distinction between two views of human nature and normative authority. Utilitarianism is associated with a reductive, egoistic view of human psychology and with the denial that some principles of human nature are more authoritative than others. By contrast Butler's anti-utilitarian view insists that there are higher, more authoritative elements in human nature. Second, there is the distinction between two views of what makes actions right and wrong. For Paley, rightness and wrongness are determined by consequences; for Butler by something else. That the second of these two elements should be central to a debate between utilitarians and their opponents will not surprise anyone; the prominence of the first will be more surprising to those unfamiliar with early Victorian moral philosophy.

Whewell goes on to divide moral philosophers into two schools: the proponents of "independent morality" (like Butler) and the proponents of "dependent morality" (like Paley).

> It is indeed evident that the two opposite moral schools of antiquity, the Stoical and the Epicurean, have had their antagonism prolonged into modern times; nor can it cease to subsist so long as there is a School of Independent Morality, which, like Butler, seeks the ground of virtue or moral rightness in the faculties of man and their relation to each other; and another School of Dependent Morality, which, like Paley, looks for the criterion of rightness to external things;—pleasure, utility, expediency, or by whatever name it may be called.[12]

As we will shortly see, Mill and Whewell disagreed sharply on the accuracy and usefulness of this classification scheme.

The Elements of Morality

Whewell was consistent both in admiring Butler and in seeing his work in moral philosophy as incomplete. *The Elements of Morality* was Whewell's attempt fully to develop a moral system along the lines Butler laid out. As Terence Irwin observes, "This large book contains long, elaborate, and not invariably fascinating, discussions of particular moral and legal rules and duties," but, "Whewell's position is clear enough to suggest a reasonable alternative to utilitarianism."[13]

The best overviews of that position are given in the preface to the second edition of *The Elements of Morality* and in the "Introductory Lecture" of the *Lectures on the History of Moral Philosophy in England* (1852).[14] In the preface Whewell begins with the opposition between Epicureans and Stoics. Epicureans urge "that pleasure is the proper guide of human action"; Stoics argue "that virtue is our proper guide."[15] Each view is subject to objections. The objections to the Epicurean (or utilitarian) view suggest that there is more to morality than mere pleasure or utility. The objection to the Stoic view is that the deliverances of conscience vary between persons and societies.

To overcome the objection to the Stoic position, moral principles must be shown to be the deliverances of reason, not of a variable moral sense. Whewell begins with a single "supreme rule": "The rule to do what is right and to abstain from doing what is wrong".[16] Rightness for Whewell is reason-giving, and the reasons it gives are overriding: "In . . . describing [an action as right] we render a *reason* for it, which reason is *paramount* to all other considerations."[17] This idea of the final authority of the supreme rule is, in effect, Whewell's rendering of Butler's doctrine of the authority of conscience.

A reader is apt to respond that it is all very well, and not nothing, to say that we ought to do what is right and that rightness provides overriding reasons for action, but that the supreme rule as so far articulated does not give any real guidance as to what is right. The distinctive character of Whewell's moral system becomes apparent at this point. He aims to give morality more definite content by appealing to a classification of other central elements of human nature that are to be subject to the supreme rule. In the preface he illustrates this procedure with the example of the rule of veracity:

> [The] supreme rule will separate itself into partial rules according to the faculties, powers, and impulses which it has to govern. And by the very condition that it is a supreme and absolute rule, joined with the conditions which man's constitution supplies, we see, with irresistible evidence, the authority of

certain fundamental moral truths; we thus discern the necessary existence of certain virtues as part of this supreme rule of human action.

> For instance: man . . . has the faculty of speech. . . . His whole being cannot be under a Supreme Rule . . . except the use of this faculty . . . be under such a rule. There must be, for the use of Speech, a rule of right and wrong:—a universal and supreme rule. But the ultimate and supreme distinction of the use of Speech is that of truth and falsehood. And it is plain that there can be no ultimate and supreme rule on this subject, except that rule which makes truth to be right and falsehood to be wrong. And thus, one part of the supreme rule is, that Truth is right: that it is right to speak the Truth: that Veracity is a Virtue.[18]

The pattern of argument sketched here is supposed to apply in the same way to other key aspects of human nature. Whewell develops a classification of "the leading Desires of man . . . which . . . are, *The Desire of Personal Safety, the Desire of Having, the Desire of Family Society . . . and the Desire of Civil Society*."[19] These desires can only be right if "they . . . conform to this primary and universal Condition, that they do not violate the *Rights* of others."[20] To each of the desires or springs of action there then corresponds a right, "as the primary Desires of men are the Desire of Personal Safety, of Possession, of Family, and of Civil Society; so the primary kinds of Rights among men are everywhere the Rights of the Person, the Rights of Property, the Rights of the Family, and Political Rights."[21] What are supposed to be justified by this pattern of argument are, simultaneously, claims about virtues, claims about precepts expressible as imperatives, and claims about moral principles expressible by indicatives. Whewell claims that these moral principles are properly called axioms, are self-evident, and are knowable by intuition. It is not an objection that reflection is required to apprehend the truth of these moral principles, for the same is true of the apprehension of self-evident geometrical axioms. And the moral axioms, like axioms in other areas of thought, have the feature that they cannot be derived from some more fundamental principle.[22]

The final key element of Whewell's approach is the role he gives to law in filling out the content of moral principles. In the preface he explains this as an application in moral philosophy of his distinction between idea and fact:

> When we have come to the conviction that Truth, Justice, and the like, are the rule of our being, we have to consider, *What is Truth*, and *What is Justice*, in special cases. In pursuing this inquiry, we have to attend both to the external conditions and to the internal essence of moral action; and we are thus led to perceive that between the external conditions and the internal essence, there is a kind of necessary and universal antithesis;—the antithesis which occurs

in so many forms and in so many places, of *Idea* and *Fact*. . . . There is . . . a *factual* or historical side of every moral question, as well as a purely *moral* side; there are in it external elements, given by man's history, as well as internal rules, given by man's constitution. . . . For instance, Morality must, in some measure at least, depend upon Law. It is wrong to steal, to covet, to desire what is another's. But the law alone can determine what *is* another's.[23]

Whewell versus Bentham and Mill

The catalyst for Mill's critique of Whewell in "Dr Whewell on Moral Philosophy" (1875) was Whewell's discussion of Bentham in the *Lectures on the History of Moral Philosophy in England*. Mill frames the opposition between Bentham and Whewell as a contest between progressivism and conservatism:

We do not say the intention, but certainly the tendency, of [Dr. Whewell's] efforts, is to shape the whole of philosophy . . . into a form adapted to serve as a support and a justification to any opinions which happen to be established. A writer who has gone beyond all his predecessors in the manufacture of necessary truths, that is, of propositions, which, according to him, may be known to be true, independently of proof; who ascribes this self-evidence to the larger generalities of all sciences . . . was still more certain to regard all moral propositions familiar to him from his early years as self-evident truths.[24]

As J. B. Schneewind remarks, the *Lectures* "must be the first academic text to treat Bentham seriously as a philosopher" rather than "as the ideologist of a small and rather fanatical political group."[25] Whewell regards Bentham's position as substantively the same as Paley's, but he gives much more space in the lectures to Bentham. Some of that space is devoted to criticizing Bentham's cast of mind and the "extravagant unfairness to adversaries which was habitual in him."[26] Whewell refers by way of illustration to chapter 2 of Bentham's *Introduction to the Principles of Morals and Legislation* (1789), where Bentham considers two principles "adverse to that of utility," namely the principle of ascetism and the principle of sympathy. Whewell remarks, "Now it is plain that these are not only not fair representations of any principles ever held by moralists, or by any persons speaking gravely and deliberately, but that they are too extravagant and fantastical to be accepted even as caricatures of any such principles. For who ever approved of actions because they tend to make mankind miserable? . . . Who ever asserted that he approved of actions merely because he found himself disposed to do so, and that this was reason sufficient in itself for his moral judgments?"[27] Mill defends Bentham even in this connection, objecting to some of the ways in which Whewell expresses his criticism and professing to find something worthwhile in Bentham's characterizations of his opponents. These are not the most persuasive parts of the review, which is not surprising given that elsewhere Mill himself made very similar criticisms of Bentham: "The greatest of Mr. Bentham's defects, his insufficient knowledge and appreciation of the thoughts of other men, shows itself constantly in his grappling with some delusive shadow of an adversary's opinions, and leaving the actual substance unharmed."[28] More generally, as Laura Snyder suggests,[29] Whewell and Mill share more than Mill is prepared to acknowledge. Whewell's objections to Paley are echoed in Mill's amendments to Bentham. Whewell rejects Paley's reductive, egoistic conception of human psychology. In a similar spirit, Mill departs from Bentham by introducing the distinction between higher and lower pleasures in chapter 2 of *Utilitarianism*.

There is a deeper disagreement as to how to understand and frame the options in moral philosophy. As noted above, Whewell identifies two schools, of "independent" and "dependent" morality, aligning himself with the school of independent morality. Mill, following Bentham, distinguishes those views according to which there is an external moral standard from those according to which the standard is merely internal, and he advocates external morality. Mill uses "internal" to refer to much the same set of views as Whewell refers to as "independent," and Mill uses "external" to refer to much the same set of views Whewell refers to as "dependent." But they disagree sharply as to the nature and merits of these options. After quoting Whewell's characterization of the dichotomy, Mill complains, "There is, in this mode of stating the question, great unfairness to the doctrine of 'dependent morality,' as Dr. Whewell terms it, though the word 'independent' is fully as applicable to it as to the intuition doctrine. He appropriates to his own side of the question all the expressions, such as conscience, duty, rectitude, with which the reverential feelings of mankind towards moral ideas are associated; and cries out, 'I am for these noble things: *you* are for pleasure or utility.' We cannot accept this as a description of the matter at issue."[30]

There is plenty to complain about in both Whewell's and Mill's ways of framing the alternatives. As we will see in more detail when we come to Sidgwick, one key problem is common to both: a mistaken assumption that metaethical and normative theses must go together, that intuitionists in metaethics must oppose utilitarianism. Another is that both Whewell and Mill mistakenly depict the alternative to their own favored view as necessarily involving the kind of arbitrariness that theological voluntarism exemplifies. There are further problems specific to one or the other way of framing things. Whewell's talk of "independent morality" suggests at least three separable metaethical theses that he runs together. One is a thesis about meaning, that the fundamental evaluative concepts cannot be defined in factual terms. Another is a thesis about motivation, that there is a distinctive kind of moral motivation different from motivation by ordinary desires. A third is a thesis about metaphysics, that moral truths are not made true by anyone's will. Bentham's and Mill's distinction between external and internal is equally problematic. As Sidgwick suggests, the idea that utilitarian or other principles can be established by external evidence seems unsustainable. Ordinary factual evidence does not establish that we *ought* to aim at happiness alone; the utilitarian standard, just like any other, seems to require intuitive support. And the idea that utilitarianism is in some privileged sense the only "external" standard to which the only alternative is arbitrary reliance on personal taste or social convention seems equally unsustainable. Maximizing happiness is no more an external goal than is maximizing knowledge, or minimizing injustice, or any of an indefinite range of further possible alternatives.

Like Butler, Whewell's objection to utilitarianism is not that right actions do not maximize utility.[31] He is prepared to grant "that . . . virtue always does produce an overbalance of happiness."[32] He offers instead two main objections, summarizing them in the *Lectures*: "First, we cannot calculate all the consequences of any action, and thus cannot estimate the degree in which it promotes human happiness;—second, happiness is derived from moral elements, and therefore we cannot properly derive morality from happiness."[33] Mill responds to both. Their debate over the first objection covers some of the issues later discussed under the headings of "act" and "rule" utilitarianism. Whewell grants that it is easier to know the consequences of kinds of acts than it is to know the consequences of individual acts. But he argues that to evaluate individual actions by reference to the consequences of kinds of actions is to give up on the utilitarian position. Mill argues that such evaluations are perfectly consistent with utilitarianism.

Whewell's second objection is that Bentham only gets plausible results by appealing to nonutilitarian moral sentiments of approval and disapproval: "Why should a man be truthful and just? Because acts of veracity and justice, even if they do not produce immediate gratification to him and his friends in other ways . . . at least produce pleasure in this way;—that they procure him his own approval and that of all good men."[34] Mill responds that Whewell here misreads Bentham. Insofar as Bentham appeals to public opinion and moral sentiments, he does so only to provide sanctions. Such appeals to public opinion do not, for Bentham, determine what morality requires.

Whewell adds a final, and, as he sees it, decisive objection. He quotes the now celebrated passage where Bentham writes, inter alia: "The day *may* come when the rest of the animal creation may acquire those rights which never could have been withholden from them but by the hand of tyranny."[35] Whewell regards this implication as a reductio ad absurdum of Bentham's

utilitarian position: "The Morality which depends upon the increase of pleasure alone would make it our duty to increase the pleasures of pigs or of
geese rather than those of men, if we were sure that the pleasure we could
give *them* were greater than the pleasures of men. . . . It is not only not . . .
obvious, but to most persons not a tolerable doctrine, that we may sacrifice
the happiness of men, provided we can in that way produce an overplus of
pleasure to cats, dogs, and hogs, not to say lice and fleas."[36] Mill is as confident in treating this as a strong point of Bentham's system as Whewell is in
regarding it as a fatal weakness. Mill sees Whewell's criticism as further evidence that he is committed to a conservative endorsement of mere prejudices
rather than to rational and progressive moral thinking. "Nothing is more
natural to human beings, nor, up to a certain point, more universal, than to
estimate the pleasures and pains of others as deserving of regard exactly in
proportion to their likeness to ourselves. These superstitions of selfishness
had the characteristics by which Dr. Whewell recognizes his moral rules;
and his opinion on the rights of animals shows that, in this case at least, he
is consistent. We are perfectly willing to stake the whole question on this
one issue."[37] Mill turns from defending Bentham against Whewell's objections to attacking Whewell's own alternative view. The overall charge he
makes against Whewell is similar to the charge he later will make against
Immanuel Kant:[38] that Whewell either gives no justification whatever for
the substantive moral rules he articulates, or that he justifies them by relying
on utilitarian reasons. The supreme rule of morality is a truism that does not
tell us what is right. Whewell's introduction of fundamental rights promises
something more substantive. But this promise is unfulfilled: either Whewell
simply and unacceptably reads off the list of moral rights from some current
legal system, or he has no independent basis for generating the list of moral
rights he gives.

As his problematic framing of the options in moral philosophy would
lead one to expect, Mill sees this as an inescapable defect of intuitive or internal morality: "Dr. Whewell has failed in what was impossible to succeed
in. Every attempt to dress up an appeal to intuition in the forms of reasoning must break down in the same manner. The system must, from the
conditions of the case, revolve in a circle. If morality is not to gravitate to any
end, but to hang self-balanced in space, it is useless attempting to suspend
one point of it upon another point."[39] Thus, when Whewell does provide
anything like an argument for any one of his more specific moral rules, as he
does at various points in the *Elements*, those arguments must be utilitarian:
"Almost all the *generalia* of moral philosophy prefixed to the 'Elements' are . . .
derived from utility."[40]

Whewell responds to these charges both in the supplement to *The Elements of Morality* and in the preface to the third edition. He clearly shows

that some of the details of Mill's critique do not hit their mark (as Mill shows with respect to some of the details of Whewell's critique of Bentham). Thus, for instance, Whewell does have a reasonable position with respect to the role of law in morality: the fundamental rights are determined independent of any particular legal system, but the details in any particular case have to be filled in by local legal rules. More generally, neither Whewell nor Mill is fully convincing. Mill is surely mistaken in the view he inherits from Bentham that the only possible alternative to utilitarian reasoning is no reasoning at all. But Mill is surely also right in suggesting that Whewell could have developed his alternative to utilitarianism a lot more clearly than he did.

Whewell and Sidgwick

Whatever our verdict on the debate between Mill and Whewell, it set the stage for later Victorian moral philosophy. In what is often called the "short intellectual autobiography" that his literary executor, E. E. Constance Jones, included in the preface to the sixth edition of *The Methods of Ethics*, Sidgwick begins:

> My first adhesion to a definite Ethical System was to the Utilitarianism of Mill: I found in this relief from the apparently external and arbitrary pressure of moral rules which I had been educated to obey, and which presented themselves to me as to some extent doubtful and confused; and sometimes, even when clear, as merely dogmatic, unreasoned, incoherent. My antagonism to this was intensified by the study of Whewell's *Elements of Morality* which was prescribed for the study of undergraduates in Trinity. It was from that book that I derived the impression—which long remained uneffaced—that Intuitional moralists were hopelessly loose (as compared to mathematicians) in their definitions and axioms.[41]

The autobiography then sketches the intellectual evolution that we find realized in the *Methods*: away from the early embrace of Mill toward a position that, while still critical, finds much more that is defensible in Whewell than Mill ever did.

One move that is crucial is to separate the metaethical and normative theses that are combined by both Whewell and Mill, and thus to reject the idea they share, that there is a basic dichotomy between the independent and the dependent or the internal and the external. Sidgwick makes this move by arguing that all sensible moralists, even utilitarians, must be epistemic intuitionists:

> It should, however, be observed, that the current contrast between "intuitive" or "*a priori*" and "inductive" or "*a posteriori*" morality commonly involves a certain confusion of thought. For what the "inductive" moralist professes to

know by induction is commonly not the same thing as what the "intuitive" moralist professes to know by intuition. In the former case it is the conduciveness to pleasure of certain kinds of action that is methodically ascertained: in the latter case their rightness: there is therefore no proper opposition. If Hedonism claims to give authoritative guidance, this can only be in virtue of the principle that pleasure is the only reasonable ultimate end of human action: and this principle cannot be known by induction from experience. Experience can at most tell us that all men always do seek pleasure as their ultimate end . . . it cannot tell us that anyone ought so to seek it. If this latter proposition is legitimately affirmed . . . it must either be immediately known to be true,— and therefore, we may say, a moral intuition—or be inferred ultimately from premises which include at least one such moral intuition.[42]

He thus makes room for his own distinctive position, which he sometimes characterized as "utilitarianism on an intuitional basis."[43]

Sidgwick's metaethical position is, in general, much more Whewellian than Millian. In the most important metaethical chapter in the *Methods*, book 1, chapter 3,[44] he argues both that the fundamental moral concept cannot be defined in factual terms and that there is a distinctive source of moral motivation different from ordinary desire. Though he ultimately rejects normative views like Whewell's, Sidgwick gives them a much fuller and fairer hearing than Mill did. As we saw, Mill, following Bentham, assumes that there is no non-consequentialist way to justify moral rules. Sidgwick makes no such assumption. The argument he suggests for making the fundamental moral concept, the concept expressed by "right" and "ought," rather than the concept expressed by "good," is that only thus can we give a fair initial representation of "the Intuitional view; according to which conduct is held to be right when conformed to certain precepts or principles of Duty, intuitively known to be unconditionally binding." This is the view that Sidgwick develops under the names "Dogmatic Intuitionism" or "The Morality of Common Sense." He admits that it seems "undeniable that men judge some acts to be right and wrong in themselves, without consideration of their tendency to produce happiness to the agent or others." The question is whether such judgments can be properly sustained. The dogmatic intuitionist is the philosopher who holds that they can, that we can "obtain from this fluid mass of opinion, a deposit of clear and precise principles commanding universal acceptance."[45]

An obvious question here is how far the dogmatic intuitionist Sidgwick depicts is simply Whewell. The answer, I suggest, is that the dogmatic intuitionist occupies the philosophical position Sidgwick thinks Whewell should have held, shorn of various aspects of Whewell's actual views that Sidgwick continued to find unconvincing. Like Whewell, Sidgwick's dogmatic intuitionist is committed to the existence of self-evident non-consequentialist axioms. But other aspects of *The Elements of Morality* play little role in book 3 of the *Methods*. There is no appeal to Whewell's single supreme rule of morality; Sidgwick's dogmatic intuitionist is explicitly pluralist. Sidgwick does discuss the idea that the lower parts of our nature should be governed by the higher, but for him this is an empty truism, not a crucial insight to be drawn on in articulating dogmatic intuitionism. When in his short intellectual autobiography Sidgwick identifies a philosopher as supplying the model for his procedure in book 3 of the *Methods*, that philosopher is Aristotle, not Whewell. And though Sidgwick embraces the search for genuine intuitions, Whewell is never for him a positive model of a philosopher engaged on this search. His role models there are instead Samuel Clarke and Kant.

Having spent most of book 3 developing dogmatic intuitionism, Sidgwick argues against it in chapter 11. He does so by first articulating four criteria or conditions "which would establish a significant proposition, apparently self-evident, in the highest degree of certainty attainable."[46] The conditions can be framed as follows:

1. The terms of the proposition must be clear and precise.
2. The proposition must genuinely seem self-evident.
3. The propositions accepted as self-evident must be mutually consistent.
4. Propositions accepted as self-evident should not be denied by epistemic peers.

Sidgwick goes on to argue that none of the putative axioms of the dogmatic intuitionist satisfies these conditions. They may appear to do so when left as vague generalities, but when made clear and precise so as to satisfy condition 1, they will no longer satisfy one of the other conditions, most usually conditions 4 or 2.

Despite the fact that Sidgwick found much more to sympathize with in Whewell than Mill did, it is natural to appeal to his influence to help explain Whewell's fall into obscurity in moral philosophy. Whewell is a very recognizable intellectual presence in the *Methods*, as we have seen, but he then disappears from the stage. There are no references at all to him in the indexes to G. E. Moore's *Principia Ethica* (1903), to Hastings Rashdall's *Theory of Good and Evil* (1907), to H. A. Prichard's *Moral Obligation* (1947), or to W. D. Ross's *The Right and the Good* (1930). Even discounting for the economical referencing habits of that era, it is hard to see much trace of Whewell's thought in these later works.

The obvious explanation for his disappearance is twofold. First, if Whewell had an intellectual identity for later moral philosophers, it was as the model for Sidgwick's dogmatic intuitionist. And Sidgwick's critique of dogmatic intuitionism was widely thought by his contemporaries and successors to be an unqualified success. Second, later Victorian moral philosophers who shared Whewell's orthodox Christian sympathies tended to take a very different intellectual path than he had: to embrace the Hegelianism of Sidgwick's most important contemporary rivals, T. H. Green and F. H. Bradley.[47]

Whewell's fall into obscurity remains complete enough that there is really not a current philosophical literature on his ethics in the same way in which there is a current philosophical literature on Mill's or Sidgwick's ethics: an ongoing scholarly debate whose participants refer to each other and disagree about the interpretation and evaluation of the historical philosopher's views and arguments. There is, however, a small number of particularly noteworthy philosophical treatments of Whewell's ethics from the last half-century or so.[48] The pioneer was J. B. Schneewind, who gave a sensitive reconstruction of Whewell to set the stage for his exploration of Sidgwick.[49] Alan Donagan argues for the then-novel claim that Whewell's position survives Sidgwick's critique of dogmatic intuitionism.[50] Laura Snyder explores the Mill–Whewell debate in moral philosophy in the context of a broader treatment of their philosophical disagreements.[51] And while Terence Irwin's magisterial three-volume history of ethics does not feature a chapter devoted solely to Whewell, it does contain a sympathetic treatment of Whewell in the context of the Mill–Whewell debate.[52]

CHAPTER 15

Whewell and Liberal Education

An Essay

SHELDON ROTHBLATT

No ideal of education has been more uplifting, none so much filled with aesthetic promise, none more humanistic, none more argued as the salvation of civilization and the arbiter of culture than the one denominated "liberal." But keeping the ideal constant is always a battle. As societies change, undergoing multiple transformations in their economic foundations and political constitutions, the existing content of a liberal education changes. Its ends become problematical. Whatever subject discipline is praised as the core of a liberal education will be disputed, another offered in its place. Whatever pedagogical approach is chosen, critics will raise the alarm. Ends claimed for a liberal education will be rejected by some. Flaws are found in each position. If the flaws are not egregious, then critics will rail that the benefits are maldistributed. Many cannot buy an education unrelated to immediate employment. Far from being uplifting then, liberal education is simply a validation of existing social disparities, the unworthy claiming an equally unworthy superiority.

Over time liberal education bends. Defenders answer critics by borrowing their vocational outlook. Liberal education, says Lynn Pasquerella of the American Association of Universities and Colleges, is needed for developing skills and mindsets important to employers and "aptitudes that matter for workplace success." Oral and written communication, once preeminent, slips a bit. "Critical thinking" is still on the list, digital literacy essential. A sociology of labor markets is undeniably valuable. But the grander, elevating dimensions of liberal education are largely absent from reports of this cast, although democracy and civic obligations make an appearance.[1]

Still, not all hope is lost. Even in today's digital world, where machine learning and artificial intelligence inform every aspect of national, indeed global culture; from personal connections and the organization of work to the immensely profitable entertainment industry, no other form of education carries as much weight whenever the subject turns to the arts of living. Digital technology may be the most significant invention in all of recorded history, as is commonly said in the popular media, but this ubiquitous technology does not put everyone at ease. Discussions of artificial intelligence immediately elide into science fiction scenarios: warfare at a distance, cyberattacks on financial systems, constant surveillance, robots replacing people.[2] If the ideals of a liberal education sometimes sound utopian, depictions of the digital age often sound dystopian.

Organizational and Intellectual Constraints

Digital Plus may be a context for present-day discussions of the relevance of a liberal education. Industrialism Plus was a context for William Whewell. Born in an agricultural and trading economy, he died in the world's first wholly industrial society. No intellectual concerned about the largest moral issues facing a society in transit can avoid contemplating consequences. Whewell in particular loved to argue big issues. His views on liberal education were situated within a national debate on the place of liberal education in what textbooks used to call "The Age of Reform" and that a Victorian wag might call "The Age of Steam and Cant" (a joke about Immanuel Kant's influence).[3] Furthermore, an "attached intellectual" discussing educational options shapes discussion to existing institutional structures, and for Whewell that meant Trinity College and Cambridge University. The liberal education he both amended and defended was constrained by the organizational structure of one of only two collegiate universities in England (St. Andrews in Scotland once had a collegiate structure).

Liberal education at Cambridge was designed essentially for landed and clerical elites. At least that was Whewell's inheritance. He spoke as if the "upper classes" (his oft-used phrase) would remain in charge, setting standards of education for the rest of society. But that conclusion was no longer tenable. Political influence was shared with the representatives of manufacturing, leaders like the prime minister, Sir Robert Peel. As it happens, he was also Whewell's benefactor, recommending to the queen in 1841 that he be made head of Trinity College. The appointment was in the gift of the crown.

Whewell himself was the son of a master carpenter, rose to academic heights through the old haphazard system of patronage and merit selection, an English version of the "lad o'pairts" of Scottish myth. He twice married well and came into money. How he regarded his own social status is uncertain, but as a Cambridge don and possessed of formidable intellectual gifts, he traveled in excellent company.

All writers on Whewell remark that the intellectual mix with which he worked consisted of the rationalism of the eighteenth century with its suggestions of atheism, theism, and secularism; the politically radical ideologies of late Enlightenment thought; and aspects of Romanticism, the period in which he came of age. He took this mix into a national give-and-take with English and Scottish antagonists. But a more immediate concern was the education of undergraduates at Cambridge. Whewell did not appear overly interested in Scottish universities beyond passing allusions.[4] He was well aware of the transformation of higher education in Germany and was highly regarded there by scholarly grandees. But for an English don, the behavior of undergraduates was an aspect of liberal education. As such Whewell could not condone the indifference to student carousing prevalent in Germany, even though in general he claimed to limit his discussions to only *English* liberal education.

Whewell was capable of stylish or at least interesting prose, but his writings on liberal education are rather cumbrous and rambling. Appearing as stand-alone publications in the 1830s–1850s, they resist compression into a short chapter. He discussed the role of university professors; the function of colleges, especially his own, the leading center of thought in Cambridge; and the curriculum for undergraduates. He explained the examination system at considerable length, as this was a particularly nettlesome issue for him. Examinations drove curricula. The opposite was also true, if the instructor also examined. But in the Cambridge system of examinations necessary for degrees, teaching and examining were separate, the exercises blindly marked, competitively ranked, and affected by a high turnover in examiners.[5] Whewell was concerned about "an atmosphere saturated with examinations."[6] He tried hard, possibly too hard, to separate college and university, but at the same time he also tried to imagine solutions where the two were not in conflict and spent considerable effort in addressing the matter.

The examination system at Cambridge was composed of a "Previous," a sort of mid-career test on theology; the Gospels or Acts; some classics and arithmetic, not particularly strenuous; and an honors degree examination, or "tripos," taken by the best students, usually after three years of residence. The tripos had replaced a leftover medieval exercise of oral disputations and acquired a reputation as the most rigorous and objective examination in all of Britain. Few sat for it until well into the nineteenth century. The majority, if not dropouts, read for an "Ordinary Degree" of lesser reputation. Just getting by, they were the "hoi polloi" who "went out in the poll." The earliest history of the tripos remains foggy. The first and, until 1822 when a Classical Tripos was established, the only terminal honors examination available to undergraduates was the Mathematical Tripos, indicating the importance that Cambridge still attributed to the work of the two Isaacs, Newton and

Barrow. Until 1850 those wishing to do the classics examination had first to do mathematics. Whewell found educational merit in oral debate—public speaking in effect—but disputations could not be standardized. He gave some attention to the pollmen (students who intended to read just for a pass, not honors), but he seems to have focused on the merit-inclined sector of the student population, to which he himself had always belonged. In 1816 he placed second in the Mathematical Tripos, losing out to an accomplished poseur whose abilities he underestimated, or so it was gossiped.[7]

Whewell centered his version of a liberal education on the college: pedagogy and student discipline were its two principal functions. College tutors and lecturers had enlarged responsibilities, more broadly educational and personal than the formal academic scheme. If examinations set by an independent central board were the end-all of a university education, students would pay more attention to professorial lectures and less to their college work. The loss would be, in his words, a more personal, more "active," "practical," and "direct" form of education. He designated as "permanent" the knowledge gained in college. The function of a liberal learning was to assure it.

Professors offered "progressive" knowledge that, in Whewell's opinion, was "speculative," exploratory, and in flux. Discoveries made one day would be overthrown on the next. Undergraduates might profit by attendance—bearing in mind his caveats about examinations—but if inadequately fortified by permanent knowledge, which contained all of the elements absent from progressive knowledge, their minds might be led astray by both intellectual excitement and delight in learning that professors were fallible. The thought that immature students would become puffed up with self-conceit riled him. The other educational defect was that students attending lectures were "passive" and received instruction "indirectly." There were no required assignments as in college, but he came to think that such were possible.

Whewell occasionally conceded that permanent and progressive knowledge had the habit of intruding on each other's territory.[8] Some subjects consisted of both progressive and permanent elements. Keeping one form of knowledge from bleeding into another was unlikely ever to succeed, given the very nature of cerebral inquiry. He appears to have been of two minds about progressive knowledge. He sometimes remarked that "almost all general improvements in literature and science have been introduced and diffused by professors."[9] Yet it is difficult to square such a judgment with his concerns about the indefinite character of knowledge in the making. One answer appears to be his inherent desire to protect the college, the place for liberal instruction. Historians have noticed that he was more flexible in his discussions of knowledge in the public arena than within the territory of his own university.[10]

The primary step in discussing a liberal education must always be curricula, what is to be studied, why and how. The next step is to align the means of education with some purported ends, *in-puts* and *out-puts* in today's bureaucratic jargon. This has always been a sticking point, but it was of salient importance to Victorian thinkers because they were quarreling about the moral principles and reasoning abilities graduates would bring to society. They disagreed on whether education ought to hasten or restrain social change. Whewell rarely mentioned Edmund Burke, but he appears to have been of that ilk, willing to support or even initiate specific academic reforms. He could on occasion innovate, but in general he feared (not unreasoningly) that sudden change was destabilizing, depending of course on the nature of the change.

The subjects for study at Cambridge were many. Some thirteen chaired professors lectured on the natural sciences, classical literature, and divinity, but the curriculum of the colleges was heavily concentrated on mathematics, especially Euclid, a touch or two of the New Testament, some teaching about morality (Whewell's textbook replaced William Paley's neo-utilitarian version),[11] and the time-honored instruction in the Latin and Greek languages. Professors lectured on their specialties, and such was the weak standard of professorial appointments in Whewell's heyday that chairs could be awarded on the basis of patronage. (As Lucasian Professor of Mathematics in the 1820s, Charles Babbage even failed to take up residence.)[12] Whewell's credentials were never in doubt, and he held two chairs in succession, mineralogy and moral philosophy.

The teaching of permanent knowledge, the core of a liberal education, was the responsibility of the colleges. But how precisely was this to be achieved? Whewell's approach was through epistemology—or "philosophy of knowledge" as he sometimes called it—a topic vigorously argued in late Georgian and Victorian controversies. How does the mind acquire knowledge in the first place, and once acquired, how is it to be used? The issue was obviously critical to Victorian pedagogy because so much time and effort were spent in fathoming the process of cerebral reasoning. Whewell's principal antagonists, the ones he regarded as especially dangerous to his own position, were the Benthamites, the philosophical radicals, and the utilitarians, the descendants of the John Locke and David Hartley school of association psychology, of whom the principal representative for Whewell was John Stuart Mill.

"Whether a philosopher succeeds in transforming the way people think depends in part upon his ability to convince them that what he says conforms to unanswerable principles," wrote Lord Annan.[13] It was necessary for each school of thought to advance its argument for unanswerable principles. No convincing conclusions emerged, but not for want of trying. As historians like Laura Snyder have mentioned, the arguments of each side sometimes overlapped, agreements reached that were annoying precisely because they disrupted the attack. A reader may be forgiven for thinking that in this give-and-take each side was anxious to avoid conceding a palpable thrust that might turn out to be fatal. Therefore while each side might borrow from the other, outright concession was unthinkable.

Association psychology and its antithesis have long been a staple of the literature on the history of thought in Victorian England, to which more recently Laura Snyder, Richard Yeo, Harvey Becher, Menachem Fisch, and others have contributed, extending analysis into the main body of Whewell's intellectual corpus. A summary of key points in his quarrels with utilitarians is nevertheless necessary.

How Sensible Are the Senses?

The salient issue was the role of outside stimuli—the environment, experience, knowledge gained through sensory phenomena—in determining thought. That in turn depended on the constitution of the mind itself. The utilitarians held fast to the view that ideas always originated from the outside and were "associated" once in the mind. John Stuart Mill labeled his opponents as members of an a priori German school of psychology, misleading as it turned out. The conclusion that ideas were "innate" in the mind was more relevant to the neo-Platonic form used in the eighteenth century by Bishop Berkeley who demonstrated how easy it was for the mind to be fooled by the senses. Whewell called sensory knowledge "transitory" as some of it did not last.[14] Whewell's colleague Adam Sedgwick, a popular lecturer of a more liberal persuasion, offered a third position, arguing that while the mind did not contain "innate ideas," it did contain "innate capacities" for transforming received data.[15] There is some guessing that Kant was an influence on this mode of thought, but whether he was needed is uncertain.

Why were these abstruse philosophical arguments more than "academic"? Because the outcome of the debate determined the scope and scale of possible reform in all sectors of English life. The utilitarians were social engineers armed with a complete agenda for the total transformation of English society. If they could prove that sensory stimuli created ideas, ideas to be converted into values, and values employed to modify behavior, then educational systems could be established to promote a desirable outcome. Manipulation of the environment to insinuate the right ideas into the brain was achieved with the aid of the happiness principle of pleasure and pain, carrots and sticks. Charles Dickens scuttled its dehumanizing simplicities in *Hard Times*. At its extreme, one can imagine the frightening dimensions of "brain washing" or "group think" as known in the twentieth century. Mill himself, using a phrase in circulation in the 1830s, was concerned about a "tyranny of the majority." A reading of Samuel Taylor Coleridge had first shaken his obedience to Bentham's views on personal freedom. Mill was later rumbled by Whewell to the point where he agreed that many levels of pleasure and pain existed, to include anticipatory pleasure and duration. Whewell's *Philosophy of the Inductive Sciences* (1840) "enabled me to present my ideas with greater clearness and emphasis,"[16] he confessed without conceding.

If the contrary could be proven true, that ideas existed apart from stimuli—put there by the Almighty?—or that the mind had an independent ability to organize and direct experience (those "capacities" again), then the utilitarian program would fail. Means and ends would be unaligned. The essential pedagogical element for Whewell was how to place into the mind the kinds of data amenable to its inherent natural ability to gain "permanent" knowledge, the only kind of knowledge that was true and free of error.

The mind shaped data through processes that Whewell called "Fundamental Ideas" and "necessary truths," those processes initiated by a curriculum specifically designed for them—the essential Cambridge college curriculum of axiomatic geometry and Greek and Latin language. Whewell challenged the reader to prove that sensation or "Experience" could show not only what a thing *is* "but that *it must be*." Can experience demonstrate that two straight lines cannot enclose space? "The fallacy is . . . too palpable . . . to be dwelt upon." Who imagines that Archimedes needed an experiment to conclude that "the pressure on the fulcrum is equal to the sum of the weights?" He continued in this vein, pointing out (as had Berkeley) how difficult it is to interpret any experience.[17]

The other half of Whewell's collegiate curriculum were the classical languages, the "dead languages" as opponents called them. Citing French thinkers of like mind, he went through the usual preliminaries of explaining that Greek and Latin are "indispensable elements of a liberal education." They are the common inheritance of all European nations. All who are educated agree on their importance. They represent unity—think of a world where only jargons are spoken. They join past to present. They are models of written expression because they provide a type of universal grammar. They are the basis of modern languages and literature. Without a solid grounding in the classics, modern literature is "incoherent caprice and wanton lawlessness."[18] But the caveat is that while grammar lies within the domain of permanent knowledge, other aspects of the classical heritage are progressive. He named history, philology, philosophy, again struggling to keep alive a beleaguered distinction.

Faculty Psychology

While Whewell, Sedgwick, Mill, and others argued abstractly along conventional lines, a pseudoscientific explanation for understanding the physiology of the brain was circulating internationally. Whewell used its lexicon, especially the word *faculty*, as did almost all educational thinkers, but its alignment with his epistemology is unclear.

"Faculty psychology"—or the pseudoscience of phrenology—was a primitive form of brain physiology, and as such of some interest to those coming after Whewell. The main figures were Franz Gall and his disciple Johann Gaspar Spurzheim. Gall died in 1828, Spurzheim in 1832. Not altogether incorrectly, Gall identified cortical territories, specific "organs" or "faculties" in the brain where mental operations take place, among them reasoning, memory, space, metaphysics, language, morality, or even a sense of God. He listed twenty-seven.[19] An American writer, surveying the topic a year after Whewell's death, maintained that the nervous composition of the brain consists of cells and fibers, the result being that the brain *acts* on received impressions. He went further: actual physiological changes occur.[20] Whewell was likely to include physiology in the category of progressive knowledge because that is how to describe medicine and anatomy, but he and his coevals seemed perfectly content with psychological or metaphysical arguments about faculties. At least they were comfortable using the word and spoke as if faculties ought to exist even if they could not actually be found among the brain's lobes and hemispheres.

However understood, the significance of faculty psychology—a term Whewell himself did not use—is that it provided educators with a justification for teaching texts in particular ways. Certain texts or subjects stimulated memory or helped develop reasoning and language skills. Where contemporaries vigorously disagreed was on subjects and pedagogy that best stimulate, cultivate, or discipline particular faculties. Which subjects can massage the most faculties? At what point in education are certain subjects especially suited to the task? Words, whether classical or in the vernacular, can exercise the memory, said one participant in the discussions, but the sciences are better as "instruments of culture." Reviewing the discussions, this American author went on to scorn the supposed advantages of the "dead languages," but the real folly is starting the study of languages with grammar, "the artificial production of stupidity." Hebrew had no grammar for a thousand years. He did not follow up, but his meaning is clear. Invocations of mind "gymnastics" were so commonplace, he said, that he found a large volume on heraldry palmed off as mental discipline.[21]

The famed Edinburgh professor Sir William Hamilton scoffed at the claims made for mathematics, calling into question the very foundation of Whewell's conception of a liberal education. Mathematics cultivates fewer faculties. If any are somehow stimulated, the outcomes are feeble.[22]

Sedgwick referred to a faculty of imagination (Romantic historiographers were strong on imagination),[23] also a faculty of moral judgment or conscience containing "feelings." It is one of those "capacities" independent of teaching. The brain *acts*? Mill, often in the game, spoke about higher and lower faculties. He mentioned that when a "cultivated mind" is taught to "exercise its faculties," the result is "inexhaustible interest in all that surrounds it." There are "acquired faculties" such as conscience that took him back to the theory of experience. Somewhat confusingly, he also talked about "social feelings" that are "innate," although these can be amplified by education.[24]

Civilization without Discontents

Throughout his educational writings, Whewell employed a mélange of conventional phrases such as "mental faculties," "mental discipline," and "mental culture." To these he joined other words in common circulation, such as *civilization* and *progress*. He referred to a "general progress of national culture and civilization." He stated that universities exist to provide the "general culture of all the best faculties." He was certain that a "culture of the mind . . . is the true object of a liberal education." Again, the "object" of liberal education is "to develop the whole mental system of man." A complete mental culture consists of many elements, but the one especially indispensable "is such a discipline of the reasoning power as will enable persons to proceed with certainty and facility from fundamental principles to their consequences." The object of a liberal education is "to educe all the faculties by which man shares in the highest thoughts and feeling of his species." Furthermore, our "faculties of humanity" require us to be more than "mere" mathematicians or "mere" scholars.[25]

Reconciling Whewell's frequent appeal to faculties and mental discipline as a basis for culture, civilization, and progress with his far more developed philosophical arguments regarding induction in the sciences, fundamental ideas and necessary truths, is demanding. He offered little assistance, at least in the texts on liberal education. It is quite impossible to know how many readers and auditors were able to comprehend his convoluted intellectual explanation for how the mind works. Perhaps it was unnecessary, as outcomes such as mental culture and the disciplining of the mind were part of familiar discourses. But Whewell could probably have argued that fundamental ideas act in concert with faculties stimulated in particular ways, organizing and directing the contents therein. Whether any reader was thereby enlightened is problematical, but we must give the Victorians the benefit of the doubt.

Pedagogical Bottlenecks

Apart from curricular and epistemological issues, there was a more serious institutional difficulty that Whewell never adequately addressed because it required seminal institutional reforms with which he disagreed. The main choke points in his entire conception of liberal education as permanent knowledge were the statutory requirement that college fellows take holy orders if they wanted to retain their appointments beyond a certain number of years and that, unless appointed to headships and certain professorships, they had to remain celibate. Dating back to the original Elizabethan statutes, the fellowship system had been designed for turnover. Each college also had the right of preferment to a certain number of country benefices, a further incentive to leave. Whewell defended celibacy even though he considered resigning his fellowship and leaving Cambridge when he was a young, unmarried and lonely don.

Whewell must have understood that the quality of teaching within the colleges was greatly handicapped by a hemorrhage of talent. Much of the teaching was schoolboy slogging. Sometimes, he asserted, undergraduates would simply have to be forced into "uncongenial kinds of study."[26] Much of it was uncongenial because of Whewell's insistence on highly technical teaching, usually geometry and grammar. His understanding of liberal education was constrained by the urgent necessities of his epistemological outlook.

There was also the problem of the system of degree examinations, the fear that undergraduates would seek the kind of preparation that was not provided in college, professorial or otherwise. This fear was very real. Undergraduates did go out of college to hire private tutors or "coaches." Some were fellows eager to have the substantial extra income, but others were simply graduates without gainful employment.[27] For seven years after receiving a first degree and before gaining a college appointment, Whewell had taken private students. His biographer says he was conscientious, traveling with his charges on reading and study trips to places like Wales. This was a common practice in the early years of the Romantic movement. Nature would have a role in broadening the education of youth, but Whewell was hardly enthusiastic about "drudging your intellects at 40 pounds per annum for stupid pupils," as a friend wrote to him in 1817.[28] From his days as a private tutor on, he preferred to be a scholar but yielded to circumstances.

Without mincing words, Whewell several times called "coaching" a great evil. He listed and approved of Cambridge policies intended to control coaching and tried to design an alternative system, Permanent Studies for Lower Honors, Progressive Studies for Higher Honors, viva voce, as well as written examinations. This might restrict access to coaches, "so far as it is desirable that the practice should be limited."[29] He went from topic to topic, thinking of ways in which college teaching and university examinations could dovetail, or how he could separate the better students from the "inert" and "frivolous" students. One plan was to create a "side" of three or four serious students. If they received the close attention of college tutors, discipline could be maintained, and they would not have to resort to coaches.

The larger body would be placed within another side that could then also be farmed out to private coaches. Possibly an arrangement could be worked out to retain some college authority. At one point he blithely dismissed the extra expense caused by coaching, possibly because it was generally the richer students who were the most inert. But these tangled thoughts were finally only schemes devised to work out solutions to problems arising from his conception of liberal education.

Especially interesting are the compatriot words that Whewell connected to his discussion of mental discipline, progress, culture, and civilization. He asked a very Victorian question: Is there actual progress in civilization over time: mental, moral, material, spiritual? His answer was that there is, or there could be, progress, or that there is progress in some intellectual domains, that advances in civilization and culture are possible, have in fact been achieved, but might not last. This answer is neither forthrightly positive nor forthrightly negative. Where exactly are we?

Civilization came to Whewell from the eighteenth century. The word struggled to disentangle itself from an older word, *civility*, attached to the courtier tradition.[30] Insofar as courtier ideals were pertinent to civilization—manners, politeness, self-effacement, the social graces—Whewell could respect them. He casually referred to "good breeding," "manners," "faults of grammar," their absence belonging to those "less cultured," standing low on the scale of "civilization."[31] The irony is that his own bullying persona, which detracted from his thoughtful contributions to education, did not exactly fall within the canons of politesse.

Various schools of history, economics, and thought had cataloged stages of civilization, sometimes described as stages of economic advancement, but French thinkers also thought in terms of broader levels, from primitive religion to science (Whewell found errors in and generally disagreed with the work of the Comtean "positivists"). But not all thinkers accepted progress of some kind as inevitable. The Scots had been concerned that wealth accumulated through commerce would lead to a decline in "virtue." They were possibly thinking of the old clan system and its heroic warriors. The political economist David Ricardo, whose economics Whewell faulted, thought that men of business could progress, but not the generality of mankind.[32] Thomas Malthus was unabashedly gloomy. Population growth outraced the resources needed for survival. Lord Macaulay, in the famous third chapter of his vastly popular history of England, mocked the naysayers. If there is discontent, it is because moral and material improvement heighten expectations. There never was a past golden age.[33] Incidentally, Whewell read an essay by Macaulay and found him brilliant, elegant, and superficial.

Whewell himself did not actually think in terms of specific stages of civilization. He simply recognized discreet periods, past or present, defined by the right kind of liberal education capable of realizing fundamental ideas, a civilization that kneaded the mental faculties, especially the "Faculties of Reasoning and Language," producing then an optimal condition.[34]

Like so many thinkers, historians, travelers, and writers of fiction, he distinguished between rude and refined societies. But he was generous. No society would remain in a rude state if its intellectual leaders are properly educated or come into contact with an advanced civilization from which it can learn.[35] Civilizations can reach a higher state through, for example, "great scientific advances in knowledge," but they can also fall back. Only the Governor of the universe, cautioned Whewell, can decree whether England's greatness will continue or, "before the final close of all things, the brightness of civilised England must wane and become dim." He added: like Greece and Rome.[36]

At this point Whewell became startlingly animated. With Euclid and Plato (Whewell translated Plato for the general English reader), ancient Greek civilization had reached eminence but, forgetting the reason why civilizations advance, Greece subsequently had declined. The replacement of axiomatic thinking with "the speculative study of philosophy" was the cause. "What great men had already commonly taught mankind, was perverted or forgotten by their degenerate followers." A "cumbrous apparatus of crystalline spheres" (Aristotle and Ptolemy) supplanted and debased "the simple geometrical conceptions" of the school of Plato. The "mechanical truths" of Archimedes, were, "like his tomb, overgrown with the rank and unprofitable vegetation of later days," lost for fifteen hundred years. Natural philosophy was dead, and literature mortally smitten. Speculative philosophy survived, but "light after light goes out, and all is night."[37]

This is almost Carlylean in its colorful language, apocalyptic, a self-righteous denunciation of wrong-headed decisions that he came close to considering willful. Speculative philosophy was the undoing of a great civilization, and yet at another point he referred to the "speculative faculty" as "one of the elements of civilization," the difference between rude and refined nations.[38] This is puzzling, if *faculty* and *philosophy* are equivalent. Elsewhere he seems to require a civilization to have both permanent and progressive knowledge.

What next appeared in his great episode of finger-wagging was an explanation for how the nations of Europe ultimately climbed back to the summit of civilization. Once again we are surprised. He attributed revival not to periods labeled today as modern but to those once dismissed as the dark ages. The monasteries "restored practical teaching . . . and bequeathed the results to the English universities." The consequence was a "genuine culture."[39] We recall that practical education is the function of college teaching.

But do we also notice a howling inconsistency? One scholar finds Whewell evaluating the medieval period from the perspective of rationalists, yet he denounces dogmatism, mysticism, a neglect of observation, enthusiasm, and illusion.[40] The contradiction is inexplicable, unless in some unspecified fashion the monasteries are not included in the mainstream of the medieval period.

Civilization and *culture* appear synonymous, although German thinkers were starting to separate them, and some English historians were thereby influenced.[41] Whewell correctly assumed that readers would know what he meant. We are not yet quite at that point in the nineteenth century when cultures were regarded as autonomous, no one superior to another. Racist theories, emerging at the same time, would however make the nefarious distinctions. Mill was beside himself when he mistakenly thought that Whewell approved of slavery. Whewell thought it heinous, yet, always fearing anything sudden, law was law until change could come. Mill was not exactly mollified.[42]

Whewell's commonplace Victorian reference to culture can be interestingly compared to Matthew Arnold's fully Romantic notion of culture as an abstraction. Culture for him represented the best that has been thought and said in the world. Thus far Whewell could approve. Arnold went on to equate culture with Church and State, but the Broad Church Anglicanism that represented cultural unity is no more, leaving only the State as the guardian of culture, a metonym for the departed religion. The State represents order. Synonymous with culture, it will heal the divisions in society and thereby liberate what is best in human nature.[43]

This is clearly not how Whewell approached issues of culture. He was "practical." His objection to certain kinds of mathematics was that they are abstract. His opposition to structural reforms proposed by royal commissions indicates how much he suspected the State. Universities should reform themselves. He even referred to himself as an "Old Whig,"[44] identifying with the aristocrats of yore who opposed a centralizing monarchy. How can a state that has no unity itself, Whewell and his friends wondered in 1846, possibly save society from itself?[45] Indeed.

The One Thing Needful

One point on which mid-Victorians could agree is that character formation, behavior modification generally was the outcome of a properly regulated liberal education, which is why fights over the psychology of the mind were so captious. In the history of liberal education, ideal-type characters and ideal qualities make frequent appearances. The Athenians had citizens, the Spartans had warriors, the medieval tellers of knightly tales had chivalry, and saints were holy. Romantics like Thomas Carlyle extracted heroes from the past, contrasting them with fickle representatives of the present. *Unum necessarium*: which personal quality is the one thing needful? Love of truth, said the great scholars of the Italian Renaissance, is reached via philology, the close reading of texts; *Bildung* said the Germans, *bildning* for the Swedes, aesthetic and spiritual self-realization. Civility had its day, the poise and ease of the courtier traditions, the purpose of a liberal education in Georgian England. A liberal education created the "gentleman" in John Henry, Cardinal Newman's example, someone almost "feminine" in decorum. Whewell bestowed an unflattering (by today's standards) "effeminacy of mind" on those lacking breadth in education,[46] but no commentator has described him as a mid-Victorian "Muscular Christian." But masculinity itself would become an educational quality that late Victorians and Edwardians associated with colonial proconsuls able to confront adversity in the bush.[47] Manly but also healthy, said the denizens of Harvard University circa 1900. Only sickly "grinds" burned the midnight oil, and by the way, they were likely to be the sons of pushy and uncouth immigrant families.

Yes, all of these ideal types were men, and all of the qualities regarded as needful were possessed by men. Liberal education had always been for men, with the few exceptions of high-born women in the Renaissance and in Georgian England. Newman said that a liberal education was about the "making of men."[48] Whewell posited the same end: the object of a liberal education "is to make men truly men," and, he added, "not to make them men of genius, which no education can make them."[49] At one point he suddenly launched into a series of generalizations difficult to unpack. Liberal education at Cambridge provides the graduate with the "feeling that he *is* an Englishman." Being an Englishman means knowing "the principles by which the actions of his fellow-citizens are regulated, and by reference to which his own will be judged of; a sympathy with their objects, and habit of balancing himself among their impulses." Was he talking about a sense of proportion? He stopped short to offer a caveat: no system of instruction is perfect; some graduates "will act ill," to the detriment of the English university system.[50]

These flat generalities were probably just afterthoughts, for he once said that character itself was less important than qualities such as truth, justice, humanity, and mental habits.[51] We know from his overall position that he wanted graduates formed at Cambridge who would reason their way carefully through a thicket of moral choices, respectful of tradition, exemplifying English virtues that he imprecisely listed. That was why they were instructed in reasoning controlled by axioms. Whatever they did in life would then be directed by fundamental ideas and necessary truths. It was not so much a description of formed characteristics as a defense of those aspects of mathematics that fit Whewell's epistemology. But how many graduates who took the Mathematical Tripos exemplified those requisite qualities? The vast majority of graduates were never heard from again. How much the mathematical teaching in the colleges influenced the hoi polloi, the students who did not sit for honors, is unquantifiable.

A Perpetual Search

Whewell's object was to preserve the liberal education he knew.[52] But his version has not traveled well. Who today argues about fundamental ideas and necessary truths? Who maintains that induction is superior to deduction in gaining a liberal education? The primacy of Euclid and classical languages in liberal education has gone by the wayside. In some American universities classics majors are not necessarily required to learn the languages. The knowledge base has become so vast and varied that no single discipline is sovereign. While we do credit the monks of the middle ages with preserving and adding to significant works of antiquity, no one after Whewell is likely to describe that achievement as comprising the "direct," "practical," and "active" teaching that occurs within undergraduate colleges. No one distinguishes between knowledge that is progressive or permanent. All knowledge is regarded as in flux. Whewell's collegiate pedagogical ideal, if not the college idea itself, was unsustainable as his university entered a new phase of vitality, the first of many to come.

In fairness to him, no one has ever been able to guarantee a union of means and ends in liberal education. Nature, nurture, circumstances, not just education, are part of the mix. But the story of liberal education in Whewell's Cambridge is just as much about structural impediments and vested interests as about curricula and learning theory. That is why liberal education is always searching for alternative educational systems: special colleges, interdisciplinary teaching, divisions instead of departments, internships, what Gerald Grant and David Riesman called *The Perpetual Dream*.

Still: "Mais où sont les neiges d'antan?" asked the medievalist François Villon.[53] What have we lost from what Whewell wanted? A dislike of fads and fancies, something called culture or civilization? Tradition, convention? We can well wonder if the monks of the future will come to the aid of the twenty-first century.

CHAPTER 16

"In the Chapel of Trinity College"

Whewell as Priest

MICHAEL LEDGER-LOMAS

A week before his fatal riding accident, William Whewell preached in Trinity College chapel on the end of the world. Repurposing passages from a sermon preached decades earlier, he warned that it would come not with a whimper, but a bang. He drew on his now unfashionable catastrophist geology to envision the surging seas and collapsing mountains that would herald Christ's return. Joseph Barber Lightfoot dwelled on those words when days later he preached Whewell's funeral sermon in that same chapel. They established him as "a religious teacher in the truest sense." Lightfoot preached on 1 Corinthians 15:32, in which Saint Paul's fight "with wild beasts at Ephesus" symbolized the duty of Christians to make themselves spectacles of faith. For Whewell, as for Lightfoot, the "little amphitheatre" in which his hearers were "'set forth' to slay or be slain" was Trinity College chapel, where a "silent concourse" of long dead fellows gathered to watch their combats for the faith.[1] As Hulsean Professor of Divinity, Lightfoot had a mission to defend the text of the New Testament against the quibbling of German critics at a time when the university was about to undergo decisive secularization. That might explain why he represented the chapel as central to the life of his fallen master and by extension to Cambridge. Yet Lightfoot was justified in thinking that Whewell's sermons—most of which remain, like the one he quoted, unpublished—are vital in recovering his identity as a priest. They show him to have been not just a scientist but a divine, whose faith was not a thin derivation from the study of nature, but a missionary and pastoral force. He did not just want to teach people, but to save them and to convince them to save others.

Students of Whewell have long recognized that his ordination as a priest on Trinity Sunday 1826 shaped his career, even though he continuously passed over opportunities to take livings because he was reluctant to leave Cambridge and felt ill-suited to pastoral work in a parish. As his friend Julius Charles Hare urged him to recognize just before he became master of Trinity, his scientific and philosophical work was his priestly vocation.[2] As Bernard Lightman demonstrates in this volume, Whewell dissented from William Paley's utilitarian natural theology, which apparently tried to argue people into faith by pointing out ways in which God had contrived the world to meet humanity's needs. Without underrating the role of reason, Whewell suggested that its role was simply to trace the fit between the laws of human thought and the cosmos.[3] Whewell not only redefined natural theology in this idealist manner; he also placed limits on its homiletic use. As John Hedley Brooke long ago argued, he was not Paleyan, but Pauline in his understanding of human beings as fallen souls striving for salvation through Christ's atonement. He derived knowledge of their predicament not from ratiocination but revelation: the Scriptures.

Whewell's sermons were crucial in Brooke's efforts to qualify the prominence he accorded natural theology in his apologetics. Yet Brooke remained wary of the sermons as sources, noting that their "hortatory" form might obscure the "quintessential Whewell."[4] William Ashworth's recent monograph is still more emphatic in presenting Whewell's sermons as a function of his clerical status. Yet though intermittently citing the sermons to hint at the theological "anxiety" beneath Whewell's epistemological commitments, Ashworth gives no connected account of what Whewell sought to achieve in preaching them. Like Brooke, Ashworth remains more interested in situating Whewell the scientific apologist in the histories of science and scholarship, offering only a sketch of what he did as a priest and his relationship to the Church parties of his time.[5]

Instead of suspecting the sermons for their "hortatory" character, this chapter follows historians of the sermon genre in investigating what Whewell intended their exhortations to achieve. Victorian sermons were "performances," and "confessional performances" at that: tailored to particular congregations and seeking not just to move them, but in certain directions.[6] When readers read sermons first preached in a village church, a college chapel, or a cathedral, they journeyed mentally to that place, to sit with its congregation and experience the emotional sway over them of their preacher. With the exception of significant forays to the University Church of Great St. Mary's, most of Whewell's preaching happened while he was master of Trinity and was designed to stir and mold his temporary parishioners: the undergraduates. As most of them would become clergymen, their faith would be critical in determining the Church's and the nation's future. On a rare trip to preach beneath the cavernous dome of St. Paul's Cathedral in January 1860,

Whewell mused that a preacher "cannot hope to appeal to the consciences of individuals with the same effect as if he was addressing an accustomed congregation, known to him personally."[7] At Trinity, he dealt with "fellow labourers"; here, he tentatively addressed hearers "from every quarter of this city, from every walk of life."[8] Whewell's preaching offered a vivid account of what Christianity was, but one that must be understood as conditioned by a particular institution and depending on it to achieve its effects. When he published an early selection of his sermons (1847)—which this essay takes as its main source because they reached a wide audience beyond Trinity—he claimed to do so not on account of any "theological value" they might have but to exhibit "the character of our College Chapel Sermons."[9] His book contributed to the burgeoning genre established by what his friend Julius Charles Hare called Thomas Arnold's "admirable Rugby sermons," that, in affording glimpses of an institution united under its clerical head, offered an argument about how to couple edification and education in the nation as a whole.[10]

I begin this essay by reconstructing the institutional settings in which Whewell preached. Although his Cambridge contemporaries expected sermons to be "addressed to the heart more than to the head," they wanted to be persuaded as well as moved.[11] William Gibson's point that sermons were also "critical acts" that dissected Scripture or expounded Church history for "discriminating" audiences was nowhere truer than at the universities.[12] Whewell's sermons were a running argument with and about undergraduates that promoted communal worship as an indispensable prop to their moral lives. I then look more closely at how Whewell presented Christian faith in his preaching, extending Brooke's suggestion that he was a Pauline thinker who believed that revelation surpassed natural philosophy. Although it is tempting to class Whewell as concentrating on natural theology's claims in the interval between Paley and Darwin, in his college sermons he was more interested in God's movement in history than his presence in nature.[13] He urged undergraduates to think of their college as not just a sacramental community, but one that was the product of centuries of Christian effort. For Whewell such a recognition had social implications: the young could repay their debts to those who had founded and advanced Trinity by furthering the incomplete Christianization of their society. While conceding that Whewell's later, unpublished sermons reveal a stiffening defensiveness in the face of new challenges to Christian belief, I close by demonstrating that his insistence on the master of Trinity's preaching duties had a long institutional afterlife.

"Ecclesiastical Mania"

In the decades during which Whewell established his reputation at Cambridge, sermons were central to the social and intellectual life of the university. The diaries of Joseph Romilly, a contemporary fellow of Trinity, record his fastidious attendance at Sunday sermons in the University Church of Great St. Mary's and afternoon tours of the town's churches and even its dissenting chapels in quest of edification. His criteria for good preaching were eclectic: though a Whiggish Protestant, Romilly also kindled to reactionary eloquence. On April 27, 1832, Thomas Thorp preached with "much fine feeling & beautiful writing," but in "his bitter comments on the superficiality and materialism of the times," he was "too much a prophet of ill."[14] A month later Whewell's friend Hugh James Rose preached a "beautiful & eloquent" but "intemperate uncompromising High-Church Sermon" whose "inflammatory" epithets, such as calling the king "*sacred*," savored of early modern fanaticism.[15] People like Romilly were a demanding audience. Boring or eccentric preachers in St. Mary's might be "scraped"—drowned out by the shuffling of chairs and feet—or shunned. On September 15, 1833, Romilly heard an "awful Sermon" in St. Mary's: "People showed their taste rather than their piety: 3 persons in the Pit, 2 in the Gallery, V Chancellor asleep."[16] Given the prestige attached to university sermons, Whewell made them an important part of his work after his ordination, often using them to advertise his scholarship. Though never published, his 1827 sermons on the relationship of faith to knowledge set out his stall for a divinity professorship.[17] His four sermons on the foundations of morals announced his revolt against Paleyan utilitarianism, then staked his claim to the Knightbridge Professorship of Moral and Casuistical Divinity. They were homiletic rather than systematic in approach, following St. Paul in insisting that it was not "the approval of the conscience, but the mercy of the Redeemer" that was our ultimate "stay and support."[18]

His preaching in college is easier to overlook. If services took place in chapel twice daily, then sermons were heard there only rarely, being reserved for red-letter days in the calendar: the sacrament days celebrated at the end of each term and the commemoration of benefactors service. As Whewell awkwardly explained to undergraduates in May 1861, this paucity of preaching had not reflected "any light estimation" of its importance but the trust that students would get instruction at St. Mary's instead. Over time the fellowship concluded that their "exhortations" might weigh better than the "word of a stranger" and should be given every Sunday rather than just on commemoration days.[19] But not until 1860 did the college take a step that Hare had mooted to Whewell as long ago as 1834—even then, many fellows opposed it.[20] Whewell might have told undergraduates that the fellows regarded it as a "privilege" to speak to them on matters of "unspeakable importance" in which they had a "common interest," but in earlier decades many had taken neither chapel services nor occasional preaching seriously. When called on for this duty, Romilly read out sermons written by others.[21] In the 1830s and 1840s even High Churchmen often editorialized on college life or national politics from the pulpit. In 1833 Romilly fumed as George Peacock preached a "vile sermon against the lower orders having anything to do with politics" at the commemoration service. Fellows, Whewell included, often made slipshod errors in the liturgy when taking a service.[22] Such ecclesiastical sprezzatura had reflected a divide within the Trinity fellowship between zealous churchmen and those who had sought ordination only reluctantly as a condition of retaining their fellowships. In the late 1830s Alexander Chisholm Gooden found that his coach Thomas Burcham did not "stand high with the parson-*fellows*, being a man who never appears at chapel and whose morality does not stand on the most unimpeachable foundation." In November 1839 the master and dean enraged Burcham by refusing to sign his testimonials for ordination on the grounds of those absences.[23]

The nonattendance of undergraduates, rather than of fellows, soon became a cause of controversy. During Christopher Wordsworth's mastership, the college developed an elaborate system to ensure that undergraduates went to at least five morning and evening chapels a week, or four including Sundays, as a condition of keeping term. Wordsworth gave the deans tall chairs from which they could monitor their charges. Chapel markers armed with pins and canvas rolls recorded the attendances of several hundred students in the chapel, which was from 1836 warmed with water pipes to make it more inviting in the winter.[24] These steps merely turned chapel into a roll call exercise. Even Gooden, who went to morning chapel daily to get an early start on his studies, was amused that his friend Jenkins "regularly attends Chapel *twice* a day," which caused him to be "a *little* laughed at for his ecclesiastical mania."[25] In 1834 knowledge of these attitudes had moved the classical tutor Connop Thirlwall to attack compulsory chapel in print and to argue that fellows who like himself merely offered "literary instruction" should not have to cram undergraduates with religion. Thirlwall's pamphlet caused his friend Whewell to declare his support for compulsion. Although he regretted Wordsworth's decision to force Thirlwall's resignation as a tutor, Whewell had precipitated it by making plain that neither he nor his colleague Charles Perry, a conscientious evangelical, could work with an assistant who had undermined college discipline.[26] In two open letters to Thirlwall, he argued that his friend's principled voluntarism was not just harmful to the cohesion of the college but was "altogether inconsistent with the general scheme of our national religion," which deployed regular rites to "throw a religious character over all the business of life" and make faith not just a passing sentiment but an "inviolable custom."[27]

Four years later, the senior dean of chapel, William Carus, brought to Wordsworth's attention laxity in the enforcement of the regulations. Wordsworth's move to strengthen them with penalties unleashed a charivari of undergraduate displeasure targeting tutors Whewell and Perry.[28] An Australian undergraduate was expelled for posting to Perry a "blasphemous Parody of the Litany" that included among its invocations "that Mr Whewell may learn the manners of a Gentleman."[29] A Society for the Prevention of Cruelty to Undergraduates posted class lists of attendance by fellows with suggested punishments for the laggards. Whewell, who languished in the second class, was requested to "write out the first chapter of the History of the Inductive Sciences."[30] The society's pranks were part of a satirical drive against parson fellows. William Makepeace Thackeray sent up Dean Thorp by publishing *A Few More Words to Freshmen* in his name. His Thorp boasts that "after deep research" he has found the chapel to be as full of God as "any full-sized steepled church" and implores fashionable undergraduates "that surplice[s] (washed) should be your only linen."[31] The society's antics got the attendance requirements reduced and their enforcement slackened. A mid-century American undergraduate found that the deans left alone undergraduates who attended fairly often and were *"regular in all other respects."* In this climate, he began "sleeping over morning chapels, and consuming much claret after dinners."[32]

Whewell's writings on universities in the late 1830s show that he had no illusions about undergraduate piety. He conceded that "a College chapel is not, in the sincerity and earnestness of its devotions, all that the friend of religion could wish it to be." Yet if he did not confuse compulsory observance with piety, he insisted that it instilled the habits that sustained religion. The reading of prayer books under the eye of the deans was good discipline for young men vulnerable to the "wandering gaze and mutual glance which unsettle the serious thought." While Whewell saw that it was demeaning to enforce attendance with fines, he still upheld daily morning chapel as his ideal.[33] This reflected his understanding of the college as a home. Undergraduates came fresh from family worship in their own Christian homes to *"our* family" where they could expect to do the same.[34] Revering St. Paul as he did, Whewell followed him in seeing love and coercion not as opposites but as mutually constitutive of holiness. In a sermon on Ephesians 5:1–2, he encouraged undergraduates to see that all those who labored "in the Institutions which are dedicated to his service" were no less "prisoners of the Lord" than Paul himself.[35] On another occasion he cast undergraduates as future clergymen who would soon be "diffusing the authority of the Great Head of the School."[36] As late as 1862 he found in Paul's epistle to the Philippians the words to greet freshmen who "stand to us in a certain relationship as disciples."[37] As master and preacher Whewell was part apostle, part paterfamilias, part jailer—personae that suited an authoritarian, not to say despotic personality.[38] No wonder one undergraduate scrawled in a chapel book Francis Doyle's epigram that the intergalactic traveler "will find when he reaches the verge of infinity, / That God's greatest work is the Master of Trinity."[39] Yet Whewell sincerely aimed to wield his power for the Church. Immediately after his ceremonial reentry into Trinity as its master, he had closeted himself with Carus and Thorp, the senior and junior deans of chapel, and urged them to persist with an evening service after their celebratory feast that night, so "earnest" was he "to have the going to chapel attended to."[40]

Whewell's alliance with Carus reveals much about his ambitions for chapel. A weighty Churchman within and beyond the college, Carus had started his career as curate to the Greek professor James Scholefield, who had been Charles Simeon's curate.[41] He had written himself into this evangelical line of succession, inheriting Simeon's position at Holy Trinity Church and editing his memoirs. Cherubic, if corpulent, and as yet unmarried, he drew female admirers to Holy Trinity, who doted on his every sermon. Carus made Holy Trinity the center of an international web of American Episcopalians, Prussian conservatives, and evangelical missionaries to the Jews.[42] He also had a following among godly undergraduates. When he married and left Cambridge to take up a living, some of the hundreds who had attended Sunday evening conversaziones in his Trinity rooms founded a Greek Testament prize (1854) in his honor, which is still awarded today. Carus was a pronounced Tory, so much so that he felt it prudent to clear up doubts over his loyalty to the Whiggish Victoria after her accession. Although historians have lingered over Whewell's ties with Coleridgean Romantics such as Hare and Thirlwall, it was with this staunch evangelical that he collaborated most closely on chapel discipline.[43] Dedicating his book of sermons to Carus, Whewell suggested that they could testify to the "devout demeanour of the Congregation, and of the religious thoughtfulness with which, so far as the human eye can judge, young and old alike, bring to such acts of worship" and hail the "long course of improvement in such matters" that they had "endeavoured to forward, with feelings of gratitude to God."[44] Carus's collaboration extended to Whewell's pious first wife Cordelia. He revered her for her attendance at chapel and collaborated with her on the Sabbath-keeping of college servants.[45]

Twenty years later Whewell would be even more confident of success, telling the undergraduates of 1861 that he detected in them a "temper of manly self-respect and of thoughtful regard for their position," a recognition that "life is a serious and mighty business," for which college was vital preparation.[46] In reality, the godliness of undergraduates was less a settled achievement than a problem requiring watchfulness. Whewell reproved undergraduates who persistently skipped Sunday services or committed such enormities as bringing a lit cigar into chapel.[47] Yet he viewed the chapel less as a disciplinary engine than a sacramental space whose power he conjured up in his preaching. In keeping with that attitude, he refused a donor's request to house Bertel Thorvaldsen's statue of the dissolute Lord Byron in the antechapel and sought to block the installation of the skeptical Lord Macaulay there too.[48] For both the Whewells, the chapel mattered most as the place in which they took Communion. In December 1841 Romilly remarked on the curiosity that Cordelia Whewell "staid [for] the Sacrament . . . I never before remember any Lady being present." She gave a cue to the college, with attendance on sacrament day high when the Whewells were present and dwindling otherwise. Whewell's priestly celebration of the sacrament was scrupulous, Romilly noting that he "observed the Rubric & consecrated the Elements "standing *before* the Table" with his back to the congregation."[49] That reflected his careful churchmanship. Sticklers been horrified when on their first visit to Cambridge Victoria and Albert sat in King's College Chapel with their back to the Communion table. Under Whewell, the sacrament was detached from rote discipline. From 1843 sacrament day was moved from the last Thursday of term to a Sunday and made a distinct rite, separate from the compulsory service that preceded it.[50] This reflected his loyalty to traditional churchmanship, for Trinity continued to limit the sacrament to once a term rather than offering it every Sunday, as real high flyers advocated.[51]

Because many of Whewell's sermons and certainly the majority of his published ones prefaced the celebration of the Eucharist, they repeatedly articulated its meaning, revealing his passionate conviction of Christ's real presence in the sacrament. In February 1842 he remarked that whereas "angels' food" had sustained the Israelites in the wilderness, they came to "feed upon that more precious manna by which, we trust, our bodies and souls are preserved even to everlasting life."[52] Participation in the Eucharist was a source of moral strength. On Michaelmas 1841 Whewell offered undergraduates—tyro Christians who came to college with "drops of baptismal dew . . . not yet dry upon your faces"—a "most thrilling call . . . to a more intimate union with Him, the great Head of our Christian body."[53] During Easter term 1842 he reminded them that through participation in the sacrament, "our hearts [were] purified, our souls elevated, our minds enlightened and our strength supported" for the performance of duty.[54] This was an abiding concern for him. In a sermon of April 1864 he rebuked undergraduates who "reject[ed] the invitation to the Table of the Lord," turning their backs on it and leaving chapel before Communion. In doing so there "arrives a slight put upon this Holy Ordinance, and a disturbance upon the solemn calm with which this service has hitherto been attended."[55] In conformity with the Prayer Book, Whewell directed the congregation to polish off any bread and wine that remained after Communion. That practice offended ritualist undergraduates, who regarded it as sacrilege against the real presence of Christ, and they responded by skipping Communion altogether—a display of priggishness that countered Whewell's vision of the Eucharist as a sacrament of togetherness.[56]

"Active, Diffusive Christianizing Charity"

Whewell accordingly entered his chapel as a priest rather than a natural philosopher. His sermons seem less interested in reconciling science and religion than in emphasizing their separate spheres. "There is a providential history of the world and a natural history of the world," he argued in one late sermon, epitomizing a lifetime's teaching. The "survey of nature" and the reading of human records allowed partial reconstruction of the latter, but the "providential history of the world we have given to us in our Bibles" shed a "light which reason alone, drawing her knowledge from the contemplation of nature, cannot give."[57] Whewell's sermons therefore devoted themselves primarily to the exposition of Scripture. Yet while insisting on its revealed truth, they also inserted its meaning into the bigger and deeper sweep of human history, reflecting Trinity's liberal Anglican understanding of past ages as a unity "full of the hand of God."[58]

Whewell embedded Scripture in history by asking how far the truths declared in the New Testament had been anticipated in the literatures of Greece and Rome. In doing so he voiced the understanding of classical philology developed by his Trinity friends Thirlwall and Hare. As William Thompson, the Greek professor and Whewell's successor as master, put it in a mid-century commemoration sermon, they had turned Cambridge classics from verse making into a searching branch of philology, which used "German materials" to recover the "religious mind of antiquity." Thompson put a conservative slant on their Romantic explorations, observing that the point of studying the encounter of the "unaided Grecian mind" with the "great problems of human nature" was to underline the limits of human reason.[59] Whewell did the same, musing in Easter term 1842 that while Greek literature did not "*answer* the mighty questions," it had offered "penetrating guesses" about the "Providential sway" of the world. He developed the point through an exposition of *Oedipus Colonnus*, a text chosen "for the benefit of the Freshmen"—even if Romilly, who heard the sermon, judged it too "hard and too metaphysical" for them.[60] In describing how Oedipus sought absolution for his crimes, Sophocles had shown how humans craved "renovation and purification" and were in the grip of "a superintending Providence, which holds in its hands the issues of life and death." His Eumenides, the "*All-seeing Gracious Ones*," were "images of a higher reality, the shadows of an eternal substance" that the Greeks could not name. Enter Paul, who had scriptural answers to Greek questions. In a flight of fancy Whewell imagined him stopping in the Grove of the Eumenides on his visit to Athens. He had evoked it in his speech at the nearby Areopagus, when he declared to the Athenians that they remained, notwithstanding their piety, ignorant of the God he now unveiled to them. Paul explained that what Sophocles had described as a "vague supernatural influence" was in reality the Living God of the Psalms and Proverbs.[61]

Paul knew that the Hebrew Scriptures stood to Greek thought as revelation did to reason, completing while surpassing it. This opposition, insistent in the published sermons, confirms Brooke's insistence that Whewell was a Pauline thinker for whom revelation surpassed all other paths to truth. While his exegesis demonstrated Hare and Thirlwall's interest in historicizing ancient texts, his anxiety to exhort his hearers caused him to argue that the force of any scriptural passage was not limited by settling what it originally meant to its writer and readers. In the late 1820s, Whewell had sympathized with Hugh James Rose's allegation that German biblical critics mangled revelation to fit their pinched sense of what was rational.[62] His sermons proclaimed the hermeneutic principle that Christ's sayings applied "not in the times of the Apostles alone, but in all times" until His second coming.[63] Whewell was always keen to understand a parable, such as the Prodigal Son, not in historical critical terms but as an elastic "narrative" or "story" that stretched to comprehend "human life as it has existed in all ages and countries": "Who does not know families in which such events have occurred?"[64] Whewell was just as adamant in closing the cultural or temporal gap between undergraduates and the Old Testament. He believed that Christ's appeal to the Old Testament gave it the same force for Christians as the New. Yet this "ray cast from the Gospel" not only lit up the Hebrew Scriptures but altered them, making them speak of Christ and his Church. Sometimes divines had dealt with objectionable passages in the Old Testament, such as the curses in David's Psalms, by dismissing them as remnants of Jewish tribalism. Whewell preferred to argue that after Christ's coming these verses became condemnations of Jews who rejected the Messiah. The argument had a homiletic rationale: his hearers should reflect that their unregenerate natures were just as "busy in the work of the crucifixion of Christ" as contumacious Jews had once been.[65] Whewell always subscribed to what he took to be Paul's teaching, which represented Christianity as "superseding" Judaism and transferring God's promises from the "Israel of men ... the collective body of those who were Jews outwardly" to the "Israel of God."[66] The spiritual Israel was easily confused with Britain. On Michaelmas 1843 he mused in preaching on Isaiah that, to modern Christians like the British, the nation was now their "terrestrial Zion." Rightly so: in coming to save all humanity, Christ did away with the need for a "separate people." The "earthly nationality of the Jews was rendered void," and their spiritual privileges were "repudiated, removed, destroyed," just as their temple had fallen.[67]

If Whewell's jealousy for the unity and authority of Scripture was rem-iniscent of Rose, a godfather to the Tractarian movement, then Ashworth goes too far in suggesting that he drifted into sympathy with his friend's sub-ordination of reason to Church authority.[68] As Tractarianism encroached on Cambridge, Trinity chapel remained a bastion of Protestant rationality. Indeed, Wordsworth's only reservation in recommending that Whewell re-place him as master was whether he had "sufficient staple of divinity" to fight Tractarianism.[69] Preaching at the first commemoration after Whewell's in-auguration, J. W. Blakesley recalled that he had sworn to be "ever be guid-ed by the essential principles of Protestantism," which involved testing all creeds and documents against "universal reason" and ranging "wherever the spirit of Man has left traces of itself" rather than fetishizing one "spiritu-al ancestry" as Tractarians did.[70] Whewell's preaching developed a similar position, warning that the defense of the "Eternal Truth" was incompatible with subjugation to a spiritual conqueror like the pope. While opposing any High Church embellishment of the chapel, Whewell paid in the mid-1840s for statues of Francis Bacon and Isaac Barrow in its antechapel, memorial-izing an "ancestry" for the use of reason in clarifying the grounds of faith that he and other fellows had long celebrated in commemoration sermons.[71]

Whewell's preaching became more outspoken once John Henry New-man's 1845 conversion made Tractarianism look like a stalking horse for Rome. In November 1846 there was "much hubbub" when a "weak minded" Trinity undergraduate converted to Roman Catholicism. In 1850 Cardinal Wiseman's declaration that England was returning to Rome provoked anger in the university and inflamed anxiety about Romish sympathies among the undergraduates. Carus, who had launched "one fling at the Puseyites" after another in Holy Trinity, now preached against "Jezebel," while Vice-Chan-cellor George Corrie masterminded the dispatch to the queen of a "dull and violent address" against the pope.[72] At this heated moment Whewell pub-lished a college sermon against the "Egyptian pomp and splendour" of Rome. At the Reformation the English had undergone an exodus to "a region of pure worship and reasonable service." Now was not the time to fall back into "slavery."[73] His contrast between the "ancient paths" of prayer and Scripture and Romish wiles was conventional.[74] But Whewell's battle against youthful vice meant that his polemic also had a pastoral dimension. Just as the young felt "something of a bewildering and intoxicating efficacy, which stimulates them, when gathered together, to especial acts of levity and extravagance," so those whose faith had warped "are vehemently bent upon drawing others into the state of mind in which they themselves are."[75]

In preaching that confidence in Scripture rested on a deeper commit-ment to truth itself as "fixed on a basis of rock; adamantine," Whewell could sound static. His insistence that truth "in its essential nature, is always the same" caused him to extol such superannuated disciplines as Euclidean geometry and to defend their teaching in the university.[76] As another Trinity fellow fretted in an 1857 commemoration sermon, outsiders thought of Trinity's commitments to its chapel and to Euclid as "old-world arrangements."[77] Yet if Whewell's epistemological commitments ossified, then his vision of Christianity in his later sermons remained in certain respects progressive. He continually defined the Church not as a fixed doctrinal deposit, but as a vehicle for "active, diffusive Christianizing charity" that had created the institutions he and his hearers inhabited and that needed to be reinvigorated to face contemporary challenges. "They not only preached Christ," Whewell liked to say of the apostles, "they Christianized the world."[78] The sermons showed little interest in defending or even exploring the creedal technicalities that impassioned Rose. In a late sermon on the Trinity, Whewell offered only lumpy circumlocutions of the dogma—a "compendious and significant definition of the Divine Nature"—before assuring undergraduates that its "practical application" was "what concerns us."[79] A practical application was a social one. In two occasional sermons to maritime and prison charities, Whewell urged their officers to stay true to the religious motives of their founders, which alone provided a "permanent" impulse to philanthropy.[80] The apostles had taught the first Christians to love one another as members of a "little household of faith" that held aloof from "a nation of strangers"— the pagan world around them. Over the centuries that had changed: "society itself now claims to be a Christian society," which meant that Christians answered for the treatment of its most marginal members.[81]

Whewell's definition of criminals in 1847 was as compassionate as it was verbose: "those carried to the edge of perdition by the working of the machinery of our society."[82] The sermons showed a nagging awareness that Christianity's prospects relied on the zeal that his undergraduates could show for those who fell between the grinding gears. When they left college and took orders, they must join the "company of amenders of the world," for as Romans 14:7 said, "None of us liveth to himself."[83] Whewell's preaching admittedly represented socialism as an anti-Christian yelp of rage. He was privately equally contemptuous of Thomas Carlyle's jeremiads on the condition of England and of Sir Robert Peel's efforts to link repeal of the Corn Laws to the Irish famine.[84] But he had long shared Hare's Romantic aversion to materialism and uttered full-throated denunciations of the rich who "pile up wealth in their storehouses, while the multitudes of the poor pine in nakedness and want, in filth and darkness." The inequalities of the market economy were a "new evil under the sun" for Christians to fight, for history showed that "Christianity is constantly labouring to bring the world into obedience to God and the world is constantly changing the mode of its resistance to this attempt."[85] As he twice married into money and speculated in railway shares, Whewell was an unlikely Jeremiah against capitalism. He was stronger on exhortation than solutions, putting his faith in a revival of church building to quell greed and class hatred by once more erecting tangible pointers to heaven.[86] Yet they must not wait passively for that heaven. His sermons preached a postmillennial eschatology, in which "thy will be done" should always be preceded by "thy kingdom come." Providence would collaborate with their striving for improvement, until the "whole earth" became a "shining passage to the eternal gates."[87]

"A Great Spiritual Power"

This concern with social fracture faded as the hungry 1840s receded. It is true that whenever the aging Whewell felt that society was relaxing its commitment to Christianization, he was stirred to eloquence. At St. Paul's Cathedral in January 1860, his exegesis of an innocuous passage from Luke turned into an attack on the cruelty with which the Indian rebellion had been suppressed and the use by young soldiers of racial slurs to describe their "fellow subjects, members of a race whose intellectual acuteness is as good as our own" and who had been civilized "at the time of Alexander of Macedon."[88] Yet in the next breath he urged the British to behave to "heathen people" as "older and stronger scholars" might do to the "younger and weaker" in a college—complacently modeling the world on the institution he headed. A year later Whewell encouraged his undergraduates to think of Britain's abolition of slavery as the "noblest national act of which history contains a record." The "philanthropic feelings" of the British needed defending against "sophists," just as Paul had exhorted the Galatians to "not be weary in noble doing," but their virtue was not in doubt.[89] Whewell's later preaching displays not just a complacency about institutions but the tendency to spy out dangers to belief that characterized his later publications in natural theology.[90] His sermons frothed against agnosticism, the Darwinian quest for human origins, and the authors of *Essays and Reviews* (1860). Their efforts reminded him of Zacchaeus, who had scrambled into a tree to spy on Jesus. He implored them to "come down from thy watchtower of self-complacency" and to mingle with ordinary Christians in accepting Christ as a "saviour," not just a "historical character."[91] He enjoined an awareness of life as evanescent—no less "temporary" and "short" in its way than an undergraduate degree—and warned of the need to be mindful of Christ's imminent return.[92]

Whewell's Tory vision of Trinity as a bastion of patriotic, philanthropic Christianity was too rigid to last. Yet it is important to distinguish the content of the sermons from their genre, which was enduring. Whewell's belief that undergraduates must go to chapel and that the master should preach to them was an unlikely survivor of the secularization of the university. Although reform of the college statutes in the 1880s threw open the mastership to non-clerics, the first appointee under the new statutes was Henry Montagu Butler, an undergraduate and then fellow of Trinity in Whewell's day, then a clerical headmaster of Harrow. Butler resembled Whewell and his successor William Thompson in being a godly classicist, who amused himself by translating Old Testament episodes into Homeric Greek.[93] His clerical friends urged that in Trinity the finely poised battle for "Christendom" could be won. Butler mused in accepting the offer that one of his aims must be to make the chapel "a great spiritual power."[94] He defended compulsory attendance by undergraduates—the expected level having fallen to twice on Sundays and twice in the week—down to 1913.[95] In the preface to his college sermons, he claimed that his role was to put "central Christian truths" before "young men from Christian homes and Christian ideals." A college pulpit might be no place to explore "theological or ecclesiastical difficulty," but he could "offer sympathy." Butler was wary alike of the new breed of lay fellows who were dismayed at his appointment and of the High Churchmen whose interest in liturgical innovation ran against his liberal Protestantism. He never felt that he had cracked the crust of undergraduate indifference.[96] Yet Butler's perseverance in Whewell's enterprise reminds us that Victorian faith was glued up in communities and institutions, whose power to shape habits of feeling and action could be persistent. It was the very power that Whewell's sermons in Trinity chapel had always celebrated.

CHAPTER 17

Whewell on Church and State, Science and Empire

JOHN GASCOIGNE

Reared in the traditions of the union of Church and State with which Cambridge was so deeply intertwined, William Whewell devoted himself to maintaining such an alliance in the face of rapid change.[1] His career straddled the great constitutional revolution brought about by the Great Reform Bill of 1832, and much of his energy was devoted to steering the university through the rough waters it brought in its wake. Whewell's attitude to change was similar to Robert Peel's, who spelled out his response to the post–Reform Bill world in his "Tamworth Manifesto" (addressed to his constituents at Tamworth) preparatory to becoming prime minister in 1834. While firmly opposed to change for its own sake—which he termed a "perpetual vortex of agitation"—Peel accepted the Reform Bill if it "impl[ied] merely a careful review of institutions, civil and ecclesiastical, undertaken in a friendly spirit." He did not, for example, support "the admission of Dissenters as a claim of right, into the universities." Crucially, he sought to reform the Church "to extend its sphere of usefulness, and to strengthen and confirm its just claims upon the respect and affection of the people."[2] Appropriately, it was Peel who appointed Whewell to the mastership of Trinity College (a position in the royal gift) in 1841, having been recommended by Bishop Charles James Blomfield of London, Peel's lieutenant for ecclesiastical reform, and Whewell's predecessor as master, Christopher Wordsworth. Indeed Wordsworth, as he told Whewell, deliberately deferred resigning until Peel's administration took office to increase the likelihood of Whewell being made master.[3]

Whewell was on sufficiently close terms to stay with Peel during the 1839 Birmingham meeting of the British Association for the Advancement of Science (BAAS)[4] and shared much of his political outlook. Like Peel, Whewell sought wherever possible to conserve traditional institutions by conceding moderate change. Such an approach was to color his wide activities and writings. Early in his career he had followed the Whig traditions of Trinity College in supporting limited concession to the position of the established church to reduce animus against it. Thus he had refused to sign a university address against opening the franchise to Roman Catholics in 1821. In the following year he complained to his sister of the election of John Bankes as an MP for the university with his "bigoted intolerance towards Papists."[5] Bankes's affirmation to follow the "most steady and decided opposition to any measure tending to undermine or alter the established church," however, won him election.[6] At the same time Whewell supported conserving reform within the university, being actively involved in 1822 in improving the examination system to "a very tolerable system of it as far as legislation is concerned."[7]

The approach of the Reform Bill further divided the university. In 1831 the senate carried by ninety-one to fifty-three a petition to the Commons to oppose the "sudden and sweeping changes" proposed in the bill. This was countered by a declaration of support for the sitting Whig members representing the university seat signed by Whewell along with his Whig scientific colleagues, Adam Sedgwick, John Stevens Henslow, and George Biddell Airy. Fear of reform, however, led to the defeat of those Whig reformers, Lord Palmerston and William Cavendish (the future Duke of Devonshire). Elected in their place in 1831 were two opponents of the Reform Bill, William Peel, Robert Peel's brother, and Henry Goulburn, a faithful lieutenant of Peel. For fifty years thereafter there were only four contested elections during which the members were Conservative.[8] Whewell's increasing wariness about political change led him largely to disengage from national politics and devote his formidable political skills to university politics.[9] His natural conservatism was evident in the comment by his friend J. W. Clark that "he had so strong a respect for existing institutions that he hesitated long before he could bring himself to sanction any change."[10]

Though he accepted cautious reform of the parliamentary system, Whewell was wary about change to the Church's position. The attempt in 1834 by his Trinity Whig colleague, Connop Thirlwall, to make attendance at the daily college chapel service voluntary he considered "inconsistent with any college management, and any religious establishment."[11] True to tradition he told Thirlwall that "more good is done by retaining our College Chapel service than would be done by abolishing it."[12] In one of his two 1834 pamphlets directed at Thirlwall, he spelled out at greater length the grounds for his opposition to the abolition of compulsory chapel—grounds that reflected his larger conception of the relations between Church and State. He saw the debate as going to the heart of the issue of "whether or not an Established Church shall be maintained"—a consideration that applied not only to England but "whether ministers shall be placed in every part of the empire." For "the tendency of such arguments [by Thirlwall], would be to make it seem, that religion ought to be entirely disconnected from all civil institutions" and "that an Established Church is contrary to the essence and nature of religion." On the contrary, asserted Whewell, an established church made religion a habit connected with the life of the nation, thus instilling religious devotion. Such considerations made Whewell cautious about allowing dissenters into Cambridge, given that "the Universities have hitherto been among the most important supports of the Church." If such a change were to occur it was vital that change be moderate and "that our commonwealth suffer no detriment in its most vital parts, its religious constitution."[13] In a subsequent pamphlet he affirmed the centrality of the union of Church and State, contending that his "predominant fears are not for the security of the Universities, but of the Church of England" believing "that Church to be the very heart of our social body."[14] He did not approve, however, of the master of Trinity, Christopher Wordsworth, demanding Thirlwall's resignation from a college tutorship[15]—a sentiment shared by many of his colleagues at that strongly Whig college.

Restoration dissenters could matriculate at Cambridge (in contrast to Oxford) but could not take degrees without subscribing to the Church of England. The resistance to changing this form of adherence to the established church was evident in the pamphlet on the subject written in the same year by the Regius Professor of Divinity Thomas Turton, whom Whewell described as an "excellent and valued friend."[16] In his view the issue of allowing dissenters to take Cambridge degrees without affirming membership in the Church of England was part of a larger conspiracy to disestablish the Church, with the universities being the place to start as a particularly vulnerable part of the constitution.[17] Such fears were magnified by the passage through the House of Commons of a bill to allow the admission of dissenters, though this was defeated in the House of Lords. Eventually parliament stepped in in 1856 and legislated to abolish religious tests for undergraduate degrees, though making concessions to the weight of clerical opinion by retaining such tests for the master's and higher degrees. Not until 1871, however, were dissenters allowed to be appointed as fellows or to hold university posts.[18] The clerically aligned universities were slow to change their practice to accommodate the post–Reform Bill order in Church and State.

Unlike many of his Cambridge allies, Whewell was hopeful that the essential bonds between Church and State could be maintained. In face of the forces of change Whewell expressed the view in 1835 that it was wise to "hope for the best. . . . Nobody can deprive us of the Church, if they would, for it has the affections of the people in its favour, and, even if it had no sup-port from Government, would be supported by the conviction of truth and the good it does to all."[19] When supporting Lord Lytteton for the position of high steward of the university in 1840, he reaffirmed his conviction that Church and State should continue to form a unity glorying in "the bless-ing of living in a Church such as ours, so interwoven with the constitution, so beneficial in its influence upon religion, morals, manners and politics." Such unity, he continued, should even extend to the politically fraught area of the Church's role in education.[20] In any event, Lytteton was defeated by Lord Lyndhurst, a friend of Sir Robert Peel.[21] Whewell more often took the view that change should be organic, being the result of natural evolution rather than revolution or even external reform. "I believe in the National Constitution and in our National Religion," he wrote in 1842, a year after becoming master, "not as mere formulae, but as living things, as the most essential part of the social and spiritual life of the nation. . . . I cannot believe in any Church or State which is not constantly unfolding, enlarging, and renewing itself." Such a condition fostered "a formative spirit which makes *reform* unnecessary."[22] The spirit of promoting cautious change led Whewell to lobby for the election of Prince Albert as chancellor in 1847. Keen to pro-mote the sciences, the prince actively supported Whewell's introduction of new triposes in the natural sciences and moral sciences in 1848. The oration marking the prince's election underlined the uneasy outlook of many mem-bers of the Church in urging him to "look carefully after the interests of the Church." Failure to elect the prince, wrote Whewell, would have "mark[ed] the resolution of the University and the Church to assume an attitude of suspicion and hostility towards the State."[23]

Science, Religion, the State, and the Empire

In Whewell's scheme of things science would play its part in maintaining the union of Church and State by illustrating the conformity of rational inquiry with religion. When defending Cambridge's emphasis on pure and applied mathematics in 1826, he remarked, "I am persuaded that there is not in the nature of science anything unfavourable to religious feelings."[24] In 1833 he had produced the first of the much acclaimed series of the Bridge-water Treatises, *Astronomy and General Physics Considered with Reference to Natural Theology* (1833), which followed in the tradition of Trinity's most famous son, Isaac Newton, in using current science to defend the religious establishment. So strong was the association of Newton with the defense of Christianity that Whewell sprang to his defense when he was accused of having wronged Flamsteed, decrying the way in which readers were called upon "to cast away all their reverence for the most revered name of our na-tion."[25] He was urged on by Stephen Rigaud, Savilian Professor of Astronomy at Oxford, who wrote to Whewell, "If Newton's character is lowered, the character of England is lowered and the cause of religion is injured."[26] The Bridgewater Treatises represented a high point in the history of natural theology that went back to Isaac Newton, Richard Bentley (a former master of Trinity), Robert Boyle, and beyond. The choice of Whewell as a contributor was made by the reformist bishop Blomfield of London, a former fellow of Trinity and Peel's ecclesiastical lieutenant. As Morrell and Thackray have shown, the broad church theology, fostered by works such as the Bridgewater Treatises, permeated the early BAAS.[27]

Whewell did, however, differ from some of his colleagues in the BAAS in his attitude to government involvement in the promotion of science. For Whewell traditional liberal ideology about maintaining limited powers of the State extended to the interplay between it and science. The effective founder of the association, William Harcourt, remarked in 1831, the year of the association's founding, that Whewell and those who thought like him were "crazy about the decline of science and hostile to Government doing anything about it."[28] Such a response was in tune with the reply that Harcourt received from Whewell to a proposal to lobby government for aid: that he would not support "any Association which had for one of its objects to influence Government in its proceedings with regard to science and its cultivators. I believe that, in England at least, men of science, as a body, will secure their dignity and utility best by abstaining from any systematic connection or relation with the government of the country, and depending on their own exertions."[29]

Such a conception of the State's role in the promotion of science was reflected in his monumental *History of the Inductive Sciences, from the Earliest to the Present Time* (1837), in which he had little to say about the role of the State apart from acknowledging the role of Louis XIV in founding the Paris observatory and the eighteenth-century sovereigns of Prussia and Russia in the founding of their scientific academies on the French model. State support for scientific expeditions also received some acknowledgment.[30] The *Edinburgh Review* drew attention to this lacuna, criticizing the way in which "Mr. Whewell has almost shunned the subject of scientific patronage and of national endowments for science; yet his peculiar opinions are betrayed even by his science." "The chivalry of science," the reviewer, David Brewster, continued, "must be incorporated by the State, and patronised by the Sovereign."[31] Whewell's suspicion of the hand of government extended to his opposition to the founding of the statistical branch of the BAAS, since such studies often involved political considerations.[32] Fear of being drawn into party politics lay at the root of his suspicion of a close alliance between science and the State, a fear that he shared with John Herschel. "One great danger in the present excited state of politics," Herschel wrote to Whewell about the association in 1831, when politics was overshadowed by the Reform Bill, "is its assumption of a political character, and the avowed object of 'influencing government' will of necessity set this string vibrating."[33]

But Whewell's attempt to draw a line between the workings of government and the goals of the BAAS was not altogether consistent. The proposal by David Brewster to provide "direct national provision for men of science" Whewell characterized in 1835 as an attempt "to get pensions from the government for our friends."[34] In the same letter he ridiculed the possibility of giving direct service to the government in the form of advice of the placement of railways or fisheries. Yet in the following year, at the request of a parliamentary inquiry, he gave advice on appropriate experiments to determine the best route for a railway.[35]

As president of the BAAS in 1841 Whewell compared the association with Francis Bacon's Salomon's House but emphasized the important difference that the association was not dependent on the State. He also urged the association to avoid political controversies.[36] Yet elsewhere he lauded the role of the State as a patron of science. As part of his attempt to persuade foreign states to cooperate in his study of tides, he urged a Belgian acquaintance to involve his own government. Such largesse would realize "patronage of science, and especially of astronomy, which is now the pride and the glory of the most civilized nations."[37] Later in his *Elements of Morality, Including Polity* (1845) he argued that the State had "the Duty of assisting and rewarding the progress of science and literature, as for instance by means of Universities, Observatories, Voyages and the like, a Duty of Intellectual Culture."[38] A number of Whewell's close friends were to benefit from patronage from a government that was becoming more sympathetic to the claims of science. Sir John Herschel became Master of the Mint in 1850, and when George Airy was made astronomer royal in 1835, Peel granted him a state pension acknowledging his services of science to the State. The pension, wrote Peel, was "dictated exclusively by public considerations . . . and will enable the King to give some slight encouragement to Science, by proving to those who may be disposed to follow your bright Example, that Devotion to the highest branches of Mathematical and Astronomical Knowledge shall not necessarily involve them in constant solicitude."[39] Another indication of the changing relations between science and the State was the conferring in 1845 on Whewell's close friend James David Forbes, professor of natural philosophy at Edinburgh University, a pension of two hundred pounds a year for the services he had rendered to science.[40]

Whewell's membership in the BAAS led to further involvement with government. In 1835, the same year he critiqued a proposal for government aid for scientists, Whewell was involved in a petition from the association to the State for support for a scientific expedition to the Antarctic. Whewell himself had an interview with the first lord of the Admiralty, who was not encouraging. Persisting with this call on the State, Whewell suggested that the government might be swayed by arguments about the way in which such expeditions increased naval skills "so that all the discoveries are pure gain."[41] He cautioned, however, that it should be a private approach lest the association be embarrassed by a public refusal. He himself was to sound out a former Trinity student, Thomas Spring-Rice, chancellor of the exchequer. Such lobbying was eventually successful when in 1839 the James Ross's expedition set off for the Antarctic.[42] The concentration of this expedition on magnetic variation led in turn to a chain of State-sponsored magnetic observatories around the globe.[43] The growing rapport between the association and the State was underlined in a letter of 1844 to Whewell from the great geologist Sir Roderick Murchison. Murchison rejoiced at the way in which the Association had provided seed money for projects later taken up by government. "This very year," Murchison wrote, "the government have taken up works *begun* by us to an extent of £1500."[44]

Whewell's involvement with the BAAS therefore drew him closer to the workings of the State. More and more the armed forces, the State's chief expense, depended on the advice of scientists with a consequent growth in bureaucracies such as the Admiralty's Hydrographic Service. It was this state agency that gave support to Whewell's main claim to original science, his work on tides between about 1833 and 1850.[45] This branch of science had already been opened up by John Lubbock, a former student of Whewell, who had focused on the movement of the tides in the Thames River and who continued to work with him. Whewell aimed at a more generalized project that would seek to arrive at general laws on the basis of an array of localized studies—a case study of the inductive method that Whewell saw as the heart of true science.[46] Government agencies such as the Navy and the Coast Guard provided much of the basic information, and the work of calculation to put such data in manageable form was largely conducted by officials from the Admiralty. Some of the funds for this were provided by the BAAS, but the Admiralty itself covered much of the cost.

Ties with the British State provided a platform for more international connections. Simultaneous observations around the British coast in 1835 were matched by similar observations on the coasts of North America, Spain, Portugal, France, Belgium, the Netherlands, Denmark and Norway.[47] When reporting this in his paper for the Royal Society's *Transactions*, Whewell gave thanks to the Board of the Admiralty that had "promoted [the project] with great zeal" as well as for the good offices of the Duke of Wellington as foreign secretary. The first lord of the Admiralty was given particular thanks for employing two extra clerks to undertake the calculations.[48]

The reach of the British State was lengthened by the growing scale of empire. The Cape Colony provided the site for an observatory for the southern skies. This was to be used by Whewell's close friend John Herschel during his time in South Africa, where he was on close terms with the astronomer royal there, Thomas Maclear. The year after his arrival in 1834 Herschel undertook tidal observations for Whewell, widening the scope of his study.[49] Whewell hoped that one informal arm of empire, the missionaries, might provide scientific data for the metropolis. In 1835 he voiced the hope that he could glean information on the scantily chartered Pacific and its tides and "I think the persons most likely to be able and willing to make such observations are the missionaries." In this instance such hopes were not realized since "our friends in New Zealand [by whom he probably meant the Church Missionary Society] will hardly begin our operations in time."[50] Though direct collection of results by missionaries was limited, Whewell did benefit from the close study of missionary publications provided by his sister.[51] Another arm of empire was the East India Company, which furnished its tidal records.[52]

In general, however, the empire does not loom large in Whewell's work, focused as he was on the traditions and connections provided by a largely "little England"—Cambridge. At the founding of the BAAS, he would have been party to the declaration that one of the goals of the association was "to promote the intercourse of those who cultivate science in different parts of the British empire" but this was slow to bear fruit.[53] On the few occasions he alluded to the subject of the empire, he was surprisingly critical of British rule. In the wake of the Indian uprising of 1847, he alluded in a sermon at St. Paul's in 1850 to the national hardness of heart evident in its treatment of India. The origins of such brutality, he suggested, lay in the public schools with their "demeanour of the stronger to the weaker."[54] Comments such as these led his friend J. W. Clark to reminisce that "he was more frequently heard to deplore the severity dealt out to the natives than to admire the heroism of their victims."[55] On the other hand, in an unpublished sermon of 1862, he saw some good in imperial policy as manifested in the abolition of slavery, describing it as the "noblest national act of which history contains a record."[56] In his *Elements of Morality* he appears to provide some encouragement for imperial settler societies in writing that, since "individuals cannot acquire Property in Land, except by the derivation from the State," it followed that "a Civilized State, in discovering a country of Savages, may take possession of it" but then dismisses the issue as something to be considered in the context of international law.[57]

Moral Philosophy and Defense of the Church

Preoccupation with the morals of Britain helps to explain his increasing interest in moral philosophy. The bulk of Whewell's tidology work was completed by 1840, as was his *Philosophy of the Inductive Sciences*, and thereafter his preoccupation with science loomed less large. True, in 1841 he served as president of the BAAS, but his waning interest was evident in his remark to Murchison in 1840 that he was gradually withdrawing "from the engagements of the material sciences."[58] After the expiration of his presidency in 1842, his role in the BAAS was less and less active.[59] His resignation of the chair in mineralogy in 1832 and election to the Knightbridge Chair of Moral Philosophy in 1838 had marked a change in the direction of his interests. He did, however, maintain some contact with the State's scientific establishment through his appointment as one of the visitors of the Royal Observatory at Greenwich in 1847. But along with moral philosophy, university matters and the defense of the curriculum absorbed more of his time. Consistent with his greater focus on the affairs of the Church after his appointment as master of Trinity in 1841, he devoted more time to philosophical issues that bore on the defense of Christianity. Natural theology still played a role with the publication of his *Indications of the Creator* (1845) as a counterblast to the evolutionary theorizing of Robert Chambers's anonymous and very controversial *Vestiges of the Natural History of Creation* (1841).[60] *Indications*, however, was not an original work, consisting as it did of extracts from his previous writings. His *Of the Plurality of Worlds* (1853) did use science to argue for the uniqueness of God's creation of humanity in the face of arguments for the peopling of other worlds in the vastness of the universe. But the publication of Charles Darwin's *On the Origin of Species* (1859) weakened some of the assumptions of a Paley-style natural theology, leading Whewell to remark in 1864 about arguments concerning the antiquity of man: "It is true that a reconciliation of the scientific with the religious views is still possible, but it is not as clear and striking as it was."[61]

Though Whewell's works continued to assume a divine designer, as Brooke has argued, he turned more to the book of Revelation than the book of Nature. His lectures and reflections on moral philosophy were chiefly embodied in his *Elements of Morality, Lectures on Systematic Morality* (1846), and *Lectures on the History of Moral Philosophy in England* (1852). These sought to explore "the consideration of God's workmanship as seen in our souls."[62] The human conscience became the terrain of the philosopher and the divine.[63] The need to emphasize the Christian dimension and reject the utilitarianism of William Paley, a standard author in the Cambridge curriculum, provided a particular impetus for such works. The study of philosophy as well as the study of nature, argued Whewell, was a support to religion since "no philosophy could be at all satisfactory to thoughtful men which has not a theological bearing."[64] As well as shoring up the philosophical bases of Christian morality, Whewell was active in defending the Church's temporal interests. As vice-chancellor in 1842–1843 he forwarded to the government a petition to keep the existing number of Welsh bishops in the face of proposed reforms to the Welsh Church.[65]

Moral philosophy, he asserted in his opening lecture, was "a true science" and for the moral philosopher the "whole history of the world is his books, his apparatus, his laboratory."[66] Exposition of moral philosophy included what he called "polity," the study of the State, hence the title of his main treatise on moral philosophy, *The Elements of Morality*. For Whewell the key point, which had been glossed over by previous English moral philosophers, was that the State "is necessarily conceived of as a Moral Agent." The State, then, needed to inculcate morality in its citizens, and since it could not do that without religion, "a recognition of it [religion] by the State is requisite."[67] This did not necessarily demand an established church, but Whewell saw such an institution as being the ideal form of a partnership with the State: "Where an Established Church exists it must be looked upon as one of the greatest of national blessings," for an established church brought with it "a position of equilibrium for the Relations of Church and State, in proportion as it is fully and completely established." Religion, for Whewell, meant revealed religion, since human reason could not arrive at the moral law that "is his [God's] command. Conscience is his voice." The need for revelation brought with it an explicitly Christian belief, since "the central point of these Revelations is the coming of Jesus Christ upon Earth."[68]

His emphasis on the merits of an established church led to criticism from those who wished to dismantle the Church of Ireland that had been imposed on the largely Catholic population. Hence one critic responded to his work by asserting that his "arguments go to defend the Church of Ireland, the most unjust of all the unjust establishments in the world."[69] The critic had certainly recognized the importance of an established church in Whewell's overall system of polity. In 1845 Whewell acknowledged in a letter to the Reverend Frederic Myers that the defense of an established church was "one of my primary objects in writing the book," while recognizing that "the union is now less close than it was." As in his *Elements of Morality* he concluded that "the polity of an established Church, when it is possible is far the happiest and the best fitted for all the noblest ends of a State's existence."[70] In his natural habitat as master of Trinity he was even more emphatic about the benefits of an established church. In a sermon in Trinity chapel in 1847 he dwelt on its benefits to the State, portraying it as the "direst of national calamities, any event which should shake these foundations."[71]

Whewell's emphasis on a God-given code of morality was deliberately intended to counter the utilitarian arguments of William Paley, whose textbook on morality was part of the curriculum. In response to Paley's moral criterion of what served "to promote Human Happiness," Whewell urged the need to "ask what is *right?*, not, what is *useful?*"[72] The same quest to promote Christian morality underlay his writings on political economy and its gloomy depiction of the lot of the bulk of the population in a free market economy. As early as 1830 Whewell published a paper that sought to discredit classical economics by treating it mathematically.[73] His critique of classical economics, and especially Ricardo, was most systematically laid out in his *Lectures on Political Economy* (1862). Such an attack, writes Yeo, was "part of Whewell's affirmation of the existing nexus between Church and University and the importance of a Christian framework for knowledge."[74]

It was within Cambridge, his natural base, that Whewell was most vigilant about preventing the hand of the State from disturbing the established church and its age-old alliance with the university. Seeing the probability of a royal commission as early as 1845, five years before it was established, he wrote that "such an interference from without, with the legislation of the universities would, I am fully persuaded, be productive of immense harm."[75] In the same year he stated that he could see "tolerably plain indications that the old Universities are not to expect a continuance of the protection that they have been accustomed to receive at the hands of the Government."[76]

Not that Whewell was totally against change in the university where it did not weaken its bonds with the Church and if it arose from within. What he favored was incremental change of the kind he had himself advocated. "During the whole of my long residence in the University," he wrote to the geologist Charles Lyell, "I have been constantly engaged in urging and attempting reforms."[77] As he wrote to government commissioners in 1858 defending the requirement that fellows take orders: "Constitutions are not made, but grow."[78] One change that Whewell successfully introduced was intended to make the university more useful to the Church by introducing the Voluntary Theological Examination in 1843. Given that the university provided little formal theological instruction, it was intended that this examination would be of service to future clergymen, and Whewell urged its merits to the bishops.[79] On the other hand, he was against more fundamental changes that reduced the clerical character of the university; relaxing the requirement for most fellows to be ordained, he argued, would result in losing that "reverent and religious tone which is of inestimable value to the nation."[80] In this instance, as in others, while reluctantly tolerating the establishment of the 1850 Royal Commission on Cambridge and the ensuing Parliamentary Commission of 1855, Whewell sought to prevent fundamental change. He was particularly sensitive to change in the college as opposed to the university. He had already shown how reluctant he was to alter the college's practices since, at the revision of Trinity's statutes in 1842, he had advocated only minimal change.[81] The commission's attentions to Trinity prompted him to fight back with a resolution that was backed by the other colleges, stating that they rejected any "measures which would tend to impair the existing connection between the colleges and the Church of England."[82]

As it happened, the Royal Commission's reforms were such as Whewell could accept: the requirements about fellows remaining celibate and taking orders remained, except for a few specified college tutors and university professors.[83] Other changes were the revision of the curriculum and the taxing of the colleges to support the central university and its professors. Whewell's grudging acceptance of the outcome was evident in a letter to James Forbes in 1863: "I do not think our Commissioners have done us much harm; but they done us some and they have failed altogether of the good they might have done."[84]

Whewell had been a leader in Cambridge University's holding operation to ensure most of its traditions were maintained—for the present at least. Concessions had to be made, but these only minimally affected what was of prime importance to Whewell and many of his colleagues: the maintenance of the links between the university and the Church. Whewell's activities were so various and prolific that it is easy to forget that his primary identity was that shaped by his Cambridge setting, with its defense of the constituted order in Church and State. Whewell could accept some of the changes wrought in Church and State by the post–Reform Bill polity provided these could be accommodated within the traditional matrix of a union of Church and State. Much of his writing on diverse topics had at its foundation the defense of a religious order, as was the case with his use of science to shore up natural theology or his exposition of the basic principles of morality, polity, or economics. Whewell also engaged with the growing role of the State in the post–Reform Bill era, using its expanded reach to support some aspects of science, even though this went against his traditional belief that science should make its own way in the world. Such support for science could be seen as the study of the book of Nature, which complemented God's revelation through the book of Revelation.

It was Whewell's achievement to make sufficient compromise to play a part in passing on in recognizable form the world in which he had been formed. Further concessions to the forces of modernization were largely evolutionary rather than revolutionary. As at Oxford clerical tutors were metamorphosed into dons,[85] the range of the curriculum expanded considerably as more triposes were offered, and the central university grew in significance, thus enabling the erection of facilities such as laboratories. But not all changes would have been to Whewell's liking: the 1871 University Tests Act meant that, apart from a handful of clerical positions, all university and college posts were open without having to take orders, and the late nineteenth century saw the coming of married fellows, thus weakening the established practice of surrendering a fellowship on marriage and, where possible, taking up a parish living. Such incremental changes meant that the university was less of an arm of the established church. But compromise yielded dividends: the Church survives to today, and Cambridge maintained its collegiate character and many of its traditions. Whewell also personified the changing relation between science and the State with his tentative calls on the State to promote research expanding in the second half of the nineteenth century. The many-faceted Whewell was to be both a champion of the old regime in Church and State and a promoter of conserving change. Janus-like he looked both to the past and the future.

CHAPTER 18

The Whewell Papers at Trinity College Library

DIANA SMITH

Trinity College, Cambridge, played a central part in William Whewell's life; in turn, Whewell ensured he would be remembered for playing a central part in the life of Trinity College. Most prominently, he designed and built a new court bearing his name, his coat of arms displayed on the outer walls, echoing the royal coats of arms on the Great Gate just across the lane. His will, which specified the donation of his bust and portrait that he had used as decoration in the Master's Lodge, also directed that one thousand volumes from his library be given to Trinity College Library.[1] What he may not have anticipated was that his personal papers would follow the volumes to the library and that they would form the nucleus of an impressive collection of personal archives of Trinity members. These range across many fields of study, with J. J. and G. P. Thomson's scientific archives, Srinivasa Ramanujan's mathematics notebooks, Ludwig Wittgenstein's philosophical papers, Lord Tennyson's poetry notebooks, and the political archives of Lord Houghton and Rab Butler. The William Whewell papers are one of the larger collections at Trinity, with over one hundred boxes of correspondence, writings, and subject files spanning the length of his career at Trinity and the breadth of his interests. The acquisition of the collection appears to be the first large collection of personal papers obtained by the library and represented a significant step forward in terms of depth of coverage and consequent interconnections to other papers and books already there. It was not a smooth path, however. The acquisition, care, and cataloging of a large collection of personal papers posed challenges that are reflected in the way the collection is arranged and presented to this day.

There is, disappointingly, no specific mention of Whewell's papers in his will. His bequests to Trinity College included books to be left on special shelves in the Master's Lodge, the bust and painted portrait, and the thousand volumes to the library. All other personal property was to be left to his sister Ann Whewell (later known as Ann Newton).[2] It seems likely that the papers joined the books as his sister's gift to Trinity; this argument is made stronger by the fact that Trinity librarian William Aldis Wright was in the post at that time. Wright served as librarian from 1863 to 1870 and was active in acquiring collections for the library during his tenure and after, while serving as Trinity's senior bursar, maintaining a strong interest in the library and its collections for the rest of his life. At his death in 1914 he bequeathed £5,000 to the library in addition to a collection of early printed books and over 150 Hebrew manuscripts,[3] his collection of Edward FitzGerald's papers, and his own papers.[4] His interest in the Whewell papers is mentioned in a letter from Trinity master W. H. Thompson to Trinity fellow and Whewell executor James Lempriere Hammond in November 1872: "The Seniority . . . would prefer that the papers should be examined before they leave the Library & this I hope would be found an easier task in consequence of the trouble the then Librarian, W Wright, took in classifying them when they were first placed there."[5] The date at which the papers were placed in the library is not known, but a letter from William Mathison, one of Whewell's executors, to Ann Newton's husband William in December 1867 includes his apology for not having yet undertaken "a series of good long searches in the Library among the papers of the late Master" for some letters written by the queen and Prince Albert that had been sent to Whewell during Lady Affleck's final illness. The letters from Thompson and Mathison raise the question of the attitude to the letters even while they were treated as Trinity property. Rare books and manuscripts were loaned to certain eminent scholars at the time, most famously Theodor Mommsen, whose house fire in November 1880 destroyed Trinity manuscript O.4.36.[6]

Ann Newton made it her mission to find an author for her brother's biography,[7] but was unsuccessful in finding members of Trinity College who were both available and healthy enough for such a task.[8] In the end the biography was divided into three parts, with the correspondence and literary remains in one part, a discussion of Whewell's academic career and Trinity College work in another, and a personal memoir as a third work. Isaac Todhunter, a mathematician and historian of mathematics, was asked to write a memoir about Whewell's writings.[9] Todhunter took pains in the preface to his *William Whewell, D.D., Master of Trinity College, Cambridge: An Account of His Writings with Selections from His Literary and Scientific Correspondence* (1876) to make clear his awareness of his lack of Trinity credentials and the difficulties of editing the works of a polymath: "I engaged in the work without any presumptuous hope that I was competent to appreciate the wide extent of learning for which Dr Whewell is so justly famous."[10] Todhunter's finest contribution to Whewell scholarship, however, had less to do with his analysis of Whewell's works and more with the arrangement and description of the Whewell papers in preparation for this work.

The preface proves to be a gold mine of information about the state of the papers, which were sent to Todhunter's house in Cambridge. He describes the arrival by installments of "a mass formidable on account of its extent and the confusion into which it was thrown," and comments, "It would be difficult to convey an idea of the hopeless disorder in which the papers were involved."[11] It is not clear what happened to Aldis Wright's "classification": whether it was lost in the packing or if it wasn't a sturdy enough framework to cover the extent of the papers. Todhunter describes Whewell's habit of pinning leaves of manuscripts relating to the same subject and his work in preserving and improving these groupings: "The manuscripts are now carefully sorted and catalogued; so that it will be easy henceforward for any specialist, if necessary, to consult all those belonging to the matter in which he may be interested."[12] Todhunter's respect for original order was ahead of its time. The archival profession today values original order as the key to much that is important: establishing provenance, patterns of use, significance of the material to the owner. The first handbook of archival practice would not be published for another twenty-five years, the *Handleiding voor het ordenen en beschrijven van archieven* (*Manual for the Arrangement and Description of Archives*), written by three Dutch archivists, Samuel Muller, Johan Feith, and Robert Fruin.[13] With another instinctual archival touch, Todhunter created a catalog of the collection, now shelved in the library.[14] Related lists in his hand are housed with groups of papers throughout the collection.

Whewell's niece, Janet Mary Douglas, the daughter of his brother-in-law John Marshall, agreed to write a personal memoir, and William Aldis Wright initially agreed to write the work covering Whewell's academic career.[15] The "pressure of other engagements" caused Wright to abandon this project, so this part of the biography was also given to Janet Douglas, with help from executor James Lempriere Hammond, who unfortunately died before the book was finished.[16] Douglas appears to have been a reluctant memoirist, referring in her introductory remarks to the proposal to publish a personal memoir from Whewell's letters: "The responsibility of making this selection was pressed upon me by Mrs. Newton in a manner which made me feel it impossible to refuse."[17] She also notes that Todhunter left materials collected for the "Academic Life" untouched, as they would be used for another publication.[18] A list in Douglas's hand indicates that William Aldis Wright sent her material, mostly subject files, in London.[19]

Another large group of materials Douglas worked with were letters written by Whewell to his father, aunt, and sisters, which were loaned to her by

Ann Newton.[20] Douglas printed many of them in her memoir, *The Life and Selections from the Correspondence of William Whewell, D.D., Late Master of Trinity College, Cambridge* (1881), published two years after Ann Newton's death. Douglas may well have kept the letters: she was Whewell's niece and one of his nearest surviving relations. The letters were out of view for some time, until they were discovered by W. G. Rimmer while researching the Marshall family, apparently in an outbuilding of the Marshall house in the Lake District.[21] As Douglas was a member of the Marshall family, this may not be entirely unexpected. There are mysteries, however: another box of early letters written by Whewell to his family dated 1815–1823 was apparently separated from the other family letters and offered to Trinity by bookseller Maurice Dobbs in 1982,[22] and a box of letters from William Whewell to Ann Whewell was found among the papers of Henry Montagu Butler, who became master twenty years after Whewell's death.[23] It is possible that the papers were given to the college while Butler was master and became mixed with his papers. Whewell's papers turned up in hidden places even at Trinity: on August 27, 1870, W. H. Thompson wrote to J. L. Hammond: "An unexpected treasure has revealed itself to one of our maids at the Lodge. She accidentally discovered the existence of two secret drawers, opening by a spring in a large sideboard or chiffonier which stands in the large drawing room." He describes the contents as a group of unbound copies of Whewell's translations with some sheet music and some lithographs.[24]

Finding Aids

Given the variety of sizes of the papers in the collection and their bulk, the Whewell papers may never have been stored together on the shelves in one dedicated space in the library, where it was the custom to mount or affix guard papers to individual manuscripts and small groups of material and bind them in book format for easy placement on the shelves. What is notable is that the material does not appear to have been deaccessioned or "weeded" after arrival; the papers Todhunter described are still in the library. Only Todhunter appears to have considered the papers in their entirety: once they returned to the library, boxes were separated from one another, perhaps because of the practicalities of storing a large collection and inexperience in handling such a large group of papers. They were gradually moved about the library and given new classification marks over the years, including marks for classes O and R in the Wren Library. In general, Whewell's writings and subject files were placed in Wren class R, while the correspondence files were placed in Wren class O. Eventually, most of the papers in class O were removed to the newer class of Additional Manuscripts,[25] although it is hard to determine whether this was driven by patterns of use or available storage space.

Todhunter's order and storage solutions were used for a considerable time after their return to the library. His handwritten "Catalogue of the Papers of William Whewell"[26] provided general descriptions of groups of material in a sequence running from A to Q. The first few groups were in boxes: "Dr Whewell's tin box," two tea chests, and a larger and a smaller wooden bin. The list appears to have followed Whewell's organization, rather than one imposed later. Over the years successive generations of finding aids were written, all essentially following Todhunter's initial arrangement. An omission in Todhunter's catalog became apparent in an undated index from the early twentieth century[27] that refers not only to material listed by Todhunter but mentions "inedited" correspondence filed in boxes in the Adversaria Class in the Lower Library. It is possible that these were the letters of academic life passed over deliberately by Todhunter because of his understanding that another volume would treat this aspect of Whewell's life. These letters were described in 1961 by Walter (later Susan Faye) Cannon,[28] who tellingly referred to "Box A" as an open packing crate. Three years later, possibly as a result of this report, these papers were rehoused and assigned to Additional Manuscripts a class. In 1973 the Royal Commission on Historical Manuscripts produced a printed finding aid of the Whewell papers at Trinity that included a list of correspondents in alphabetical order, pointing to their various locations. This listing does not make a distinction between Whewell's own papers and those with a different provenance.

Detailed cataloging was given priority twenty years later when a project to create summaries for each letter was undertaken by William Ashworth. This project was unfortunately never completed, unsurprising given the very large number of letters in the collection. While these summaries are doubtless valuable to researchers, it is possible that they give unintended weight to those records that carry them, in that it implies selection due to significance, which is not necessarily the case.

The papers may now be found in four different classes: Additional Manuscripts a, Additional Manuscripts c, and Wren Library classes O and R, with the result that the unity of the collection as a traditional personal archive has been obscured. It is hoped that the advent of the online archival catalog will alleviate some of these issues, with the ability to search across shelf marks by name, date, and keyword. A collection-level record has been made to bring together all the materials now cataloged in different locations with links to them for easier access.[29]

The Correspondence

The papers are housed in 114 boxes and contain material spanning from the sixteenth to the nineteenth centuries, with the bulk of the papers dating from Whewell's adult life, 1812–1866. They consist of correspondence, subject files, writings, diaries, other Whewell papers, family papers, and later papers of others that provide information about Whewell's life and work in a variety of different fields, including the history and philosophy of science, education, theology, mathematics, astronomy, etymology, politics, poetry, architecture, and Trinity College and Cambridge University business.

The correspondence, which is dispersed throughout the collection, consists of at least 7,000 letters, most of them received rather than written by Whewell. The 1973 finding aid[30] lists an estimated 1,850 correspondents, although this included Whewell material from all sources in the library at the time, not just those in Whewell's personal papers. Over 4,500 letters dated 1814–1866 appear in two main correspondence runs, and there are many smaller groups of letters as well. The subject files include another 1,000 letters arranged by topic, the family papers include another 1,460, and another 100 may be identified as the later papers of others. Not all of the letters in the last two groups were addressed to him. There are a few groups of Whewell's outgoing letters: some of these Whewell had retrieved during his lifetime; a very few letters from different sources were added to the papers after their arrival at the library. Papers by or relating to Whewell that were acquired from other sources are now cataloged separately; this category includes another large group of nearly 1,000 letters, most of them written by Whewell to his family and found on the Marshall family property in 1970.[31]

The correspondence provides a richly informative and often entertaining window into Whewell's varied mathematical, philosophical, scientific, and literary career. In addition, many of his correspondents were some of the most prominent and intellectually active men and women of his day. Michael Faraday appears, as does Mary Somerville, Henry Hallam, Sidney Smith, Robert Peel, Thomas Babington Macaulay, James Challis, James Stephen, Edward Everett, Heinrich Fick, David Livingstone, Francis Beaufort, David Brewster, W. H. Smyth, Charles Darwin, Richarda Airy, Charles Babbage, Alexander Beresford-Hope, David Brewster, Henry Goulburn, Ada Lovelace, and Prince Albert. The letters are a mix of everyday business and deep dives into mathematics and logic, inductive science, moral philosophy, political economy, the construction of hexameters, and many other topics. What may be less expected are the small evidences of ordinary life, and from Whewell himself: confiding to a friend in 1819, "Marianne is going to be married to that porpoise looking man the landlord of the Red Lion and I never trouble myself about her,"[32] and John Ruskin's dismissal of the plot of Goethe's *Hermann und Dorothea* that Whewell had translated: "The story is a sufficiently absurd one—for this or the last century. Prudent young men do not bring home a strolling girl and betroth themselves to her 24 hours after seeing her."[33] Interesting moments are captured: a letter from Michael Faraday approving the word *scientist* but disapproving of *physicist* ("The equivalent of three aspirate sounds of s in one word is too much"),[34] and the anonymous letter signed by [Captain] "Swing": "If you do not call in all the copies of your Mechanics which have promoted the building of Machines you shall hear further from Swing."[35]

Nearly 1,000 letters in the papers were written by seven of Whewell's close friends, relationships characterized by friendships formed by common intellectual endeavor. George Biddell Airy is represented by the largest number of letters, with another large group from Richard Jones, followed by over 150 letters each from James David Forbes and John Herschel, and smaller groups of letters from Julius Charles Hare, Adam Sedgwick, and Augustus De Morgan. These friends were both supportive and free with their opinions when they disagreed, for instance George Airy's bluntness in a letter of February 14, 1863: "When you say that Laplace's theory gives us no light which the [equilibrium] theory had not given before, it seems to me that there is a moral perversion; you think that success founded on false principles is at least as good as failure founded on true principles which are imperfect. . . . I must protest against such a judgement, in toto."[36]

Whewell appears to have been a deliberate collector of letters, perhaps for practical reasons: when he wrote memoirs of his friends, he was able to consult his own letters received and actively sought their papers to help him in his work. In these collecting habits he was not alone, as evidenced by the 285 letters kept by Richard Jones, the largest number of letters sent by Whewell to a friend to be found in the collection.[37] Whewell also acquired a group of his outgoing letters to Julius Charles Hare, dated 1818–1854,[38] but his correspondence with Michael Faraday was reconstructed after the fact, with a bound volume put together in 1932,[39] taking letters from Faraday in the Whewell papers and adding to them a separate acquisition of Whewell's letters to Faraday that had been donated to the library in the early twentieth century.[40] These contain much discussion of Faraday's work, in particular concepts he was struggling to find appropriate words to describe, and thanking Whewell for constructing words for him, such as *anode, cathode,* and *ion.*[41]

Trinity College and Cambridge University business abounds in the collection. A journal kept by Whewell recording his work as master includes drafts of over seventy letters from 1841 to 1853 and includes a draft of a letter for the librarian to send to J. O. Halliwell asking how he came to possess several manuscripts missing from Trinity Library to then sell them to the British Museum.[42] There are many letters relating to reforms, graces, revisions of statutes, as well as glimpses of student life. Projects undertaken by Whewell independent of his work in Cambridge and that required the input of others are well represented in the collection, particularly his work in tidal studies, or tidology. Nearly three hundred letters concern Whewell's experiments in tidal observations worldwide; these consist of discussions of arrangements for and interpretation of reports and include correspondence with naval officers and local officials.[43]

There are a very few letters received by Whewell's first wife Cordelia in the collection, and no letters written by William to her, but her collections of autographs and franks are present. These include the earliest dated item in the Whewell papers, a sixteenth-century letter from William Cecil, 1st Baron Burghley, to Robert Petre.[44] These autographs were often gathered from letters addressed to her parents, the linen manufacturer John Marshall and his wife Jane, the latter a childhood friend of Dorothy Wordsworth. Whewell's second wife, Lady Affleck, is better represented, with about one thousand incoming letters, many of them written by other women: her relatives and the wives and daughters of William's friends, including Richarda Airy, Frances Trench, Margaret Herschel, and Susan Myers. This is the largest group of letters written by women in the collection; the project in the 1990s to create summaries of letters did not reach these, and some of the correspondents in this group are not yet identified. Over two hundred letters were written by Lady Affleck's brother Robert Leslie Ellis, whose papers form a subset within the Whewell papers. These papers are discussed in *A Prodigy of Universal Genius: Robert Leslie Ellis, 1817–1859*, edited by Lukas M. Verburgt.[45] This correspondence is accompanied by letters written to other members of the Ellis family, including Henry Ellis, explorer and colonial governor of Georgia and Nova Scotia.[46]

As mentioned in the introduction above, there are nearly one thousand letters that are related to the Whewell papers but were not part of his personal collection of papers. These include over seven hundred William wrote to his family dating from 1811 to 1866.[47] This is by far the largest group of letters written by Whewell in the library, and while many of them were published in Douglas's memoir, not all were. His care for his aunt and sister Elizabeth is clear in the letters written at the time of their final illnesses and deaths, and his correspondence with his sister Ann continued to his death. The frequency and constancy of the letters provide a continual commentary on his personal and professional life that serves as a frame for the letters in his personal papers. This collection also includes letters written to Ann Newton by both of his wives, with over one hundred letters from Lady Affleck and nearly fifty from Cordelia Whewell.[48]

There is evidence that papers were removed from the collection before it reached Todhunter, most clearly the letters from the British royal family that were found and given to Ann Newton in 1873;[49] these letters are still missing from the papers today. There is also evidence of additions: letters not originally part of the papers, including a series of letters written by John Willis Clark describing Whewell's fall in March 1866, the progress of his illness, and death;[50] and a small group of letters written by J. L. Hammond, Janet Douglas, and Isaac Todhunter after Whewell's death, including a group of sixty-six letters received by Hammond in response to his appeal for letters from Whewell[51] that matches a group of letters given separately to the library; Isaac Todhunter's letters to Hammond that discuss letters sought and received from Whewell's friends.[52]

Writings, Subject Files, and Other Papers

Isaac Todhunter's memoir provides details on the writings to be found in the papers and is a good record of the collection at that time. The papers contain drafts of some but not all of Whewell's works, and some of these include associated material. *The Elements of Morality, Including Polity* (1845) is represented particularly well, with notes and letters, drafts and proof sheets.[53] *The History of the Inductive Sciences* (1837)[54] and *The Philosophy of the Inductive Sciences* (1840)[55] are represented by drafts and notes. Drafts of over one hundred sermons are housed together,[56] with a separate group of drafts and page proofs for *Sermons Preached in the Chapel of Trinity College, Cambridge* (1847).[57] There is a variety of other writings in the papers, including drafts of the *Platonic Dialogues for English Readers*[58] and a draft of a paper on the subject of communication between the earth and the moon that Todhunter published in his memoir.[59] Hexameter poetry gathered for *English Hexameter Translations from Schiller, Göthe, Homer, Callinus, and Meleager* (1847) includes work by George Biddell Airy and John F. Herschel,[60] and a group of miscellaneous poems includes a copy of "The Worship of This Sabbath Morn" in Dorothy Wordsworth's hand[61] and an "Account of a Charade on the Name of the Learned Professor Whewell as Represented at Castle Ashby December 29 1837."[62]

Researchers investigating the writings should also consult the online book catalog at Trinity College Library, as the distinction between what was considered printed material and manuscript is often blurred: proof sheets of books may be found cataloged as if they are in their final form. These may be found by searching for Whewell as Provenance in the Advanced Search of the book catalog.[63]

The collection also contains subject files comprised of letters, notes, writings, and printed matter covering twenty different subjects. Not all materials of a similar nature were kept together, and so some subject files are now spread across multiple boxes. As mentioned above, there are over one thousand letters in these files, most of them incoming, but with some drafts of Whewell's letters as well. The biggest files relate to Whewell's work on etymology,[64] tidology, hexameters, and royal visits to the University of Cambridge. The etymology papers date from early in Whewell's career and may have been kept with an intention to return to the subject later in life. The tidology papers include a large file of tide tables and notes from all over the world, dating from the 1830s and 1840s, and is accompanied by another file of drafts of writings on tides, notes, and letters.[65] Such materials may be expected to comprise his papers, but less expected may be the presence of his vice-chancellor's notebook and cash book for 1842–1843,[66] which he appears to have considered personal property, much as tutorial files were considered the property of tutors.

The definition of *diary* is stretched in this collection, for beside six diaries dated 1820–1840[67] and a journal of Whewell's daily work as master of Trinity, 1841–1853,[68] there are also notes on books read with a record of the date the notes were taken. These were previously identified solely as diaries, and while they may be used by researchers to understand work done on a certain date, there is nothing about his movements and appointments. These are dated 1817–1830 and 1841–1853.[69] Other Whewell papers include account books, memorandum books, documents, sketches and sketchbooks, and printed material. Many of the thirty-three sketchbooks are pocket-size with accomplished drawings and sketches of architectural details, with extensive notes on the architecture sketched.[70] The account and memorandum books date from the 1820s and 1830s.[71]

The modern manuscripts and archives at Trinity have benefited from an on-line archival catalog, and the ability to see what used to be handwritten and printed finding aids online makes clearer what work is needed to improve access. In recent years some boxes of correspondence have been digitized with links to these provided in the archival catalog, and it is possible that this will continue. More work is planned to further catalog other collections at the library, and it is logical to predict that more Whewell material will be found in those collections. In addition, the Trinity College Archive will soon be given its own online archival catalog, and this will doubtless turn up unexpected treasures, given Whewell's long association with the college.

Notes

List of Abbreviations

The following abbreviations are used in the notes.

BAAS	British Association for the Advancement of Science
HJR	Hugh James Rose, Anglican priest and theologian.
JCH	Julius Charles Hare, theological writer and archdeacon
JDF	James David Forbes, Scottish physicist and glaciologist
JH	John F. W. Herschel, mathematician and astronomer
OED	*Oxford English Dictionary*
ODNB	*Oxford Dictionary of National Biography*
RJ	Richard Jones, political economist
JDFP, SAUL	James David Forbes Papers, St. Andrews University Library
TCL	Trinity College Library, Cambridge
WW	William Whewell, polymath

Introduction. William Whewell, Victorian Polymath

I would like to thank Christopher Stray and Bernard Lightman for their valuable comments on earlier versions of this introduction.

1. In this introduction, references to primary and secondary sources are kept to a minimum, as fuller references to the relevant literature will be provided in the notes of individual chapters. The story it tells of Whewell's life and work is based largely on two sources, in addition to archival and primary materials: Becher, "William Whewell's Odyssey"; and *ODNB*, s.v. "Whewell, William," by Richard Yeo, ed. Lawrence Goldman, accessed May 1, 2023, https://doi.org/10.1093/

ref:odnb/29200. Unlike in previous Whewell scholarship, in the present volume the abbreviation *WP* is not used to refer to items from Whewell's papers, held in Trinity College Library, Cambridge. Normally within collections of papers the shelfmark is dependent on the collection it is in. That is not the case here, where the shelfmark is an independent number, not a subsidiary number within Whewell's papers. I am thankful to Diana Smith for pointing this out to me.

2. There currently exists no full-length intellectual biography of Whewell's life and work. See Christopher Stray's chapter in the present volume for an overview of the available literature.

3. WW to John Whewell, June 26, 1814, quoted in Douglas, *Whewell*, 12; emphasis added.

4. WW to Elizabeth Whewell, October 1, 1817, Add.Ms.a.273.16, TCL.

5. The classic study of Whewell as a pillar of the early Victorian scientific community and a "gentleman of science" is Jack Morrell and Anthony Thackray, *Gentlemen of Science: Early Years of the British Association for the Advancement of Science* (Oxford: Oxford University Press, 1981). See Heather Ellis's chapter in the present volume for the gendered nature of this scientific body.

6. WW to James Garth Marshall, December 27, 1842, quoted in Douglas, *Whewell*, 281, 282–83.

7. WW to JCH, October 15, 1838, quoted in Todhunter, *Whewell*, 2:271.

8. WW to James Garth Marshall, December 27, 1842, quoted in Douglas, *Whewell*, 280, 282; emphasis added.

9. Michael Ruse, "William Whewell: Omniscientist," in Fisch and Schaffer, *Whewell*, 87.

10. See Diana Smith's chapter in the present volume for these numbers.

11. WW to Rev. G. Morland, December 1, 1815, quoted in Todhunter, *Whewell*, 2:10.

12. See, for instance, William Whewell, notes on books read, 1817–1840, R.18.9; and "Notes on books read, Oct–Dec, 1817," R.18.16/1, both in TCL.

13. See WW to JCH, October 15, 1838, quoted in Todhunter, *Whewell*, 2:271.

14. This is part of the title of Becher's chapter in Fisch and Schaffer, *Whewell*.

15. See Tony Crilly's and Ben Marsden's chapters in the present volume.

16. See Harro Maas's chapter in the present volume.

17. See Edward Gillin's chapter in the present volume.

18. See Michael Reidy's chapter in the present volume.

19. WW to JDF, April 2, 1838, quoted in Todhunter, *Whewell*, 2:269.

20. WW to R. I. Murchison, September 18, 1840, quoted in Todhunter, *Whewell*, 2:286.

21. WW to JH, November 1, 1818, quoted in Todhunter, *Whewell*, 2:29.

22. WW to RJ, August 16, 1822, quoted in Todhunter, *Whewell*, 2:48; emphasis added.

23. WW to JH, April 9, 1836, quoted in Todhunter, *Whewell*, 2:235.

24. The word *metascience* was not used by Whewell himself, who in this period variously referred to himself as a "system-builder" and "critic." It was introduced by Richard Yeo in *Defining Science*. For *induction* and *theology*, see, for instance, WW to RJ, July 23, 1831, quoted in Todhunter, *Whewell*, 2:124.

25. See Bernard Lightman's chapter in the present volume.

26. This is the twentieth-century term used by Jonathan R. Topham to describe Whewell's 1833 book; see Topham, *Reading the Book of Nature: How Eight Best Sellers Reconnected Christianity and the Sciences on the Eve of the Victorian Age* (Chicago: University of Chicago Press, 2022).

27. *Astronomy and General Physics Considered with Reference to Natural Theology* (London: William Pickering, 1833), 324, 340.

28. Whewell's sermons are the subject of Michael Ledger-Lomas's chapter in the present volume.

29. See WW to JH, March 6, 1817; and WW to RJ, February [?], 1831, both quoted in Todhunter, *Whewell*, 2:16, 115; emphasis added.

30. WW to RJ, October 6, 1834, quoted in Todhunter, *Whewell*, 2:193. See the items under William Whewell, "Notebooks and drafts of a history of the philosophy of science, 1820–1860," R./18.17, TCL.

31. WW to RJ, October [?], 1825, quoted in Todhunter, *Whewell*, 2:61.

32. WW to JH, April 9, 1836, quoted in Todhunter, *Whewell*, 2:234.

33. WW to RJ, October [?], 1826; WW to RJ, October [?], 1825, both quoted in Todhunter, *Whewell*, 2:72, 61.

34. For this terminology, see Peter Burke, *The Polymath: A Cultural History from Leonardo Da Vinci to Susan Sontag* (New Haven, CT: Yale University Press, 2020), 6, 148, 6. For "fiddle faddle," see WW to RJ, February 4, 1829, quoted in Todhunter, *Whewell*, 2:97. Burke defines the "centripetal" polymath as one "who has a vision of the unity of knowledge and tries to fit its different parts together in a grand system," and the "clustered" polymath as someone who juggles "several subjects more or less simultaneously."

35. William Whewell, *The History of the Inductive Sciences, from the Earliest to the Present Time*, 3 vols. (London: John W. Parker, 1837), 1:xii.

36. WW to RJ, November 13, 1833, quoted in Todhunter, *Whewell*, 2:172.

37. WW to RJ, October 21, 1833, quoted in Todhunter, *Whewell*, 2:171.

38. WW to JH, April 9, 1836, quoted in Todhunter, *Whewell*, 2:235; emphasis added.

39. WW to RJ, October 21, 1833, quoted in Todhunter, *Whewell*, 2:171.

40. WW to RJ, July 27, 1834, quoted in Todhunter, *Whewell*, 1:90.

41. Whewell, *History*, 1:viii.

42. See Aleta Quinn's chapter in the present volume.

43. See my chapter on Whewell's history and historiography of science in the present volume.

44. Whewell, *History*, 1:viii.

45. William Whewell, *The Philosophy of the Inductive Sciences, Founded upon Their History*, 2 vols. (London: John W. Parker, 1840), 1:48.

46. Whewell, *History*, 1:ix.

47. WW to Sir M. I. Murchison, September 18, 1840, quoted in Todhunter, *Whewell*, 2:286; original emphases.

48. Whewell, *Philosophy*, 1:xii; WW to JH, April 9, 1836, quoted in Todhunter, *Whewell*, 2:235; original emphasis.

49. WW to JDF, February 20, 1837, quoted in Todhunter, *Whewell*, 2:251; WW to RJ, July 27, 1834, quoted in Todhunter, *Whewell*, 1:90.

50. Todhunter, who first made this claim, put forward the evidence in chapters 8 and 10 of his *Whewell*, vol. 1.

51. Whewell, *History*, 1:xi.

52. See Sheldon Rothblatt's chapter in the present volume.

53. WW to Rev. H. Wilkinson, December 26, 1837, quoted in Todhunter, *Whewell*, 2:266.

54. WW to JH, April 22, 1841, quoted in Todhunter, *Whewell*, 2:298.

55. On the Mill–Whewell debate, see Snyder, *Reforming Philosophy*.

56. John Stuart Mill, "Dr. Whewell's Moral Philosophy," in *Collected Writings of John Stuart Mill*, ed. J. M. Robson (Toronto: University of Toronto Press, 1969 [1852]), 10:168; emphasis added.

57. John Stuart Mill, "Autobiography," in *Collected Writings of John Stuart Mill*, ed. J. M. Robson (Toronto: University of Toronto Press, 1981), 1:269–70.

58. WW to JH, April 8, 1843, quoted in Todhunter, *Whewell*, 2:315.

59. WW to RJ, December [?], 1826; WW to RJ, August 16, 1822, both quoted in Todhunter, *Whewell*, 2:81, 49.

60. Whewell, *Philosophy*, 2:iv, 297. The reference is to Adam Sedgwick's famous *A Discourse on the Studies of the University of Cambridge* (Cambridge: Cambridge University Press, 1833).

61. William Whewell, "Notebook," n.d., 12–13, R.18.17/8, TCL.

62. See David Phillips's chapter in the present volume.

63. William Whewell, *Lectures on the History of Moral Philosophy in England* (Cambridge: Deighton, Bell, 1852), xviii.

64. Compare William Whewell, *On the Philosophy of Discovery: Chapters Historical and Critical. Including the Completion of the Third Edition of the Philosophy of the Inductive Sciences* (London: John W. Parker, 1860), 344; Whewell, *The Elements of Morality, Including Polity*, 2 vols. (Cambridge: J. J. Deighton, 1845), 1:307.

65. Burke defines the "centrifugal" polymath as one who accumulates knowledge without worrying too much about connections. See Burke, *Polymath*, 6.

66. On these and other aspects of Whewell's classificatory efforts, see Aleta Quinn's chapter in the present volume.

67. See James Clackson's chapter in the present volume.

68. Todhunter devoted an entire chapter to Whewell's work on English hexameters. See Todhunter, *Whewell*, vol. 1, chap. 15. For a more recent discussion of the English hexameter movement, see, for instance, Daniel Brown, "Mathematics and Poetry in the Nineteenth Century," in *The Palgrave Handbook for Literature and Mathematics*, ed. Robert Tubbs, Alice Jenkins, and Nina Engelhardt (Cham, Switzerland: Palgrave Macmillan, 2021).

69. See William Whewell, "Review of Henry Wadsworth Longfellow, *Evangeline, A Tale of Acadie*," *Fraser's Magazine* 37 (1848): 298.

70. William Whewell, *The Platonic Dialogues for English Readers*, 3 vols. (Cambridge: Macmillan, 1861), 3:v; Whewell, "Of the Platonic Theory of Ideas: Paper Presented to the Cambridge Philosophical Society, November 10, 1856," *Transactions of the Cambridge Philosophical Society* 10 (1857): 94.

71. WW to JDF, February 29, 1860, quoted in Todhunter, *Whewell*, 2:419.

72. On Whewell's engagements with geology and palaetiological sciences, see Max Dresow's chapter in the present volume.

73. Samuel Taylor Coleridge, *On the Constitution of the Church and State*, 2nd ed. (London: Hurst, Chance, 1830), 7.

74. J. B. Lightfoot, *In Memory of William Whewell, D.D. Master of Trinity College, Cambridge. A Sermon Preached in the College Chapel, on Sunday, March 18th, 1866* (London: Macmillan, 1866), 18.

75. J. W. Clark, "Half a Century of Cambridge Life," *Church Quarterly Review* (April 1882): 169.

76. Menachem Fisch and Simon Schaffer, "Preface," in Fisch and Schaffer, *Whewell*, vi.

77. Fisch and Schaffer, "Preface," vii.

78. See Susan F. Cannon, *Science in Culture: The Early Victorian Period* (New York: Dawson and Science History Publications, 1978); Laura J. Snyder, *The Philosophical Breakfast Club: Four Remarkable Friends Who Transformed Science and Changed the World* (New York: Broadway, Crown, 2011).

79. Ashworth, *Trinity Circle*. See also Lukas M. Verburgt, ed. *A Prodigy of Universal Genius: Robert Leslie Ellis, 1817–1859* (Cham, Switzerland: Springer Nature, 2022).

80. See Jack B. Morrell, "Individualism and the Structure of British Science in 1830," *Historical Studies in the Physical Sciences* 3 (1971).

81. During the writing and editing of this introduction, the situation in Herschel and Babbage scholarship has been changing. On Herschel, see Stephen Case and Lukas M. Verburgt, eds., *The Cambridge Companion to John Herschel* (Cambridge: Cambridge University Press, 2024); and Stephen Case, *Creatures of Reason: John Herschel and the Invention of Modern Science* (Pittsburgh: University of Pittsburgh Press, 2024). *The Cambridge Companion to Charles Babbage* is forthcoming in 2025.

82. See John Gascoigne's and Heather Ellis's chapters in the present volume.

1. Whewell's Early Life and Education

This chapter has benefited from the advice of Rosalind Eyben (Brighton), Lyda Fens-de Zeeuw (Leiden), Paul Andrew, Gordon Clark, Jenny Cornell, Keri Nicholson, Andrew White and Michael Winstanley (Lancaster), Christopher Langmuir (Seville), Simon Hornblower and Chris Pelling (Oxford), James Clackson, Diana Smith, Jonathan Smith, and Gill Sutherland (Cambridge).

1. Douglas, *Whewell*; Todhunter, *Whewell*. As their prefaces show, the prehistory of the biographizing of Whewell was littered with deaths and refusals. For the tripartite plan, see "William Whewell," *Saturday Review*, May 28, 1881, 690–91, and J. W. Clark, *Old Friends in Cambridge and Elsewhere* (London: Macmillan, 1900), 1–2, who commented, "we assert most distinctly that Dr Whewell was the last man whose biography should have been so treated. His life . . . presented a singular unity."

2. Menachem Fisch and Simon Schaffer, "Preface," in Fisch and Schaffer, *Whewell*, x–xi. "Literary . . . science(s)" is a curious phrase, perhaps formed on the model of the Cantabrigian title "moral sciences."

3. Harvey Becher, "Whewell's Odyssey: From Mathematics to Moral Philosophy," in Fisch and Schaffer, *Whewell*.

4. "Haversham" for "Heversham": Becher, "Whewell's Odyssey," 2.

5. Since then the situation has changed; see, for example, Gillian Beer, *Darwin's Plots: Evolutionary Narrative in Darwin, George Eliot and Nineteenth-Century Fiction* (Cambridge: Cambridge University Press, 1983); Alice Jenkins, *Space and the "March of Mind": Literature and the Physical Sciences in Britain, 1815–1850* (Oxford: Oxford University Press, 2007); Ben Marsden, Hazel Hutchinson, and Ralph O'Conner, eds., *Uncommon Contexts: Encounters between Science and Literature, 1800–1914* (London: Pickering & Chatto, 2013).

6. The implication was that the paper came from sewage; it was probably torn from wrappings. Newspapers, or books (toilet paper was invented only in 1857, in the United States). For the Whewell anecdote, see Gwen Raverat, *Period Piece: A Cambridge Childhood* (London: Faber and Faber, 1952), 34, quoted in Yeo, *Defining Science*, 18; and in Leah Price, *How to Do Things with Books in Victorian Britain* (Princeton, NJ: Princeton University Press, 2012), 9. For most of the nineteenth century the River Cam was an open sewer, and undergraduates with rooms next to the river kept their windows closed to keep out the stink. Drainage mains were not installed until 1894, when a pumping station was built. The account of Victoria's visit in J. W. Clark and T. M. Hughes, *The Life and Letters of Reverend Adam Sedgwick*, 2 vols. (Cambridge: Cambridge University Press, 1890), 2:59–64, does not mention the story.

7. Charles A. Bristed, *Five Years at an English University*, 2 vols. (New York: G. Putnam, 1852), 1:99; compare the annotated edition: Christopher A. Stray,

ed., *An American in Victorian Cambridge: Charles Astor Bristed's "Five Years in an English University"* (Exeter, UK: University of Exeter Press, 2008), 73; F. Pollock, *The Land Laws* (London: Macmillan, 1883), 13. Both Bristed and Pollock date the anecdote to Whewell's time as tutor at Trinity, 1823–1839.

8. For example, see the story of his confounding a group of Trinity colleagues who attempted to crush him by reading up on Chinese music in an encyclopedia, only to find that he had written the relevant article: F. Espinasse, *Lancashire Worthies*, 2nd series (London: Simpkin Marshall, 1877), 368–69. J. M. F. Wright, who was at Trinity from 1814 to 1819, recalled that Whewell had "gone through the Encyclopedia Britannica, so as to have the whole at his 'fingers' ends'"; Wright, *Alma Mater, or, Five Years at the University of Cambridge*, 2 vols. (London: Black, Black and Young, 1827), 1:212. An annotated edition edited by Christopher A. Stray is: *Student Life in Nineteenth-Century Cambridge: John Wright's* Alma Mater (Exeter: University of Exeter Press, 2023).

9. Whewell's parents were married in Lancaster on August 10, 1793. Of his three brothers, two died in infancy and a third, born in 1803, died in 1812. His eldest sister, Elizabeth, born in 1797, died in 1821. Two other sisters, Martha and Ann, both married. Douglas, *Whewell*, gives Martha's husband's name first as (Rev. J.) Staller (2), then as "Statter" (591); in fact she married "James Slatter" on October 28, 1824 (https://www.lan-opc.org.uk/Lancaster/stmary/marriages_1824–1827.html). The Lancashire parish register has "Statter:" James Statter was as ordained deacon in 1830 and made stipendiary curate in the same year, and priest in 1831. Ann married William Newton, a widower, in 1865 after the death of his first wife, her cousin Sarah Bennison. See letters from William #Whewell to Ann Newton, 1845–1850, Add.Ms.c.190/1–58, TCL.

10. "In a court lower down Brock Street" according to Richard Owen (Douglas, *Whewell*, 3); Lucy Street according to A. L. Murray, *A Biographical Register of the Royal Grammar School, Lancaster* (Cambridge: W. Heffer, 1955), 19. A 1909 photograph by William Sumner of what was identified as Whewell's birthplace is held in DDX/2743/11/3, Lancashire Archives; it shows a terraced house with what may be a workshop next to it. In his *Historic Notes on the Ancient Borough of Lancaster* (Lancaster: Eaton and Bulfield, 1891), 241, "Cross Fleury" [R. E. K. Rigbye] gave Whewell's birthplace as 16 Lucy Street. John Whewell's will mentions property he owned on Brock, Lucy, and George Streets, all three being parallel and adjacent (Act upon probate, July 2, 1816, WRW/A/R140/35, Lancashire Archives). A lot in Lucy Street had been bought by William Whewell (presumably our William's grandfather) in 1783.

11. For "carpenter and builder," see the entry on Whewell in Anthony C. Grayling, Andrew Pyle, and Naomi Goulder, eds., *The Continuum Encyclopedia of British Philosophy*, 4 vols. (Bristol: Thoemmes Continuum, 2006), 4:3393. The Whewells' marriage record refers to him as a "House Carpenter," a phrase that then referred to a carpenter who built the woodwork of a house, especially its

timber framework (*OED*, s.v. "house," https//:www.oed.com, C10). This helps to explain "carpenter and builder": in the late eighteenth century many houses were built with timber frames.

12. See the entry on Whewell in Grayling, Pyle, and Goulder, *Continuum Encyclopedia*, 4:3393. Laura J. Snyder calls him "a house carpenter and joiner with a workshop employing one or two journeymen" (Snyder, *The Philosophical Breakfast Club: Four Remarkable Friends Who Transformed Science and Changed the World* [New York: Broadway Books, 2011], 11). This may rely on Sheldon Rothblatt's description of John Whewell as "a master carpenter . . . who very likely employed a journeyman or two"; Sheldon Rothblatt, *The Revolution of the Dons: Cambridge and Society in Victorian England* (London: Faber & Faber, 1968), 35. If so, it exemplifies the transformation of possibility into fact to make a better story.

13. W. G. Clark, "William Whewell. In Memoriam," *Macmillan's Magazine*, April, 1866, 545.

14. W. G. Clark, "William Whewell," 545. The *Ladies' Diary* appeared annually from 1704 to 1840. It became well known for its mathematical problems, contributed by women and men alike. See J. Albree and S. H. Brown, "'A Valuable Monument of Mathematical Genius': The Ladies' Diary (1704–1840)," *Historia Mathematica* 36 (2009). Snyder stated, in her *Philosophical Breakfast Club*, that Elizabeth Whewell had published poems in the *Lancaster Gazette*. She was perhaps following Clark, *Old Friends*, 18. I have been unable to find any trace of these poems.

15. Unitarians denied the divinity of Jesus, seeing him simply as an outstanding individual.

16. The Independents changed their name during the nineteenth century to Congregationalists; their mission arm was the London Missionary Society.

17. J. Price, "Lancaster Sunday School 1796–1900," *Contrebis* [Lancaster Archaeological and Historical Society] 27 (2002); compare J. Price, "William Whewell 1794–1866, Victorian Polymath," *Contrebis* 28 (2003).

18. About sixty of these schools were founded between the sixteenth and eighteenth centuries. The generic naming was retrospective, most of them taking their names from local institutions.

19. *Lancaster Records, or Leaves from Local History. Comprising an Authentic Account of the Progress of the Borough of Lancaster during the Period of Half a Century, 1801–1850* (Lancaster, UK: G. C. Clark, 1869), vii. The "hospital" was a set of twelve alms-houses with a chapel attached. It was also referred to as "the Old Hospital" (10).

20. In 1803 fifteen boys were chosen to attend the school: *Lancaster Records*, 10.

21. Snyder, *Philosophical Breakfast Club*, 13, states, but without providing evidence, that he was at school in the mornings and worked with his father in the afternoons. Blue Coat schools appear to have held classes all day: the Birmingham school, for example, operated from 7 a.m. to noon and 2 to 5 p.m. *A Short Account*

of the Blue Coat School in St Philip's Church Yard, from its Foundation in 1734 to 1817 (Birmingham, UK: R. Jabet, 1817), 44.

22. The school received a royal warrant from Queen Victoria in 1851; see Murray, *Lancaster Royal Grammar School*.

23. Douglas, *Whewell*, 2–5.

24. The Owens had lived in a more prosperous part of the city, but Owen's father's finances had declined, and so they moved into the Castle Hill area where the Whewells lived. Richard Owen was a biologist, comparative anatomist, and paleontologist, and an outspoken critic of Darwin's theory of evolution by natural selection. A controversial figure in many respects, Owen was respected as a naturalist and achieved success with a campaign that led to the establishment, in 1881, of the Natural History Museum in London. See N. A. Rupke, *Richard Owen: Victorian Naturalist* (New Haven, CT: Yale University Press, 1994).

25. The meeting with Rowley was probably in the summer of 1808; William had reached the age of fourteen, the usual age for apprenticeship, on May 24.

26. Murray, *Biographical Register*, 19. Whewell's younger brother John (see note 9) was admitted in 1810 but died in 1812 (22).

27. See Murray, *Biographical Register*; A. F. Leach, "Schools," in *The Victoria County History of Lancaster*, ed. W. Farrer and J. Brownbill (London: Constable, 1908).

28. N. Carlisle, *Concise Description of the Endowed Grammar Schools of England and Wales*, 2 vols. (London: printed for Baldwin, Cradock and Joy, 1818), 1:665–69. For Carlisle's survey questionnaire and methodology, see Christopher A. Stray, "Introduction," in Carlisle, *Concise Description of the Endowed Schools of England and Wales*, 6 vols. (Bristol, UK: Thoemmes, 2002 [1818]), 1:v–xiii.

29. Carlisle, *Concise Description*, reported that the Eton Latin and Greek grammars were used at the school. These books, published in 1758 and 1768, respectively, were used in most grammar schools in this period.

30. The two boys have been identified as Richard Owen and his elder brother James, whom Whewell gave two black eyes: Richard Owen, *The Life of Richard Owen*, 2 vols. (London: John Murray, 1894), 1:8–9. Compare Q. Wessels and A. M. Taylor, "Anecdotes to the Life and Times of Sir Richard Owen (1804–1892) in Lancaster," *Journal of Medical Biography* 25 (2017).

31. The boy was the son of the local gravedigger Billy Bindloss: R. D. Humber, *Heversham: The Story of a Westmorland School and Village* (N.p.: printed by Titus Wilson and Son, 1969), 29.

32. Stray, *American in Victorian Cambridge*, 86. Whewell was ordained as deacon in 1824 and priest in 1826.

33. Clark, *Old Friends*, 55.

34. Letters from Whewell to Morland are quoted in Douglas, *Whewell*, 10–11, 13–14, 24–25, 26–27; and in Todhunter, *Whewell*, 2:1–8, 10–15.

35. Thomas Satterthwaite to John Whittaker, 13 November 1809, DDX

2743/11/3/6695, Lancashire Archives. In fact Whittaker went to St. John's College in 1810; he was a fellow in 1814–1825. He corresponded with Whewell from 1814 to 1844: Add.Ms.a.214/77–86, TCL.

36. Humber, *Heversham*, 1–4.

37. In 1822 the church commissioners valued the scholarship at £44.5s: *Report of the Schools Inquiry Commission* (London: Her Majesty's Stationery Office, 1868–1870), 19, 350. The Dallam Tower account books record that in 1812 Whewell was paid £40 4s 9½d: Humber, *Heversham*, 29.

38. *A Digest of Parochial Returns Made to the Select Committee Appointed to Inquire into the Education of the Poor, House of Commons, 1 April 1819*, Parliamentary Papers, 1819, 2:1101.

39. W. W. Rouse Ball and J. A. Venn, *Admissions to Trinity College Cambridge* (Cambridge: Macmillan, 1916), 1:56.

40. The Bridge Inn may have been what is now the Gilpin Bridge Inn in the village of Levens, between Lancaster and Kendal.

41. Gough was celebrated by both Wordsworth and Coleridge; see Edward Larrissy, *The Blind and Blindness in the Literature of the Romantic Period* (Edinburgh: Edinburgh University Press, 2007). Gough also coached Joshua King, master of Queens' College, Cambridge, and Thomas Gaskin, tutor of Jesus College, Cambridge, and was much in demand as an examiner for the Mathematical Tripos. His last pupil was John Dalton, famous for originating the atomic theory and also known for his research on color blindness. Gough's manuscript autobiography, "The Dark Path to Knowledge," survives in WDX 935/1, Cumbria Archive Centre, Kendal.

42. WW to John Whewell, March 6, 1812, quoted in Douglas, *Whewell*, 6.

43. W. G. Clark's obituary called him a "sub-sizar," that is, a poor boy who could be upgraded to a sizar on the basis of the first-year college examination. The term *subsizar* occurs in Venns's *Alumni Cantabrigienses* fifty times, all with reference to entrants to Trinity from 1832 on; in using it to describe Whewell, Clark may have read back from later practice to 1811. J. and J. A. Venn, *Alumni Cantabrigienses: A Biographical List of All Known Students, Graduates and Holders of Office at the University of Cambridge, from the Earliest Times to 1900* (Cambridge: Cambridge University Press, 1922–1954). Venn's is eccentric, indeed almost Gollumesque.

44. Most of the Trinity men who came from Heversham were sizars. On undergraduate expenses, see Rothblatt, *Revolution of the Dons*, 66–68; Snyder, *Philosophical Breakfast Club*, 15–17. On John Whewell's money, see his will, proved on July 2, 1816, WRW/A/R140/35, Lancashire Archives.

45. WW to his father, February 17, 1813; WW to his aunt Alice Lyon, November 2, 1815, both in Add.MS.a.301/2, TCL. It is of course possible that some of the money John Whewell sent his son was borrowed from Alice Lyon.

46. Whewell to Ann Newton , January 11, 1815, Add.MS.a.273/1; Whewell

to Alice Lyon, September 9, 1815, Add.MS.c.191/5, both in TCL. These figures cast doubt on Snyder's view (*Philosophical Breakfast Club*, 16) that John Whewell earned less than £50 a year.

47. Matriculation consisted of registering with the university as a student; this was done after admission to a college, sometimes months after (in Whewell's case, over a year).

48. Murray, *Biographical Register*, 91; compare Rothblatt, *Revolution of the Dons*, 34–35.

49. At this point, third-year Trinity undergraduates ("senior sophs") were not examined; such an examination was introduced in 1818. The first-year examinations were mostly classical; those of the second year were dominated by mathematics.

50. The examination was later known as the Mathematical Tripos, to distinguish it from the Classical Tripos, which was first examined in 1824.

51. Sophia De Morgan, *Memoir of Augustus De Morgan, with Selections from His Letters* (London: Longmans, Green, 1882), 17–18.

52. *ODNB*, s.v. "William Whewell," by Richard Yeo, accessed January 9, 2024, https://doi.org/10.1093/ref:odnb/29200.

53. Seventy-five men were admitted in 1812, but this soon rose to over 100, and in 1823 to over 150. A new court was opened in 1824 to cope with the increased numbers: Rouse-Ball and Venn, *Admissions to Trinity College*, 1:9.

54. Whewell to his father, October 17, 1817, quoted in Douglas, *Whewell*, 8. At that point dinner was at 2.30 p.m.

55. The Cambridge Union Society, which provided an extra-collegiate meeting place for undergraduates, was not founded until 1815.

56. See [John Willis Clark], "The Rev. William Whewell, D.D.," *The Athenaeum* 2002 (March 10, 1866).

57. Yeo, *Defining Science*, 18. Yeo was followed by Susannah Gibson, *The Spirit of Enquiry: How One Extraordinary Society Shaped Modern Science* (Oxford: Oxford University Press, 2019), 42, who added that Whewell's statement "caused much amusement among his fellow students." This is reading too much into the *Athenaeum* text.

58. Mrs. Andrew Crosse, "Hours Counted on the Sundial," *Temple Bar* 95 (1892): 389. Cornelia Crosse was the widow of Andrew Crosse, "the thunder and lightning man," an early experimenter with electricity. "Worthless publication" was Whewell's favorite phrase in rejecting books proposed for the college library, including, J. W. Clark thought, Darwin's *On the Origin of Species*. See Clark, *Old Friends*, 71.

59. Whewell to RJ, August 17, 1829, quoted in Todhunter, *Whewell*, 2:103. Whewell apparently felt no need to explain the word to Jones.

60. He suffered from a digestive illness, which was apparently cured by a Cambridge doctor.

61. Harvey Carlisle [Goodwin], review of Janet Stair Douglas, *William*

Whewell, Macmillan's Magazine, December 1881, 139. Harvey Goodwin had been second wrangler in 1840. Since 1869 he had been bishop of Carlisle, and signed his article "Harvey Carlisle" according to ecclesiastical convention.

62. My thanks for advice on this matter to Joan Beal, emeritus professor of English language, University of Sheffield; Richard Coates, emeritus professor of onomastics, University of the West of England; and Laura Wright, reader in English language, University of Cambridge. Nowadays the name is usually pronounced "Hew-ell," though some of its bearers use "Wee-well."

63. "William Whewell," *Saturday Review,* 691.

64. M. E. Bury and J. D. Pickles, *Romilly's Cambridge Diary 1842–1847* (Cambridge: Cambridgeshire Records Society, 1994), 41. For the history of such undergraduate misbehavior, see Christopher A. Stray, "Rank (Dis)order in Cambridge 1753–1909: The Wooden Spoon," in *History of Universities,* vol. 26, pt. 1, ed. Mordechai Feingold (Oxford: Oxford University Press, 2012).

65. [George Biddell] Airy to his wife, September 17, 1845, quoted in Wilfred Airy, ed. *Autobiography of Sir George Biddell Airy* (Cambridge: Cambridge University Press, 1897), 117.

66. Clark, "William Whewell. In Memoriam," 545. Clark had entered Trinity in 1840, and had been a fellow since 1844.

67. Watson was born in Heversham in 1737; his father had been headmaster of the school from 1698 to 1737. See R. Watson, *Anecdotes of the Life of Richard Watson, Bishop of Llandaff: Written by Himself at Different Intervals, and Revised in 1814* (London: printed for T. Cadell and W Davies, 1817); Humber, *Heversham,* 14–21.

68. "A Lay-Member of Merton College Oxford" [R. Hodgson], *The Life of Dr Beilby Porteus* (London: J. Davis, 1810), 72.

69. P. J. Mannex, *History, Topography and Directory of Westmorland* (London: Simpkin Marshall, 1849), 270.

70. Ramsden was first chancellor's medalist in 1786, and later fellow and senior dean. Bell was senior wrangler in the same year; see H. Gunning, *Reminiscences of the Town, County and University of Cambridge, from the Year 1780,* 2nd ed. (London: G. Bell, 1855), 2:105–6. Compare B. R. Schneider, *Wordsworth's Cambridge Education* (Cambridge: Cambridge University Press, 1957), 40–41. Ramsden was also remembered for a sermon in which he pronounced *judges* as "joodges": G. Pryme, *Autobiographic Recollections* (Cambridge: Deighton, Bell, 1870), 88.

71. [Clark], "Rev. William Whewell," 333–34.

72. *Surrey Comet,* August 18, 1855.

73. L. P. Wenham, ed. *Letters of James Tate* (Leeds, UK: Yorkshire Archaeological Society, 1966), 27–28. Compare the Johnian William Makepeace Thackeray's description of the college as "the lowest, most childish, piggish, punning place"; quoted in G. N. Ray, ed. *The Letters and Private Papers of William Makepeace Thackeray: 1817–40* (London: Oxford University Press, 1945), 107.

74. Henry Jackson to G. O. Trevelyan, March 19, 1916, quoted in R. St. J. Parry, *Henry Jackson OM* (Cambridge: Cambridge University Press, 1926), 241.

75. Yeo, *Defining Science*, 18.

2. Whewell and Cambridge Mathematics

1. Whewell's role in the teaching of mathematics at Cambridge is treated in A. Warwick, *Masters of Theory: Cambridge and the Rise of Mathematical Physics* (Chicago: University of Chicago Press, 2003), 94–101.

2. Douglas, *Whewell*, 3. Whewell expressed his debt to Joseph Rowley "who was one main cause of my coming to [Trinity] College"; WW to Dr Mackreth (Vicar of Lancaster), February 2, 1851, quoted in Douglas, *Whewell*, 415, 532.

3. The grammar school at Heversham was founded in 1613 and could award an exhibition of £50 to fund an entrant to Trinity College, Cambridge. The award of the exhibition in 1811 was a walkover for Whewell, as no parishioner of Heversham applied for it.

4. Douglas, *Whewell*, 6.

5. John Hudson became a fellow of Trinity College in 1798 and a tutor. He was Charles Babbage's tutor when he arrived in Cambridge in 1810. In this role he advised Babbage not to waste his time reading Lacroix's *Traité* since it would not be examined in the Senate House Examination; such activity did not "pay."

6. John Gough was from a Quaker background. A contemporary taught by Gough was Richard Dawes, who followed Whewell to Trinity College in 1813. He was fourth wrangler (4W) in 1817 and mathematics tutor (1818–1836) at the newly founded Downing College.

7. Whewell to his father, March 6, 1812, quoted in Douglas, *Whewell*, 6. Whewell developed widespread interests in emerging sciences (he coined the term *scientist*). At the beginning of the nineteenth century the age of specialization had yet to arrive, so Whewell was noted for his polymathy—and the butt of the offhand remark that "omniscience was his foible." See the introduction to the present volume for context.

8. Douglas, *Whewell*, 6. Gough was well known as a teacher. See T. T. Wilkinson, "Biographical Notices of Some Liverpool Mathematicians," *Transactions of the Historic Society of Lancashire and Cheshire* 14 (1861–1862).

9. Isaac Newton had also been a subsizar at Trinity College.

10. Since the 1770s the examinations for a BA degree at Cambridge were traditionally known as the "Senate House Examination," named after the building in central Cambridge where the examinations took place. At that time questions were conducted orally but written answers were expected. Printed examination papers were introduced only in 1827. The *Mathematical Tripos* terminology was introduced in 1824 to distinguish the mathematics degree from the newly created Classical Tripos. See J. W. L. Glaisher, "The Mathematical Tripos," *Proceedings of the London Mathematical Society* 18 (1886).

11. For the evolution of the Mathematical Tripos and the place of mathematics in Cambridge education preceding Whewell's time, see, for instance, John Gascoigne, "Mathematics and Meritocracy: The Emergence of the Cambridge Mathematical Tripos," *Social Studies of Science* 14 (1984).

12. On the Classical Tripos, see Christopher Stray, "The First Century of the Classical Tripos (1822–1922): High Culture and the Politics of Curriculum," in *Classics in 19th and 20th Century Cambridge: Curriculum, Culture and Community*, ed. Stray (Cambridge: Cambridge Philological Society, 1999). On the Moral and Natural Sciences Triposes, see Peter Searby, *A History of the University of Cambridge*, vol. 3, *1750–1870* (Cambridge: Cambridge University Press, 1997), chaps. 5–6; and David Palfrey, "The Moral Sciences Tripos at Cambridge University, 1848–1860" (PhD diss., Cambridge University, 2002).

13. On the history of Newtonian fluxions at Cambridge, see, for instance, Niccolò Guicciardini, *The Development of Newtonian Calculus in Britain, 1700–1800* (Cambridge: Cambridge University Press, 2003 [1989]). On the cultural significance of Newtonian natural philosophy and mathematics for Cambridge in the eighteenth and nineteenth centuries, see John Gascoigne, *Cambridge in the Age of Enlightenment: Science, Religion and Politics from the Restoration to the French Revolution* (Cambridge: Cambridge University Press, 1988); Iwan Rhys Morus, *When Physics Became King* (Chicago: University of Chicago Press, 2009), chap. 2. In Britain the division of geometry into synthetic geometry and analytical geometry in library catalogs lasted to the 1970s. Analytical geometry is the way of studying geometry through algebraic operations and the algebra generated from the use of coordinates. Synthetic geometry (often referred to as *pure* geometry) refers to properties gained directly from geometrical figures without recourse to algebra. This is the geometry as conducted by Euclid directly from axioms and figures, and in the Cambridge context the way geometry was studied by Newton. For an account of eighteenth- and nineteenth-century Cambridge mathematics textbooks, see Kevin Lambert, *Symbols and Things: Material Mathematics in the Eighteenth and Nineteenth Centuries* (Pittsburgh: University of Pittsburgh Press, 2022), chap. 1.

14. WW to Richard Gwatkin, August 10, 1815, quoted in Todhunter, *Whewell*, 2:9.

15. Douglas, *Whewell*, 10.

16. Edward Jacob was elected a fellow of Gonville and Caius College, called to the bar at Lincoln's Inn, and became an editor and writer on legal matters. Whewell joined the list of famous second wranglers (including W. Thomson [1845], W. K. Clifford [1867], and J. J. Thomson [1880]). Those who stayed on at Cambridge tended to be from the wrangler class.

17. WW to his father, February 6, 1816, quoted in Douglas, *Whewell*, 21–22.

18. Ashworth, *Trinity Circle*, 24.

19. See, for instance, Harvey Becher, "Woodhouse, Babbage, Peacock and Modern Algebra," *Historia Mathematica* 7 (1980); Christopher Phillips, "Robert Woodhouse and the Evolution of Cambridge Mathematics," *History of Science* 44, no. 1 (2006); and Helena M. Pycior, "George Peacock and the British Origins of Symbolical Algebra," *Historia Mathematica* 8, no. 1 (1981).

20. See P. C. Enros, "The Analytical Society (1812–1813): Precursor of the Renewal of Cambridge Mathematics," *Historia Mathematica* 10 (1983); Harvey Becher, "Radicals, Whigs and Conservatives: The Middle and Lower Classes in the Analytical Revolution at Cambridge in the Age of Aristocracy," *British Journal for the History of Science* 28 (1995).

21. Silvestre-François Lacroix, *Traité du calcul différentiel et du calcul integral*, 3 vols. (Paris, 1797, 1798, 1800). Encyclopedic in nature, it comprised 1,790 quarto pages. When Lacroix was appointed to the École Polytechnique, he distilled this into a textbook: *Traité élémentaire du calcul différetiel et du calcul integral* (Paris: Duprat, 1802), frequently referred to as the "small Lacroix." This textbook achieved widespread popularity and in the period 1802–1881 went through nine editions. See the J. M. CdM. Domingues, "The Calculus According to S. F. Lacroix (1765–1843)" (PhD diss., Middlesex University, 2007).

22. Enros, "Analytical Society," 40. While a student at Cambridge, Sir Edward Thomas Ffrench Bromhead, FRS, was a central member of the Analytical Society. Bromhead encouraged both George Boole and George Green by allowing them to make use of his library at Thurlby Hall, Lincolnshire.

23. It has not been established that Whewell actually joined the Analytical Society. His position within the "Cambridge network" and the Analytical Society is critically discussed, for instance, in Menachem Fisch, "'The Emergency Which Has Arrived,'" 266–76.

24. See WW to JH, March 6, 1817, quoted in Todhunter, *Whewell*, 2:15–17.

25. D. M. Peacock, *A Comparative View of the Principles of the Fluxional and Differential Calculus* (Cambridge: J. Smith, 1819), 85; Becher, "Radicals, Whigs and Conservatives," 416–20; Enros, "Analytical Society," 37–38.

26. J. M. F. Wright, *Alma Mater, or Seven Years at the University of Cambridge*, 2 vols. (London: Black and Young, 1827), 2:28. William Dealtry, FRS, graduated 2W and 2SP (1796) and also wrote *The Principles of Fluxions* (1810, 1816). This is an example of a book attacked by members of the Analytical Society. It follows Newton's scheme for calculus: fluxions (derivatives) and fluents (integrals) and Newton's dot-age.

27. J. Challis, *Remarks on Cambridge Mathematical Studies and Their Relation to Modern Physical Science* (Cambridge: Deighton & Bell, 1875), 5. James Challis became a prominent Cambridge astronomer.

28. WW to JH, March 6, 1817, quoted in Todhunter, *Whewell*, 2:16.

29. [William Whewell], "Science and the English Universities," *British Critic*

and *Quarterly Theological Review and Ecclesiological Record* 9 (1831): 85; also quoted in Enros, "Analytical Society," 37, where this author identifies the anonymous reviewer as Whewell.

30. WW to JDF, June 9, 1831, quoted in Todhunter, *Whewell*, 2:119. Whewell may have thought *Traité* of little use because in it the derivative was treated as part of algebra in the Lagrangian tradition based on a series expansion, whereas what was needed in applications to mechanics was the derivative defined in terms of limits.

31. WW to JH, November 1, 1818, quoted in Todhunter, *Whewell*, 2:30. In the summer of 1816 Whewell took pupils to Bridlington on the coast of East Yorkshire. In 1817 he prepared himself for the fellowship examinations. In 1818 he took pupils to North Wales.

32. WW to JH, November 1, 1818, quoted in Todhunter, 2:30.

33. WW to HJR, November 1, 1818, R./2.99/12, TCL. James Wood (1760–1839) was SW and 1SP in 1782. He spent his entire life at St. John's College writing mathematical textbooks for students and serving as master (1815–1839), a pathway Whewell was to follow at Trinity College.

34. William Whewell, *Elementary Treatise on Mechanics Vol. I, Containing Statics and Part of Dynamics* (Cambridge: J. Deighton & Sons; London: G. & W. B. Whittaker, 1819), iii. For an overview of further editions see the "List of Whewell's Publications" online.

35. William Whewell, *Analytical Statics: A Supplement to the Fourth Edition of an Elementary Treatise on Mechanics* (Cambridge: J. & J. J. Deightons; London; Whittaker & Arnot, 1833), iv, 105. For a discussion of Whewell's mechanics textbooks, see Ben Marsden's chapter in the present volume.

36. William Whewell, *The Doctrine of Limits, with Its Applications; Namely Conic Sections, the First Three Sections of Newton, the Differential Calculus. A Portion of a Course of University Education* (Cambridge: J. and J. J. Deighton; London: John W. Parker, 1838), 18. The opening lines of book 2 are symptomatic of Whewell's intuitive interpretation: "The Limit of a proposed magnitude is that to which the magnitude can approach indefinitely near." In this he appealed to Newton's *Principia* (book 1, sec. 1). The rigorous definition of a limit due to Augustin-Louis Cauchy was not picked up at Cambridge until the twentieth century by G. H. Hardy and J. E. Littlewood. Many Victorians at Cambridge knew Cauchy as "Corky." See R. Kanigel, *The Man Who Knew Infinity: A Life of the Genius Ramanujan* (New York: Scribners, 1991), 150. According to Todhunter, Whewell's *Doctrine of Limits* seemed "to have achieved less popularity" than his other books, and this was the only edition published (Todhunter, *Whewell*, 1:121).

37. William Whewell, *A Treatise on Dynamics: Containing a Considerable Collection of Mechanical Problems* (Cambridge: J. Deighton & Sons; London: G. & W. B. Whittaker, 1823), x.

38. William Whewell, *Analytical Statics: A Supplement to the Fourth Edition of*

an Elementary Treatise on Mechanics (Cambridge: J. and J. J. Deighton; London: Whittaker & Arnot, 1833).

39. In 1822 the number of sides at Trinity College was increased to three, the tutors being Whewell, George Peacock, and John Higman, FRS. Higman graduated 3W in 1816, the same year that Whewell was 2W. He was a tutor at Cambridge (1822–1834) before accepting a church appointment. He authored *A Syllabus of the Differential and Integral Calculus* (1826) in which he set out to show how calculus may be applied to geometry via a rule-based approach based on the Leibniz notation for derivatives and differentials.

40. The details of Airy's account of his student days have to be treated with caution. See Wilfred Airy, ed., *Autobiography of Sir George Biddell Airy* (Cambridge: Cambridge University Press, 1896), 48.

41. A discussion of Airy's attitudes to teaching and his debate (of 1867) with Arthur Cayley on "pure vs applied" mathematics can be found in Tony Crilly, *Arthur Cayley: Mathematician Laureate of the Victorian Age* (Baltimore: Johns Hopkins University Press, 2008), 302–9; D. A. Winstanley, *Later Victorian Cambridge* (Cambridge: Cambridge University Press, 1947), 223–24.

42. See D. B. Wilson, "The Educational Matrix: Physics Education at Early-Victorian Cambridge, Edinburgh and Glasgow Universities," in *Wranglers and Physicists: Studies on Cambridge Mathematical Physics in the Nineteenth Century*, ed. Peter M. Harman (Manchester: Manchester University Press, 1985), 15.

43. WW to JDF, May 28, 1836, quoted in Todhunter, *Whewell*, 2:240. Forbes supported the ideals of the Analytical Society and looked to Whewell and Cambridge as having the correct educational system for teaching mathematics, as opposed to the system in Scotland, tied as it was to philosophical underpinnings. See George E. Davie, *The Democratic Intellect*, 3rd ed. (Edinburgh: Edinburgh University Press, 2013 [1961]), 116–19; Douglas, *Whewell*, 184; and Wilson, "Educational Matrix," 15.

44. Whewell's metascientific views, formulated in the 1830s and 1840s, is discussed in Fisch *William Whewell*.

45. WW to JCH, December 13, 1840, quoted in Douglas, *Whewell*, 207.

46. Frederick Myers to WW, July 31, 1845, quoted in Douglas, *Whewell*, 323.

47. WW to his sister, July 2, 1841, quoted in Douglas, *Whewell*, 223.

48. WW to JCH, July 25, 1841, quoted in Douglas, *Whewell*, 223.

49. Whewell married in October 1841. This allowance was conferred on the master, the only member of the Trinity Foundation allowed to marry.

50. WW to JCH, March 13, 1842, quoted in Douglas, *Whewell*, 264.

51. Whewell to Kate Marshall, 20 March 1856, quoted in Douglas, *Whewell*, 465.

52. Christopher A. Stray, ed. *An American in Victorian Cambridge: Charles Astor Bristed's "Five Years in an English University"* (Exeter, UK: University of Ex-

eter Press. 2008), 35, 79. Bristed has left us an estimate of Whewell's character, especially at a time when he was appointed master of Trinity. Whewell's peremptory treatment of undergraduates and dons contributed to his unpopularity.

53. Whewell's role in the Mathematical Tripos is analyzed in detail in A. D. D. Craik, *Mr Hopkins' Men: Cambridge Reform and British Mathematics in the 19th Century* (London: Springer, 2008), 66–77.

54. William Whewell, *Of a Liberal Education in General, and with Particular Reference to the Leading Studies of the University of Cambridge* (London: John W. Parker, 1845), 203, 62, 204. See also Becher, "William Whewell and Cambridge Mathematics," 24, 31.

55. WW to JH, August 20, 1845, quoted in Todhunter, *Whewell*, 2:328.

56. In the 1820s Airy wrote his *Tracts* on applied subjects: *Mathematical Tracts on Physical Astronomy, the Figure of the Earth, Precession and Nutation, and the Calculus of Variations. Designed for the Use of Students in the University* (Cambridge: J. Smith, 1826); WW to JH, August 20, 1845, quoted in Todhunter, *Whewell*, 2:329. It seems clear that Whewell had in his sights to exclude the "progressive" papers published in the *Cambridge Mathematical Journal*, the local journal founded in 1837. On this younger generation of Cambridge mathematicians, see, for instance, Lukas M. Verburgt, "Duncan F. Gregory, William Walton and the Development of British Algebra: 'Algebraical Geometry,' 'Geometrical Algebra,' Abstraction," *Annals of Science* 73, no. 1 (2016).

57. Craik, *Mr Hopkins' Men*, 74.

58. Becher, "William Whewell and Cambridge Mathematics," 39; Glaisher, "Mathematical Tripos," 20–23.

59. Whewell, *Of a Liberal Education*, 184. An exception was made for astronomy, and parts of this subject were brought back in the 1860s.

60. Whewell made no claim to be an original mathematician, and he acknowledged his erstwhile student George Biddell Airy as his superior. See Douglas, *Whewell*, 101, 107.

61. See Tony Crilly, "The *Cambridge Mathematical Journal* and Its Descendants: The Linchpin of a Research Community in the Early and Mid-Victorian Age," *Historia Mathematica* 31 (2004).

62. "Preface," *Cambridge Mathematical Journal* 1, no. 1 (1837): 1–2; emphasis added.

63. Harvey Goodwin, quoted in H. D. Rawnsley, *Harvey Goodwin, Bishop of Carlisle* (London: John Murray, 1896), 214–15. See also R. Robson and W. Cannon, "William Whewell, F. R. S. (1794–1866)," *Notes and Records of the Royal Society* 19, no. 2 (1964).

3. Whewell and Mechanics

Thanks to Isobel Falconer; Ralph O'Connor; Ellen Packman; librarians, archivists, and catalogers at Trinity College Library, Cambridge, and University of St. Andrews Library; and the editor for his patience.

1. Isaac Todhunter, "Publications Relating to Mechanics," in Todhunter, *Whewell*, 1:xi, 13–28.

2. Todhunter, *Whewell*, 1:5. The standard study of Whewell's mathematics in the Cambridge context is Becher, "William Whewell and Cambridge Mathematics." See also Tony Crilly's chapter in the present volume. For reprints of Whewell's works on mechanics, see Yeo, *Collected Works of William Whewell*, vol. 8, *Mechanics and Architecture*.

3. For discussion, see Harvey W. Becher, "Woodhouse, Babbage, Peacock and Modern Algebra," *Historia Mathematica* 7, no. 4 (1980); Becher, "Radicals, Whigs and Conservatives: The Middle and Lower Classes in the Analytical Revolution at Cambridge in the Age of Aristocracy," *British Journal for the History of Science* 28, no. 4 (1995); M. V. Wilkes, "Herschel, Peacock, Babbage and the Development of the Cambridge Curriculum," *Notes and Records of the Royal Society* 44 (1990); A. D. D. Craik, *Mr Hopkins's Men: Cambridge Reform and British Mathematics in the 19th Century* (London: Springer, 2008).

4. WW to HJR, September 14, 1817, quoted in Todhunter, *Whewell*, 2:17.

5. Jonathan R. Topham, "A Textbook Revolution," in *Books and the Sciences in History*, ed. M. Frasca-Spada and N. Jardine (Cambridge: Cambridge University Press, 2000).

6. WW to RJ, October 16, 1817, quoted in Todhunter, *Whewell*, 2:23.

7. Todhunter, *Whewell*, 1:11.

8. WW to Herschel, June 19, 1818, quoted in Todhunter, *Whewell*, 1:11.

9. WW to RJ, August 21, 1818, quoted in Todhunter, *Whewell*, 2:26–28.

10. WW to JH, November 1, 1818, quoted in Todhunter, *Whewell*, 2:28–30.

11. William Whewell's early reading notes, quoted in Todhunter, *Whewell*, 1:355.

12. WW to JCH, September 26, 1819, Add.Ms.a.215/3, TCL; see also Todhunter, *Whewell*, 2:33–34.

13. Todhunter, *Whewell*, 1:12. For bibliographic details, I have consulted Todhunter and the accumulated bibliographic resources available via the Jisc Library Hub Discovery (formerly Copac). Where possible, I consulted original copies, though the publication history of Whewell's works is complex. See the "List of Whewell's Published Works" online.

14. JCH to WW, December 21, 1819, Add.Ms.a.216/25, TCL.

15. JH to WW, December 1, 1819, Add.Ms.a.207/5, TCL.

16. See Todhunter, *Whewell*, 1:13–28.

17. Todhunter, *Whewell*, 1:128.

18. Elizabeth Garner, *The Language of Physics: The Calculus and the Development of Theoretical Physics in Europe, 1750–1914* (Boston: Birkhäuser, 2001), 229. For further context, see the chapters by Tony Crilly and Sheldon Rothblatt in this volume.

19. The firm traded variously as G. B. Whittaker, Whittaker, Treacher & Arnot, Whittaker & Arnot, and Whittaker & Co.

20. Todhunter, *Whewell*, 1:13.

21. JH to WW, December 1, 1819, Add.Ms.a.207/5, TCL. On attempts to reform the Society after Banks's period of office, see David P. Miller, "Between Hostile Camps: Sir Humphry Davy's Presidency of the Royal Society of London, 1820–1827," *British Journal for the History of Science* 16 (1983).

22. See William Whewell's election certificate, EC/1819/37, Royal Society of London Archives.

23. See William Whewell, *Syllabus of an Elementary Treatise of Mechanics. With Corrections and Additions* (Cambridge, 1821), Cam.e.821.22, Cambridge University Library; another copy is in the Trinity College Library.

24. Todhunter, *Whewell*, 1:14.

25. See a copy of the syllabus, originally owned by William John Speed, in RB.23.a.26629(1), British Library.

26. William Whewell, *A Treatise on Dynamics: Containing a Considerable Collection of Mechanical Problems* (Cambridge: J. Deighton & Sons; London: G. & W. B. Whittaker, 1823), 1.

27. WW to RJ, September 23, 1822, quoted in Todhunter, *Whewell*, 1:31.

28. Whewell, *Treatise on Dynamics*, iii, v.

29. Whewell, *Treatise on Dynamics* (1823), vi.

30. Whewell, *Treatise on Dynamics* (1823), xi.

31. Whewell, *Treatise on Dynamics* (1823), xii.

32. The Library of Trinity College, Cambridge, has a seven-page list of *Corrections of Some Passages in Professor Whewell's Dynamics* (Cambridge, n.d.).

33. H. Wilkinson to WW, November 21, 1824, quoted in Todhunter, *Whewell*, 1:32.

34. William Whewell, *An Elementary Treatise on Mechanics: Designed for the Use of Students in the University. Second Edition, with Numerous Improvements and Additions* (Cambridge: J. Deighton & Sons; London: G. B. Whittaker, 1824).

35. RJ to WW, [n.d.], quoted in Todhunter, *Whewell*, 1:14.

36. William Whewell, *An Elementary Treatise on Mechanics: Designed for the Use of Students in the University*, 3rd ed. (Cambridge: J. & J. J. Deighton; London: G. B. Whittaker, 1828).

37. Whewell, preface to *Elementary Treatise*, 1828, xiii.

38. On Whewell's role as censor of science, often attracting intense exchanges in the nineteenth-century periodicals, see Yeo, *Defining Science*.

39. William Whewell, "Observations on Some Passages of Dr Lardner's *Treatise on Mechanics*," *Edinburgh Journal of Science* 3 (1830): 149. See also Todhunter, *Whewell*, 1:45.

40. David Brewster to WW, May 30, 1830, Add.Ms.a.201/78, TCL. See Henry Kater and Dionysius Lardner, *Mechanics* (London: Longman, Rees, Orme, Brown, and Green, 1830).

41. Whewell, "Observations," 149, 150, 154.

42. See Forbes's 1831 journal, msdep7—Journals, Box 14, no. 1/10, JDFP, SAUL.

43. JDF to WW, May [?], 1831, msdep7—Incoming Letters 1831, no. 15; WW to JDF, July 14, 1831, msdep7—Incoming letters 1831, no. 25, both in JDFP, SAUL.

44. David Wilson, "The Educational Matrix," in *Wranglers and Physicists: Studies on Cambridge Physics in the Nineteenth Century*, ed. Peter Harman (Manchester: Manchester University Press, 1985), compares the physical milieu of Cambridge (Whewell), Edinburgh (Forbes), and Glasgow (William Thomson).

45. See Jack Morrell, "Practical Chemistry in the University of Edinburgh, 1799–1843," *Ambix* 16, no. 1–2 (1969).

46. George Birkbeck to WW, April 16, 1833, Add.MS.a.201/31, TCL.

47. Eaton Hodgkinson to WW, June 2, 1832, Add.Ms.a.206/88, TCL. Todhunter dates the commencement of Whewell's friendship with Hodgkinson to this time; Todhunter, *Whewell*, 1:60. The reference must be to Whewell's *Elementary Treatise*, rather than the *First Principles*, which did not appear until October 1832.

48. WW to H. Wilkinson, October 2, 1832, quoted in Todhunter, *Whewell*, 2:146–48.

49. William Whewell, *The First Principles of Mechanics: With Historical and Practical illustrations* (Cambridge: J. & J. J. Deighton; London: Whittaker, Treacher & Arnot, 1832). The copy in the Michigan University Library is inscribed "Edward Hill . . . from the author, Oct 20 1832."

50. William Whewell, *The Mechanical Euclid* . . . (Cambridge: J. & J. J. Deighton; London: John W. Parker, 1837), viii.

51. JCH to WW, December 21, 1819, Add.MS.a.216/25, TCL.

52. The anonymous review is "Whewell's First Principles of Mechanics," *Westminster Review* 19, no. 37 (July 1833).

53. Augustus De Morgan to WW, November 12, 1832, Add.Ms.a.202/96, TCL.

54. Todhunter, *Whewell*, 1:109. The *Philosophy* has been dubbed the "moral" of the *History*. On Whewell's history and philosophy of science, see Verburgt's and Cristalli's chapters in the present volume.

55. William Whewell, *An Introduction to Dynamics: Containing the Laws of Motion and the First Three Sections of the* Principia (Cambridge: J. & J. J. Deighton; London: Whittaker, Treacher & Arnot, 1832). For a rare copy see Tracts 34/8, Royal Society of London Archives.

56. WW to H. Wilkinson, October 2, 1832, quoted in Todhunter, *Whewell*, 2:146–48.

57. Later editions include: William Whewell, *On the Free Motion of Points, and on Universal Gravitation, Including the Principal Propositions of Books I. and III. of the Principia; The First Part of* A Treatise on Dynamics. *Third Edition* (Cambridge: J. & J. J. Deighton; London: Whittaker & Arnot, 1836).

58. JDF to WW, March 31, 1833, Add.Ms.a.204/9; JDF to WW, April 13, 1833, Add.Ms.a.204/10, both in TCL.

59. WW to JDF, April 27, 1833, quoted in Todhunter, *Whewell*, 2:164–65.

60. WW to RJ, October 6, 1834, quoted in Todhunter, *Whewell*, 2:192–93.

61. JDF to WW, May 10, 1833, Add.Ms.a.204/12, TCL.

62. The original is JDF to WW, July 20, 1833, Add.Ms.a.204/14, TCL; Forbes's copy is msdep7—Letterbook II, 35–38, JDFP, SAUL.

63. Eaton Hodgkinson to WW, September 7, 1833, Add.MS.a.206/90, TCL.

64. Eaton Hodgkinson to WW, September 7, 1833.

65. William Whewell, *An Elementary Treatise on Mechanics: Designed for the Use of Students in the University. Fourth Edition, with Improvements and Additions* (Cambridge: J. & J. J. Deighton, 1833).

66. Todhunter, *Whewell*, 1:15–16.

67. WW to JDF, November 13, 1833, msdep7—Incoming Letters 1833, no. 52, JDFP, SAUL.

68. William Whewell, *Analytical Statics: A Supplement to the Fourth Edition of an Elementary Treatise on Mechanics* (Cambridge: J. & J. J. Deighton; London: Whittaker & Arnot, 1833). The final edition was published in 1847.

69. Whewell, *Analytical Statics* (1833), iii–iv.

70. William Whewell, *Additions in the Fourth Edition of An Elementary Treatise on Mechanics* (Cambridge: J. & J. J. Deighton; London: Whittaker & Arnot, 1833).

71. WW to JDF, October 12, 1834, msdep7—Incoming Letters 1834, no. 46, JDFP, SAUL. See William Whewell, *On the Free Motion of Points, and on Universal Gravitation: Including the Principal Propositions of Books I and II of the Principia: The First Part of a New Edition of a Treatise on Dynamics* (Cambridge: J. & J. J. Deighton, 1832) (a copy Whewell sent Forbes is at ForQA845.W5, JDFP, SAUL); and Whewell, *On the Motion of Points Constrained and Resisted, and on the Motion of a Rigid Body. The Second Part of a New Edition of a Treatise on Dynamics* (Cambridge: J. & J. J. Deighton; London: Whittaker & Arnot, 1834).

72. JDF to WW, October 24, 1834, Add.MS.a.204/19, TCL.

73. JDF to WW, January 7, 1836, Add.MS.a.204/25, TCL.

74. WW to JDF, May 28, 1836, quoted in Todhunter, *Whewell*, 2:239–40. See William Whewell, *An Elementary Treatise on Mechanics: Designed for the Use of Students in the University. Fifth Edition, with Considerable Improvements and Additions* (Cambridge: J. & J. J. Deighton; London: Whittaker & Arnot, 1836).

75. The copy held in the Graves Collection, University College London, contains the book plate of John T. Graves; and see "A table of corrections to Whewell's *Mechanics*, 5th edition in an unidentified hand," Add.MS.a.221/1–13, TCL.

76. Whewell, *Mechanical Euclid* (1837). See W. Johnson, "The Curious Mechanical Euclid of William Whewell, F. R. S. (1774–1866)," *International Journal*

of Mechanical Sciences 38 (1996): 1151; and Benjamin Wardhaugh, *Encounters with Euclid: How an Ancient Greek Geometry Text Shaped the World* (Princeton, NJ: Princeton University Press, 2021), 273.

77. Whewell, *Mechanical Euclid* (1837), v.

78. Whewell, *Mechanical Euclid* (1837), viii.

79. Whewell, *Mechanical Euclid* (1837), vii.

80. Thomas Turton to WW, May 13, 1837, Add.MS.a.213/167, TCL.

81. John Moore Heath [Assistant Tutor at Trinity College, Cambridge] to WW, February 1,183[7?], Add.Ms.a.206/58, TCL. Heath may have been commenting on a manuscript, since the book appeared in April 1837. A lengthy discussion in the *Edinburgh Review* 67, no. 135 (1838): 81 and elsewhere, focused on Whewell's claims about correct mathematical reasoning.

82. *ODNB*, s.v. "Willis, Robert," by Ben Marsden, ed. H. C. G Matthew and Brian Harrison, accessed January 19, 2024, https://doi.org/10.1093/ref:odnb/29200.

83. See Harvey W. Becher, "Voluntary Science in Nineteenth-Century Cambridge University to the 1850s," *British Journal for the History of Science* 19, no. 1 (1986).

84. See *Minutes of Proceedings of the Institution of Civil Engineers* and the *Transactions* of the Institution of Civil Engineers, to which all three contributed.

85. JDF to WW, February 8, 1840, msdep7—Letterbook III, pp. 50–53, JDFP, SAUL.

86. WW to JDF, September 30, 1840, quoted in Todhunter, *Whewell*, 2:289–90; and the original: msdep7—Incoming Letters 1840, no. 52, JDFP, SAUL.

87. WW to JDF, October 25, 1840, quoted in Todhunter, *Whewell*, 2:304–5; see also WW to JDF, October 31, 1840, msdep—Incoming Letters 1840, no. 62, JDFP, SAUL.

88. See Ben Marsden, "'A Most Important Trespass': Lewis Gordon and the Glasgow Chair of Civil Engineering and Mechanics, 1840–55," in *Making Space for Science: Territorial Themes in the Shaping of Knowledge*, ed. Crosbie Smith and Jon Agar (London: Palgrave Macmillan, 1998).

89. See WW to JDF, January 3, 1841, msdep7—Incoming Letters 1841, no. 1 (a,b), JDFP, SAUL.

90. WW to JDF, March 12, 1841, quoted in Todhunter, *Whewell*, 2:295–96. Forbes acknowledged receipt on March 17, 1841. His copy of the sixth edition of the *Elementary Treatise*, inscribed "from the author," is at ForQA807.W5E32, JDFP, SAUL. The book about to appear was William Whewell, *The Mechanics of Engineering, Intended for Use in Universities, and in Colleges of Engineers* (London: John W. Parker; Cambridge: J. & J. J. Deighton, 1841).

91. WW to RJ, February 14, 1843, quoted in Todhunter, *Whewell*, 2:312–13.

92. See Marsden, "'Most Important Trespass'"; and Marsden, "'Progeny of These Two Fellows.'"

93. Review of Willis's and Whewell's books, *Athenaeum*, January 1, 1842, 7–8, quoted in Marsden, "'Progeny of These Two Fellows,'" 429.

94. Letters from Henry Moseley to WW, October 3–15, 1842, Add. MS.a.216/1–4, TCL. For Moseley's work, see Stephen Timoshenko, *History of Strength of Materials* (New York: Dover, 1983 [1953]), esp. 212–13.

95. William Whewell, *An Elementary Treatise on Mechanics: Intended for the Use of Colleges and Universities. Seventh Edition, with Extensive Corrections and Additions* (Cambridge: Deightons; London: Whittaker, 1847).

96. See Whewell, *Elementary Treatise on Mechanics* (1847), preface.

97. I have not here considered the broader appropriation and international consumption of Whewell's works. However, a Chinese translation of Whewell's *Elementary Treatise on Mechanics* was completed by missionary Joseph Edkins and Li Shanlan in 1859. See John Bowring to WW, June 27, 1855, Add.MS.a.201/63, TCL; and Fuling Nie, "The Translation of the Mechanical Terms in *Zhongxue*," *Chinese Journal for the History of Science and Technology* 33 (2012).

4. Whewell and Architecture

1. *ODNB*, s.v. "Whewell, William," by Richard Yeo, accessed March 28, 2021, https://doi.org/10.1093/ref:odnb/29200. On Whewell's early life and education, see Christopher Stray's chapter in the present volume.

2. *ODNB*, s.v. "Rickman, Thomas," by Megan Aldrich, accessed March 28, 2021. https://doi.org/10.1093/ref:odnb/23607; Thomas Rickman, *An Attempt to Discriminate the Styles of English Architecture* (London: Longman, Hurst, Rees, Orme, and Brown, 1817).

3. WW to his sister, June 1, 1841, quoted in Douglas, *Whewell*, 221.

4. William Whewell, quoted in Yanni, "On Nature and Nomenclature," 206.

5. [William Whewell], *Architectural Notes on German Churches, with Remarks on the Origin of Gothic Architecture* (Cambridge: J. & J. J. Deighton, 1830), i–iii.

6. Argued in Yanni, "On Nature and Nomenclature," 205, 209.

7. [Whewell], *Architectural Notes*, 1830, xvi.

8. [Whewell], *Architectures Notes*, 1830, xxiii.

9. Yanni, "On Nature and Nomenclature," 206–9; Yeo, *Defining Science*, 154–55; Alexandrina Buchanan, *Robert Willis (1800–1975) and the Foundation of Architectural History* (Woodbridge, UK: Boydell Press, 2013), 88–89; Quinn, "William Whewell's Philosophy of Architecture."

10. [Whewell], *Architectural Notes*, 1830, ix.

11. [Whewell], *Architectural Notes*, 1830, 66.

12. [Whewell], *Architectural Notes*, 1830, iv.

13. [Whewell], *Architectural Notes*, 1830, 4.

14. [Whewell], *Architectural Notes*, 1830, 30–32.

15. [Whewell], *Architectural Notes*, 1830, 65.

16. [Whewell], *Architectural Notes*, 1830, 32.

17. WW to Thomas Rickman, September 20, 1834, quoted in Todhunter, *Whewell*, 2:191.

18. M. H. Port, *600 New Churches: The Church Building Commission, 1818–1856* (Reading, UK: Spire Books, 2006), 45.

19. On the Whewell-Willis connection, see Ben Marsden's chapter in the present volume.

20. *ODNB*, s.v. "Willis, Robert," by Ben Marsden, accessed March 28, 2021. https://doi.org/10.1093/ref:odnb/29584; Marsden, "'Progeny of These Two Fellows'": Robert Willis, "William Whewell and the Sciences of Mechanism, Mechanics and Machinery in Early Victorian Britain," *British Journal for the History of Science* 37, no. 4 (2004): 403–6.

21. Robert Willis, *Remarks on the Architecture of the Middle Ages, Especially of Italy* (Cambridge: J. & J. J. Deighton, 1835), v–vi.

22. Willis, *Remarks on the Architecture*, 15.

23. Willis, *Remarks on the Architecture*, 17–18.

24. Willis, *Remarks on the Architecture*, 21.

25. William Whewell, *Architectural Notes on German Churches. A New Edition. To Which Is Now Added, Notes Written during an Architectural Tour in Picardy and Normandy* (Cambridge: J. & J. J. Deighton, 1835), xiii–xx.

26. Whewell, *Architectural Notes*, 1835, 136.

27. Whewell, *Architectural Notes*, 1835, 142–43; original emphasis.

28. WW to Rev. H. Wilkinson, October 2, 1832, quoted in Todhunter, *Whewell*, 2:146.

29. William Whewell, *History of the Inductive Science: From the Earliest to the Present Times*, 3 vols. (London: John W. Parker, 1837), 1:237–38.

30. Whewell, *History*, 1:247.

31. Whewell, *History*, 1:247.

32. Whewell, *History*, 1:343.

33. Whewell, *History*, 1:344–45.

34. Whewell, *History*, 1:345.

35. Whewell, *History*, 1:347.

36. Robert Mark, "Robert Willis, Viollet-le-Duc and the Structural Approach to Gothic Architecture," in *The Engineering of Medieval Cathedrals*, ed. Lynn T. Courtenay (Abingdon, UK: Routledge, 1997), 1–3.

37. A. Welby Pugin, *Contrasts: Or, a Parallel between the Noble Edifices of the Middle Ages, and Corresponding Buildings of the Present Day; Shewing the Present Decay of Taste* (London: Charles Dolman, 1841), 3.

38. Anuradha Chatterjee, *John Ruskin and the Fabric of Architecture* (London: Routledge, 2018), 106.

39. Buchanan, *Robert Willis*, 78.

40. Cynthia Gamble, "France and Belgium," in *The Cambridge Companion to John Ruskin*, ed. Francis O'Gorman (Cambridge: Cambridge University Press: 2015), 71.

41. William Whyte, "Architecture and Experience: Regimes of Materiality in the Nineteenth Century," in *Experiencing Architecture in the Nineteenth Century: Buildings and Society in the Modern Age*, ed. Edward Gillin and H. Horatio Joyce (London: Bloomsbury, 2019), 16–17; Marcel Proust, "Preface to John Ruskin's *Bible of Amiens*," in *On Reading Ruskin: Prefaces to La Bible d'Amiens and Sésame et Les Lys with Selections from the Notes to the Translated Texts*, trans. Jean Autret, William Burford, and Phillip J. Wolfe (New Haven, CT: Yale University Press, 1987), 19–20.

42. Nikolaus Pevsner, "William Whewell and His Architectural Notes on German Churches," *German Life and Letters* 22, no. 1 (October 1968): 45–46.

43. Yanni, "On Nature and Nomenclature," 210.

44. William Whewell, "Of Certain Analogies between Architecture and the Other Fine Arts. Read at the General Meeting of the Royal Institute of British Architects, 9 March, 1863," in *Papers Read at the Royal Institute of British Architects, Session 1862–63* (London: J. H. James Parker, 1863), 176.

45. Whewell, "Of Certain Analogies," 179.

46. Whewell, "Of Certain Analogies," 181–82.

47. William Whewell, quoted in Dougas, *Whewell*, 251.

48. *ODNB*, s.v. "Salvin, Anthony," by Richard, Holder, accessed March 28, 2021. https://doi.org/10.1093/ref:odnb/24585.

49. William Whewell, quoted in Douglas, *Whewell*, 251.

50. Douglas, *Whewell*, 252.

51. William Whewell, quoted in Douglas, *Whewell*, 515–18.

52. Roger Wolfe, "'Quite a Gem': An Account of the Former Mortuary Chapel at Mill Road Cemetery, Cambridge," *Proceedings of the Cambridge Antiquarian Society* 84 (1996).

53. Geoffrey Tyack, "Architecture," in O'Gorman, *Cambridge Companion to John Ruskin*, 100.

54. See "Untitled catalogue of architectural building parts," [n.d.], R.6.12.32, TCL.

5. Whewell and Political Economy

1. William Whewell, "Mathematical Exposition of Some Doctrines of Political Economy. Read March 2 and 14, 1829," *Transactions of the Cambridge Philosophical Society* 3 (1830); Whewell, "Mathematical Exposition of Some Leading Doctrines in Mr. Ricardo's Principles of Political Economy and Taxation. Read April 18 and May 2, 1831," *Transactions of the Cambridge Philosophical Society* 4 (1831); Whewell, "Mathematical Exposition of Some Doctrines of Political Economy. Second Memoir. Read April 15, 1850," *Transactions of the Cambridge Philo-*

sophical Society 9 (1856); Whewell, "Mathematical Exposition of Some Doctrines of Political Economy. Third Memoir. Read Nov. 11, 1850," *Transactions of the Cambridge Philosophical Society* 9 (1856). See James P. Henderson, *Early Mathematical Economics: William Whewell and the British Case* (Lanham, MD: Rowman & Littlefield, 1996).

2. William Whewell, *Six Lectures on Political Economy, Delivered at Cambridge in Michaelmas Term, 1861* (Cambridge: Cambridge University Press, 1862).

3. Steven Shapin and Simon Schaffer, *Leviathan and the Air-Pump: Hobbes, Boyle, and the Experimental Life* (Princeton, NJ: Princeton University Press, 2011).

4. Richard Jones, *Literary Remains: Consisting of Lectures and Tracts on Political Economy, by the Late Rev. Richard Jones. Edited, with a Prefatory Notice, by the Rev. William Whewell, D.D.* (London: J. Murray, 1859).

5. See especially Theodore M. Porter, "Rigor and Practicality: Rival Ideals of Quantification in Nineteenth-Century Economics," in *Natural Images in Economic Thought: "Markets Read in Tooth and Claw,"* ed. Philip Mirowski (Cambridge: Cambridge University Press, 1994).

6. Fisch, *William Whewell*, 75.

7. John Stuart Mill, "On the Definition of Political Economy; and On the Method of Investigation Proper to It," in *Essays on Some Unsettled Questions of Political Economy, Collected Works of John Stuart Mill*, vol. 4, ed. J. M. Robson (Toronto: University of Toronto Press, 1967 [1871]). Originally published as "On the Definition of Political Economy: and On the Philosophical Investigation in That Science," *London and Westminster Review* 4 and 26 (October 1836).

8. See also David A. Valone, "The Dark and Tangled Recesses of Knowledge: Theology and the Moral Sciences at Cambridge, 1812–1837" (PhD diss., University of Chicago, 1994), who offers an excellent close look at the interaction between Whewell and Jones; and Lawrence Goldman, *Victorians and Numbers: Statistics and Society in Nineteenth Century Britain* (Oxford: Oxford University Press, 2022), chap. 6, who nicely summarizes the main issues at stake.

9. Richard Jones, *An Essay on the Distribution of Wealth and on the Sources of Taxation* (London: J. Murray, 1831), xiii.

10. Jones speaks of the "dismal system," not of the "dismal science," a concept coined by Thomas Carlyle in the controversies following Governor Eyre's quashing of a slave revolt in Jamaica. On this context, see David M. Levy, *How the Dismal Science Got Its Name: Classical Economics and the Ur-Text of Racial Politics* (Ann Arbor: University of Michigan Press, 2001).

11. Jones, *Essay on the Distribution of Wealth*, xviii.

12. WW to RJ, September 10, 1827, Add.Ms.c.51/41, TCL.

13. Paul Oslington, "Natural Theology, Theodicy, and Political Economy in Nineteenth-Century Britain: William Whewell's Struggle," *History of Political Economy* 49, no. 4 (2017). Oslington and Snyder (*Reforming Philosophy*, 287n91) highlight the evidence of Jones's anxiety of being "scooped" by Whewell. See WW

to RJ, December 15, 1826, Add.Ms.c.51/33, TCL: "I never intended to publish, and I do not think that I shall preach anything which will brush the most delicate bloom of novelty off your plums." Oslington also signals the many examples of Whewell's frustration with Jones's delays, for instance quoting from one of Whewell's letters to Herschel, dated December 4, 1836: "I am going to stay with him [Jones] in the Christmas vacation. The only misfortune is, that he is less and less likely to write the books he owes the world. He professes that he shall still do much in that way, but I confess I doubt it: and I doubt with grief, for in certain branches of Political Economy I am persuaded he is a long way ahead of anybody else, and might give the subject a grand shove onwards"; Oslington, "Natural Theology," 589n23.

14. Oslington, "Natural Theology," 596.

15. Piero Sraffa, ed., *The Works and Correspondence of David Ricardo*, vol. 1, *The Principles of Political Economy and Taxation* (Cambridge: Cambridge University Press, 1951), 65.

16. Sraffa, *Works and Correspondence of David Ricardo*, 1:77n2.

17. Whewell, preface to Jones, *Literary Remains*, xii.

18. Jones, *Essay on the Distribution of Wealth*, 328–29.

19. Over the various editions of his essay, Malthus considerably changed his views on the unavoidability of the population trap. See Anthony Michael C. Waterman, *Revolution, Economics and Religion: Christian Political Economy, 1798–1833* (Cambridge: Cambridge University Press, 1991). See also Donald Winch, *Riches and Poverty: An Intellectual History of Political Economy in Britain, 1750–1834* (Cambridge: Cambridge University Press, 1996).

20. Neil B. De Marchi and R. P. Sturges, "Malthus and Ricardo's Inductivist Critics: Four Letters to William Whewell," *Economica* 40, no. 160 (1973).

21. Ralph S. Pomeroy, "Editor's Introduction," in Richard Whately, *Historic Doubts Relative to Napoleon Bonaparte*, ed. Pomeroy (Berkeley, CA: Scolar Press, 1985), xiv.

22. WW to RJ, July 23, 1831, Add.Ms.c.51/110, TCL.

23. Pietro Corsi, "The Heritage of Dugald Stewart: Oxford Philosophy and the Method of Political Economy," *Nuncius* 2, no. 2 (1987); Harro Maas, "'A Hard Battle to Fight': Natural Theology and the Dismal Science, 1820–50," *History of Political Economy* 40, no. 5 (2008); Oslington, "Natural Theology."

24. Nassau William Senior, appendix to Richard Whately, *Elements of Logic, Comprising the Substance of the Article in the* Encyclopaedia Metropolitana: *With Additions &c.* (London: Mewman, 1826), 383.

25. Jones, *Essay on the Distribution of Wealth*, li.

26. WW to RJ, February 24, 1831, Add.Ms.c.52/20, TCL.

27. RJ to WW, April 28, 1832, Add.Ms.c.52/53, TCL.

28. Edward Copleston, *A Second Letter to the Right Hon. Robert Peel . . . : On the Causes of the Increase of Pauperism, and on the Poor Laws* (London: J. Murray,

1819), 36–37. See Peter Mandler, "Tories and Paupers: Christian Political Economy and the Making of the New Poor Law," *Historical Journal* 33, no. 1 (1990).

29. See Nassau William Senior, *An Introductory Lecture on Political Economy: Delivered before the University of Oxford on the 6th of December, 1826* (London: J. Mawman, 1827).

30. See Corsi, "Heritage of Dugald Stewart." See also Anthony Michael C. Waterman, "Whately, Senior, and the Methodology of Classical Economics," in *Economics and Religion: Are They Distinct?*, ed. H. Geoffrey Brennan and A. M. C. Waterman (Heidelberg: Springer, 1994); and William O. Coleman, "How Theory Came to English Classical Economics," *Scottish Journal of Political Economy* 43, no. 2 (1996).

31. Dugald Stewart, *Elements of the Philosophy of the Human Mind: Collected Works*, 2 vols, ed. W. Hamilton and K. Haakonssen (Bristol, UK: Thoemmes, 1994), 2:235.

32. Richard Whately, "Lecture IX," *Introductory Lectures on Political Economy, Delivered in Easter Term, 1831. Second Edition, Including Lecture IX. and Other Additions* (London: Fellowes, 1832), 229.

33. Whately, *Introductory Lectures*, 235. Whately's extremely witty *Historic Doubts on the Existence of Napoleon Buonaparte* (1819) largely served to show that it was as easy to undermine the existence of miracles, as Hume had done, as to undermine the evidence for a historical fact or event.

34. Whately, *Introductory Lectures*, 239.

35. Whately, *Introductory Lectures*, 71–73; original emphases.

36. A. M. C. Waterman, *Revolution, Economics and Religion: Christian Political Economy, 1798–1833* (Cambridge: Cambridge University Press, 1991), 212. See Whately, *Introductory Lectures*, 60.

37. Whately, *Introductory Lectures*, 89.

38. Whately, *Introductory Lectures*, 100.

39. Whately, *Introductory Lectures*, 64.

40. Whately, *Introductory Lectures*, 68; emphasis in the original. For a detailed discussion, see Harro Maas, "Sorting Things Out: The Economist as an Armchair Observer," in *Histories of Scientific Observation*, ed. Elizabeth Lunbeck and Lorraine Daston (Chicago: University of Chicago Press, 2011).

41. Corsi, "Heritage of Dugald Stewart," 121. Hereafter I quote from a copy of Whewell's essay from his own papers, 164.c.83.5/5, TCL.

42. William Whewell, "On the Uses of Definitions," *Philological Museum* 2 (1833): 264. Emphasis in the original.

43. Whewell, "On the Uses of Definitions," 271. Emphasis in the original.

44. RJ to WW, February 24, 1831, Add.Ms.c.52/20, TCL.

45. Fisch, *William Whewell*. See also Lukas M. Verburgt's and Claudia Cristalli's chapters in the present volume.

46. WW to RJ, August 16, 1822, quoted in Todhunter, *Whewell*, 2:49.

47. WW to RJ, January 28, 1829, Add.Ms.c.51/67, TCL.

48. This table, which most likely was not based on actual statistical data, would be used throughout the nineteenth century in discussions about the possibility of mathematizing political economy. See John Creedy, "On the King-Davenant 'Law' of Demand," *Scottish Journal of Political Economy* 33, no. 2 (1986); and Stephen M. Stigler, "Jevons on the King-Davenant Law of Demand: A Simple Resolution of a Historical Puzzle," *History of Political Economy* 26, no. 2 (1994).

49. William Whewell, "Mathematical Exposition of Some Doctrines of Political Economy. Read March 2 and 14, 1829," *Transactions of the Cambridge Philosophical Society* 3 (1830): 201.

50. William Whewell, Notes and drafts of a work on the history of the philosophy of science, [1831–1832?], R.18.17/15, f. 41–42, TCL; emphasis in the original.

51. Whewell, Notes and drafts, f. 46.

52. The details of Whewell's and Jones's endeavors have been extensively examined, for instance, by Victor L. Hilts, "'Aliis Exterendum,' or, the Origins of the Statistical Society of London," *Isis* 69, no. 1 (1978); Lawrence Goldman, "The Origins of British 'Social Science': Political Economy, Natural Science and Statistics, 1830–1835," *Historical Journal* 26, no. 3 (1983). For more recent accounts see Goldman, "The Origins of the Statistical Movement 1825–1835," part 2 of *Victorians and Numbers* (Oxford: Oxford University Press, 2022); and Plamena Panayotova, *Sociology and Statistics in Britain, 1833–1979* (Cham, Switzerland: Palgrave Macmillan, 2020). These and other accounts of the actual course of affairs are not all consistent, some giving more weight to the importance of Babbage, some more to Whewell and Jones. As Babbage is not particularly known for his modesty, and the correspondence between Whewell and Jones clearly shows they discussed the matter in advance, their version is in my view the more credible.

53. Adam Sedgwick, quoted in Hilts, "'Aliis Exterendum,'" 34.

54. The prospectus was signed by Jones, Babbage, John Elliott Drinkwater, and Henry Hallam.

55. William Newmarch, "Inaugural Address on the Progress and Present Condition of Statistical Inquiry, Delivered at the Society's Rooms, 12, St. James's Square, London, on Tuesday, 16th November, 1869," *Journal of the Statistical Society of London* 32, no. 4 (1869): 386.

56. See Frederic J. Mouat, "History of the Statistical Society of London," *Journal of the Statistical Society of London*, jubilee volume (1885). Mouat wrongly ascribed Sedgwick's presidential speech of 1833 to Whewell.

57. Michael J. Cullen, *The Statistical Movement in Early Victorian Britain: The Foundations of Empirical Social Research* (N.p.: Harvester Press, 1975), 84.

58. See Mark Blaug, *Ricardian Economics: A Historical Study* (New Haven, CT: Yale University Press), 185.

59. Nassau W. Senior, "Opening Address of Nassau W. Senior, Esq., as Presi-

dent of Section F (Economic Science and Statistics), at the Meeting of the British Association, at Oxford, 28th June, 1860," *Journal of the Statistical Society of London* 23, no. 3 (1860): 357.

60. RJ to WW, March 15, 1832, Add.Ms.c.52/49, TCL.

61. WW to RJ, December 7, 1830, Add.Ms.c.51/93, TCL.

62. RJ to WW, May 10, 1844, Add.Ms.c.52/92, TCL; emphasis in the original.

63. John Stuart Mill, "On the Definition of Political Economy; and on the Appropriate Method of Its Study," in *Essays on Economics and Society*, in *Collected Works of John Stuart Mill*, ed. John M. Robson (Toronto: University of Toronto Press, 1967 [1838]), 4:322.

64. William Whewell, *On the Philosophy of Discovery: Chapters Historical and Critical. Including the Completion of the Third Edition of the Philosophy of the Inductive Sciences* (London: J. W. Parker and Son, 1860), 285.

65. John Stuart Mill, quoted in Salim Rashid, "Richard Jones and Baconianism Historicism at Cambridge," *Journal of Economic Issues* 13, no. 1 (1979): 169.

66. See Whewell, "Mathematical Exposition . . . Second Memoir," and "Mathematical Exposition . . . Third Memoir."

67. WW to RJ, April 30, 1848, quoted in Todhunter, *Whewell*, 2:346.

6. Whewell and Language

I wish to record my profound debt to Chris Stray for unstinting encouragement and advice for this paper, and to Nicolas Bell and the staff at the Wren Library, Trinity College Cambridge for assistance in viewing the Whewell archive.

1. Simon Schaffer, "The History and Geography of the Intellectual World: Whewell's Politics of Language," in Fisch and Schaffer, *Whewell*.

2. Schaffer, "History and Geography," 208–9, 218–22; original emphasis.

3. See also the chapter by John Gascoigne in this volume.

4. William Whewell, *An Essay on Mineralogical Classification and Nomenclature: With Tables of the Orders and Species of Minerals* (Cambridge: J. Smith, 1828), xvii.

5. The prescriptivist linguist H. W. Fowler (*A Dictionary of Modern English Usage* [Oxford: Oxford University Press, 1926], 42), includes *Eocene-Pleistocene* among his examples of "Barbarisms," stating, "A man of science might be expected to do on his great occasion what the ordinary man cannot do every day, ask the philologist's help; that the famous *eocene-pleistocene* names were made by 'a good classical scholar' (see Lyell in D[ictionary of] N[ational] B[iography]) shows that word-formation is a matter for the specialist." The *Oxford English Dictionary* gives Whewell the credit for *Eocene* (alongside *Miocene* and *Pliocene*), first used in a letter to the geologist Charles Lyell written January 31, 1831; *OED* s.v. "Eocene (*adj.*)," September 2023, https://doi.org/10.1093/OED/1182998798, but cites Lyell as

the first to use *Pleistocene: OED*, s.v. "Pleistocene *(adj. & n.)*," July 2023, https://doi.org/10.1093/OED/8158281430. WW to Charles Lyell, January 1831, quoted in Todhunter, *Whewell*, 2:109–11, makes clear that Lyell was already trying to "concinnate [his] nomenclature," and had suggested the first element *eo* (from the Greek word for "dawn"), but *Eocene* is wholly Whewell's.

6. Hans Aarsleff, *The Study of Language in England* (Princeton, NJ: Princeton University Press, 1967), 122.

7. William Whewell, "Letter to Edwin Guest," *Proceedings of the Philological Society* 5, no. 117 (February 20, 1852).

8. In particular, note the five boxes of material relating to the Etymological Society, classed as R.6.4–8, TCL. On September 30, 1851, Whewell wrote to Julius Hare that he had "all the papers belonging to our old Society" (quoted in Todhunter, *Whewell*, 2:368).

9. Holger Pedersen, *The Discovery of Language: Linguistic Science in the Nineteenth Century*, trans. John Webster Spargo (Bloomington: Indiana University Press, 1931), 1.

10. William Jones, quoted in Aarsleff, *Study of Language in England*, 133. There has been much debate about the importance of Sir William Jones for the foundation of historical and comparative linguistics. For a spirited rebuttal of a view that Jones's importance has been overstated, see Oswald Szemerényi, "About Unrewriting the History of Linguistics," in *Wege zur Universalienforschung: Sprachwissenschaftlichte Beiträge zum 60. Geburtstag von Hansjakob Seiler*, ed. Gunter Brettschneider and Christian Lehmann (Tübingen, Germany: Narr, 1980); for a more recent account, see M. J. Franklin, *Orientalist Jones: Sir William Jones, Poet, Lawyer, and Linguist, 1746–1794* (Oxford: Oxford University Press, 2011), 35–38.

11. Johann Christoph Adelung, *Mithridates, oder, Allgemeine Sprachenkunde mit dem Vater Unser als Sprachprobe in bey nahe fünfhundert Sprachen und Mundarten* (Berlin: Der Vossischen Buchhandlung, 1806).

12. Aarsleff, *Study of Language in England*, 153.

13. Franz Bopp, *Über das Conjugationssystem der Sanskritsprache: In Vergleichung mit jenem der griechischen, lateinischen, persischen und germanischen Sprache* (Frankfurt am Main, Germany: Andreäischen, 1816).

14. Thomas Young, "Adelung, *Mithridates, oder Allgemeine Sparchenkunde. Mithridates*, or a *General History of Languages*, with the Lord's Prayer as a Specimen, in Nearly Five Hundred Languages and Dialects," *Quarterly Review* 10, no. 19 (1813).

15. H. T. Colebrooke, *A Grammar of the Sanscrit Language*, vol. 1 (Calcutta: Honorable's, 1805); Charles Wilkins, *A Grammar of the Sanskṛita Language* (London: W. Bulmer, 1808); Colebrooke, *Cósha: Or, Dictionary of the Sanskrit Language: With an English Interpretation and Annotations* (Serampore, India: printed by Mr. Carey, 1808).

16. Nicholas Patrick Wiseman, *Twelve Lectures on the Connexion between Science and Revealed Religion. Delivered in Rome by Nicholas Wiseman* (London: Booker, 1836), 1–45; and John William Donaldson, *The New Cratylus, or Contributions towards a More Accurate Knowledge of the Greek Language* (Cambridge: J and J. J. Deighton, 1839), 16–40.

17. Donaldson, *New Cratylus*, 34.

18. James Cowles Prichard, *The Eastern Origin of the Celtic Nations: Proved by a Comparison of Their Dialects with the Sanskrit, Greek, Latin, and Teutonic Languages; Forming a Supplement to Researches into the Physical History of Mankind* (Oxford: printed by S. Collingwood, 1831). Prichard had earlier written on ethnography (Prichard, *Researches into the Physical History of Man* [London: John and Arthur Arch, 1813]), "which seems to have been the chief end of [his] studies" (Aarsleff, *Study of Language in England*, 208). In a letter to Lyell written on June 5, 1863, Whewell cites "Pritchard [*sic*] and the like" as his "masters in the science of languages"; WW to Charles Lyell, June 5, 1863, quoted in Todhunter, *Whewell*, 2:431.

19. Donaldson, *New Cratylus*, 35.

20. See James Clackson, "Dangerous Lunatics: Comparative Philology in Cambridge and Beyond," in *Classical Scholarship and Its History from the Renaissance to the Present. Essays in Honour of Christopher Stray*, ed. Stephen Harrison and Chris Pelling, Trends in Classics—Scholarship in the Making, 1 (Berlin: Walter de Gruyter, 2021), 134–35.

21. An undated letter from "JMK[emble]" is included in the bundle, R.6.5/4, WW, TCL.

22. John Mitchell Kemble to Jakob Grimm, spring 1834, quoted in R. C. Wiley, *John Mitchell Kemble and Jakob Grimm: A Correspondence* (Leiden: Brill, 1971), 57.

23. *The Anglo-Saxon Meteor, or, Letters, in Defence of Oxford, treating of the Wonderful Gothic Attainments of John M. Kemble of Trinity College, Cambridge* (N.p., 1835), cited in Haruko Momma, *From Philology to English Studies: Language and Culture in the Nineteenth Century* (Cambridge: Cambridge University Press, 2013), 90.

24. A letter from Whewell (WW to J. C. Hare, January 4, 1834, Add. Ms.a.215.30, TCL) notes that Kemble is lecturing on Anglo-Saxon in the following term. Simon Keynes, "The Compleat Anglo-Saxonist, John Mitchell Kemble (1807–57)" (unpublished working notes, dated 2013), 21n121, quotes a letter from Whewell to Kate Marshall on May 17, 1857 (not in Todhunter's *Whewell*): "I will not say, as my friend Kemble said, when a large audience assembled to hear him lecture on Anglo-Saxon, 'I'll thin them.'"

25. Momma, *From Philology to English Studies*, 78.

26. "It is all too easy to find fault with eighteenth-century attempts at the historical study of language"; R. H. Robins, *A Short History of Linguistics* (London: Longmans, 1967), 159.

27. John Horne Tooke, *Epea Pteroenta, or The Diversions of Purley* (London: Richard Taylor, 1829), 166–67.

28. Aarsleff, *Study of Language in England*, 71.

29. William Whewell, "On English Adjectives," *Philological Museum* 1 (1832).

30. Whewell, "On English Adjectives," 369.

31. Included in Whewell's papers in the Wren Library is a "List of words with Old English taken from Horne Tooke, and Horne Tooke's explanations" (R.6.5/5/22, TCL), which lists the conjunctions given in the first volume of *Epea Pteroenta*.

32. Whewell, "Letter to Edwin Guest," 133.

33. Letter discussing the Etymological Society: WW to Julius C. Hare, September 30, 1851, quoted in Todhunter, *Whewell*, 2:368. On the Etymological Society, see Aarsleff, *Study of Language in England*, 216–21; Christopher Stray, *Classics in Britain. Scholarship, Education, and Publishing 1800-2000* (Oxford: Oxford University Press, 2018), 156–64; and Ashworth, *Trinity Circle*, 76–81.

34. WW to Julius C. Hare, September 30, 1851. Stray (*Classics in Britain*, 162–64) notes how the *Philological Museum* did not long survive Hare's departure. Whewell was already trying to solicit articles to support the remaining editor, Connop Thirlwall (see below), in December 1832, when he wrote to Richard Jones, "You *once* talked of writing an article for the Philological Museum. Can't you do it now?" WW to RJ, December 27, 1832, quoted in Todhunter, *Whewell*, 2:151; original emphasis. I thank Lukas Verburgt for this reference.

35. Whewell, "Letter to Edwin Guest," 133.

36. Whewell, "Letter to Edwin Guest," 133–34.

37. Whewell, "Letter to Edwin Guest," 138–42.

38. John Mitchell Kemble, "On English Preterites," *Philological Museum* 2 (1833).

39. Whewell, "Letter to Edwin Guest," 142.

40. Roger Paulin, "Julius Hare's German Books in Trinity College Library, Cambridge," *Transactions of the Cambridge Bibliographical Society* 9, no. 2 (1987): 172, citing Henry Crabb Robinson und seine deutschen Freunde. Brücke zwischen England und Deutschland im Zeitalter der Romantik, vol. 2, ed. Hertha Marquardt and Kurt Schreinert, Palaestra, 249 (Göttingen: Vandenhoek & Ruprecht, 1967), 102. For Hare, see *ODNB*, s.v. "Hare, Julius Charles," by N. Distad, accessed January 10, 2024, https://doi.org/10.1093/ref:odnb/12304.

41. E. Bredsdorff, "Grundtvig in Cambridge," *The Norseman* 10 (1953). I am very grateful to Chris Stray for this reference. Among Whewell's papers (R.6.5, TCL), there is a copy of Grundtvig's *Bibliotheca Ango-Saxonica* (1831), dated June 19, which contains his handwritten note of thanks and a poem dedicated to Whewell.

42. An exception is some scrappy notes on adjectives and their formations: R.6.5/5/22, TCL, lists nouns where the associated adjective comes from a dif-

ferent root, such as *iron* and *ferruginous*; Whewell R.6.5/7/5 gives some nouns formed with the suffix *-th*, and adjectives with a suffix *-ical*; R.6.5/7/6, TCL, has adjectives formed with a suffix *−y*; R.6.8/4/8, TCL, apparently in Hare's hand, also gives some paired adjectives (such as *pleasing* and *pleasant*), and the comment "*Ferruginous* is a vile word, For *ferrugineus* is properly having the colour of rust, from *ferrugo*. The derivative from *ferrum* is *ferreus*."

43. Whewell, untitled document with proposals for a scheme of etymology, R.6.5/4/1b, TCL.

44. Whewell, untitled document with proposals for a scheme of etymology.

45. Whewell, untitled table with etymological notes on English, R.6.5/7/4, TCL.

46. Whewell, untitled list of names and tasks, R.6.5/4(1c), TCL.

47. Whewell, untitled list of letters and names, R.6.8/4/2, TCL.

48. In Whewell's letter to Hare of September 30, 1851, he states that he had started work with Lodge on a paper *On Words Derived from Names of Persons* "in a superficial manner" (quoted in Todhunter, *Whewell*, 2:368).

49. Whewell, untitled list of letters and names.

50. English words derived from Latin beginning with D, R.6.7/279–469; English words derived from Latin beginning with E, R.6.6/3; English words derived from Latin beginning with F, R.6.8/5/2, all in TCL.

51. English words derived from Greek beginning with D, R.6.7/216–278; English words derived from Greek beginning with E, R.6.6/2; English words derived from Greek beginning with F R.6.8/5/3, all in TCL.

52. English words derived from Saxon beginning with D, R.6.7/470–621; English words derived from Saxon beginning with E, R.6.6/4; English words derived from Saxon beginning with F, R.6.8/5(5), all in TCL.

53. English words derived from French beginning with D, R.6.7/1–215; English words derived from French beginning with E, R.6.6/1; English words derived from French beginning with F R. R.6.8/5(7), all in TCL.

54. Old English words beginning with F, R.6.8/5(9–40), TCL.

55. Junius refers here to Franciscus Junius, *Francisci Junii Francisci filii Etymologicum Anglicanum. Ex Autographo Descripsit & Accessionibus Permultis Auctum Edidit Edwardus Lye A.M. Ecclesiæ Parochialis de Yardley-Hastings in Agro Northamptoniensi Rector. Præmittuntur Vita Auctoris et Grammatica Anglo-Saxonica* (Oxford: E Theatro Sheldoniano, 1743). The other names included here refer to Stephen Skinner, *Etymologicon linguae anglicanae, seu explicatio vocum anglicarum etymologica ex propriis fontibus* (London: typis T. Roycroft; prostant venales apud H. Brome, 1671); Johann Georg Wachter, *Glossarium Germanicum continens origines et antiquitates totius linguae Germanicae et omnium pene vocabulorum vigentium et desitorum* (Leipzig: apud Joh. Frid. Gleditschii b. filium, 1737); Gerardus Joannes Vossius, *Gerardi Joannis Vossii Etymologicon Linguæ Latinæ. Præfigitur ejusdem de Literarum Permutatione Tractatus* (London: Impensis Jo. Martin & Ja. Alestry, 1662). On the

work of Skinner, Junius, and Vossius, see John Considine, "Stephen Skinner's Etymologicon and Other English Etymological Dictionaries 1650–1700," *Studia Etymologica Cracoviensia* 14 (2009).

56. English words derived from Greek beginning with D, R.6.7/216, TCL; Richardson, *New Dictionary*, 477.

57. English words derived from Latin beginning with E, R.6.6/3, TCL; Richardson, *New Dictionary*, 638.

58. English words derived from Saxon beginning with F, R.6.8/5(5), TCL; Richardson, *New Dictionary*, 768.

59. *The Times* of March 24, 1830 contains an advertisement for the 28th fascicle of the *Encyclopaedia Metropolitana* and includes the information that it extends the *Miscellaneous and Lexicographical* coverage of the *Encyclopaedia* from *investiture* to *Lahore*. I am very grateful to John Considine for this information.

60. Samuel Taylor Coleridge, "A Preliminary Treatise on Method," in *Encyclopaedia Metropolitana, or Universal Directory of Knowledge*, ed. E. Smedley and H. J. Rose (London: Fellowes, 1845), 44.

61. Charles Richardson, *A New Dictionary of the English Language* (London: William Pickering, 1837), 5.

62. Richardson, *New Dictionary*, 18.

63. M. Adams, "Reading Trench Reading Richardson," in *Historical Dictionaries in their Paratextual Context*, ed. Roderick McConchie and Jukka Tyrkkö (Berlin: De Gruyter, 2018), 5; *ODNB*, s.v. "Richardson, Charles (1775–1865), Lexicographer," by John Considine, accessed January 10, 2024, https://doi.org/10.1093/ref:odnb/37893.

64. See Keynes, "The Compleat Anglo-Saxonist"; and *ONDB*, s.v. "Kemble, John Mitchell (1807–1857), Philologist and Historian," by J. Haigh, accessed January 10, 2024, https://doi.org/10.1093/ref:odnb/15321, for Kemble's movements in 1829–1831.

65. Richardson, *New Dictionary*, 40.

66. Ashworth, *Trinity Circle*, 79.

67. R. O. Preyer, "The Language of Discovery: William Whewell and George Eliot," *Browning Institute Studies* 16 (1988): 124, 133.

68. E. W. H[ead], "On the Root of εἰλέω and Some of Its Derivatives in the Greek, Latin and Teutonic Languages," *Philological Museum* 1 (1832). See Stray, "From One Museum to Another," for the identification of Head as the author.

69. Richardson, *New Dictionary*, 797.

70. Richardson, *New Dictionary*, 824.

71. Anonymous, Old English words beginning with F, R.6.8/5/39, TCL.

72. Anonymous, Old English words beginning with F, R.6.8/5/32, TCL. Note that the Sanskrit cognate is cited here in the genitive case, where Schlegel had *podon* (Friedrich Schlegel, *Über die Sprache und die Weisheit der Indier: ein Beitrag zur Begründung der Altertumskunde* [Heidelberg: Mohr und Zimmer,

1808], 12), and Adelung *pad-* (*Mithridates*, 166), the citation form now mostly common in Western grammars. It may be significant that the three consonants of the Sanskrit genitive correspond more closely with a Hebrew triliteral root.

73. William Whewell, "On the Use of Definitions," *Philological Museum* 2 (1833): 270.

74. Aarsleff, *Study of Language in England*, 112.

75. Richardson's *New Dictionary* escaped the censure aimed at Horne Tooke from some Victorian commentators; see Adams, "Reading Trench Reading Richardson."

76. William Whewell, *The History of the Inductive Sciences, from the Earliest to the Present Time*, 3 vols. (London: John W. Parker, 1837), 3:483–84; cited in T. Craig Christy, *Uniformitarianism in Linguistics* (Philadelphia: John Benjamins, 1983), 3. See also Max Dresow's and Aleta Quinn's chapters in the present volume.

77. Donaldson, *New Cratylus*, 12. See also Christy, *Uniformitarianism in Linguistics*, 18–23, for details of other scholars who took up the metaphor. Whewell remained unconvinced by the notion of language change as uniformitarian, noting in a letter to Lyell that the "English language was formed in less than a century from the Conquest, and has undergone little essential change since . . . I think that no numerical multiplication of changes now in progress could produce the past changes; and I think I have all the philologers with me." WW to Charles Lyell, June 5, 1863, quoted in Todhunter, *Whewell*, 2:431–32.

78. See, for example, Ian Roberts, "Uniformitarianism," in *The Cambridge Handbook of Historical Syntax*, ed. Adam Ledgeway and Ian Roberts (Cambridge: Cambridge University Press, 2017); and George Walkden, "The Many Faces of Uniformitarianism in Linguistics," *Glossa: A Journal of General Linguistics* 4, no. 1 (2019), both of whom discuss uniformitarianism with consideration of Whewell's contribution to the idea. For the impact of the study of geology on Classical Studies in early Victorian England, see Louis Barry Rosenblatt, "Fossils and Myths: A Comparative Study of Geology and Classical History in Early Victorian England" (PhD diss., Johns Hopkins University, 1984) (another reference I owe to Chris Stray).

79. Whewell, *History*, 3:484.

7. Whewell and Tidology

WW to Francis Beaufort, November 2, 1836, Letter Book 6, Royal Hydrographic Office (RHO).

1. WW to RJ, November 6, 1836, Add.Ms.c.51/203 TCL.

2. Richard Yeo, *Defining Science*, 56. Yeo's sixteen-volume collection of Whewell's works makes no mention of Whewell's research on the tides. See Richard Yeo, *Collected Works of William Whewell*.

3. Fisch, *William Whewell*, 41.

4. Joan Richards, "Observing Science in Early Victorian England: Recent Scholarship on William Whewell," *Perspectives on Science* 4 (1996): 235.

5. WW to RJ, November 13, 1833, quoted in Todhunter, *Whewell*, 2:172. See Verburgt's chapter in the present volume for Whewell as a historian of science.

6. See Gascoigne's chapter in the present volume for an account of Whewell's vision on society and the State.

7. See Cristalli's chapter in the present volume for a discussion of Whewell's philosophy of science.

8. "High Tide and Inundations," *London Times*, February 14, 1825.

9. As reported in *London Times*, February 24, 1825; for Hamburg, see *London Times*, February 14, 1825.

10. F. W. Morgan, *Ports and Harbours* (London: Hutchinson's University Library, 1952), 45.

11. John McPhee, *The Control of Nature* (New York: Noonday Press, 1989); and Paul R. Josephson, *Industrialized Nature: Brute Force Technology and the Transformation of the Natural World* (Washington, DC: Island Press, 2002).

12. Vindex [pseud.], "The Almanack of the Stationers' Company," *London Times*, January 11, 1828.

13. Detector, letter to the editor, *London Times*, December 31, 1828.

14. WW to JH, October 15, 1823, O.15.47/122 TCL.

15. On Whewell's textbooks, see Crilly's and Marsden's chapters in the present volume.

16. WW to JL, December 7, 1829, 0.15.47/183, TCL, WP.

17. WW to JL, January 30, 1831, W255, Lubbock Papers, Royal Society Archives, London (hereafter Lubbock Papers).

18. WW to JL, April 1, 1832, W261, Lubbock Papers.

19. Francis Beaufort to JL, February 22, 1832, B162, Lubbock Papers.

20. William Whewell, "Essay towards a First Approximation to a Map of Co-tidal Lines," *Philosophical Transactions of the Royal Society of London* 20, no. 3 (1833): 147.

21. WW to JL, July 6, 1832, Add.Ms.a.216 TCL.

22. The first entry was William Whewell, "Memoranda and Directions for Tide Observations," *Nautical Magazine* 2 (1833): 662–65.

23. WW to JL, June 6, 1833, Add.Ms.a.216 TCL.

24. Francis Beaufort to WW, June 11, 1833, Letter Book 4, Beaufort Papers, RHO.

25. WW to his sister, March 13, 1832, quoted in Douglas, *Whewell*, 143–44.

26. WW to sister, Douglas, *Whewell*, 144. The missionary's observations were accurate and Whewell's response incorrect.

27. William Whewell, *The History of the Inductive Sciences, from the Earliest to the Present Time*, 3 vols. (London: John W. Parker, 1837), 2:248–49.

28. Whewell, "Memoranda and Directions," 664, original emphasis.

29. Whewell, "First Approximation," 148.

30. Whewell, "Memoranda and Directions," 665.

31. WW to JL, November 7, 1833, O.15.47/208 TCL.

32. Richard Spencer to WW, September 26, 1832, R.6.20/301 TCL.

33. Francis Beaufort to William Bowles, July 30, 1833, Letter Book 5, Beaufort Papers, RHO.

34. Samuel Sparshott to Francis Beaufort, October 8, 1833, S405, RHO.

35. Bowles to the Respective Inspecting Commanders, May 17, 1834, B456, RHO

36. William Whewell, "On the Results of Tide Observations Made in June 1834 at the Coast Guard Stations in Great Britain and Ireland," *Philosophical Transactions of the Royal Society of London* 125 (1835): 84.

37. Francis Beaufort to WW, February 7, 1835, Letter Book 6, Beaufort Papers, RHO.

38. Whewell, "First Approximation" (1833), 227.

39. Francis Beaufort, memorandum, "Circular to Ministers and Consuls Abroad," March 13, 1835, FO.83.84, Public Records Office, London.

40. Francis Beaufort to Lords Commissioners of the Admiralty, February 14, 1835, ADM.1.3485, Public Records Office.

41. WW to JDF, October 23, 1856, msdep7/94, JDFP, SAUL.

42. William Whewell, "Researches on the Tides—6th Series. On the Results of an Extensive System of Tide Observations Made on the Coast of Europe and America in June 1835," *Philosophical Transactions of the Royal Society of London* 126, no. 2 (1836): 291.

43. Whewell, "Research on the Tides—6th Series," 291.

44. James C. Scott, *Seeing Like a State: How Certain Schemes to Improve the Human Condition Have Failed* (New Haven, CT: Yale University Press, 1998), 11.

45. Whewell, *History*, 2:127.

46. Whewell, *History*, 2:181.

47. William Whewell, *The Philosophy of the Inductive Sciences, Founded upon Their History*, 2 vols. (London: John W. Parker, 1847), 2:100.

48. Whewell, *Philosophy*, 2:395.

49. Whewell, *Philosophy*, 2:397.

50. Whewell, *Philosophy*, 2:399.

51. Whewell, *History*, 2:246–53.

52. William Whewell, "On the Empirical Laws of the Tides in the Port of London; with Some Reflexions on the Theory," *Philosophical Transactions of the Royal Society of London* 124, no. 1 (1834).

53. William Whewell, "Researches on the Tides—4th Series. On the Empirical Laws of the Tides in the Port of Liverpool," *Philosophical Transactions of the Royal Society of London* 126 (1836): 1.

54. Whewell, "Researches on the Tides—4th Series," 3.

55. Whewell, *Philosophy*, 2:62.

56. Paul Hughes and Alan D. Wall, "The Admiralty Tidal Predictions of 1833: Their Comparison with Contemporary Observations and with a Modern Synthesis," *Journal of Navigation* 57 (2004): 213.

57. "On the Advantage Possessed by Naval Men, in Contributing to General Science," *Nautical Magazine*, 1832, 180.

58. For other aspects of Whewell's hierarchical vision of the organization of science, see Ellis's chapter in the present volume.

59. WW to John Herschel, April 3, 1836, quoted in Todhunter, *Whewell*, 2:235.

Chapter 8. Whewell on Astronomy and Natural Theology

The author would like to thank Donald McNally for his suggestions and insights into Whewell at an early stage of my research. I also appreciate Dr. Christine Luk's helpful comments on the first draft of this piece. Finally, I am indebted to Dr. Lukas M. Verburgt for his careful attention to detail in his editorial comments.

1. Rev. William Whewell, *On the Foundations of Morals: Four Sermons Preached before the University of Cambridge November 1837* (New York: E. French, 1839 [1837]), ix.

2. Ashworth, *Trinity Circle*, 8, 109.

3. John Brooke and Geoffrey Cantor, *Reconstructing Nature: The Engagement of Science and Religion* (Edinburgh: T. & T. Clarke, 1998), 142.

4. Yeo, "William Whewell, Natural Theology and the Philosophy of Science," 495.

5. Jon H. Roberts, "That Darwin Destroyed Natural Theology," in *Galileo Goes to Jail*, ed. Ronald L. Numbers (Cambridge, MA: Harvard University Press, 2009), 162.

6. Jonathan R. Topham, *Reading the Book of Nature: How Eight Best Sellers Reconnected Christianity and the Sciences on the Eve of the Victorian Age* (Chicago: University of Chicago Press, 2022), 3, 11, 14, 77.

7. Rev. William Whewell, *Astronomy and General Physics Considered with Reference to Natural Theology* (London: William Pickering, 1833), 2–3, 9, 11–12.

8. Whewell, *Astronomy and General Physics*, 141, 144–45.

9. Whewell, *Astronomy and General Physics*, 148–49.

10. Topham, *Reading the Book of Nature*, 112. John Brooke has also pointed out that the main thrust of Whewell's natural theology was directed against the calculus of pleasures and pains deriving from Bentham and Paley. See Brooke, "Indications of a Creator," 155.

11. Whewell, *Astronomy and General Physics*, 156, 158–59, 181, 184, 189.

12. Whewell, *Astronomy and General Physics*, 251, 254–55.

13. Whewell, *Astronomy and General Physics*, 322, 324, 328, 334, 349.

14. Yeo, *Defining Science*, 118, 124.

15. Whewell, *Astronomy and General Physics*, 360–61, 372.

16. John M. Robson, "The Fiat and Finger of God: The Bridgewater Treatises," in *Victorian Faith in Crisis: Essays on Continuity and Change in Nineteenth-Century Religious Belief*, ed. Richard J. Helmstadter and Bernard Lightman (Houndmills, UK: Macmillan, 1990), 77.

17. Whewell, *Astronomy and General Physics*, 356.

18. William Whewell, *Indications of the Creator. Extracts, Bearing upon Theology, from the History and the Philosophy of the Inductive Sciences* (London: John W. Parker, 1845), viii.

19. Whewell, *Indications of the Creator*, viii–ix.

20. James A. Secord, *Victorian Sensation: The Extraordinary Publication, Reception, and Secret Authorship of* Vestiges of the Natural History of Creation (Chicago: University of Chicago Press, 2000).

21. WW to RJ, July 18, 1845, quoted in Todhunter, *Whewell*, 2:327.

22. WW to JH, March 12, 1845, quoted in Todhunter, *Whewell*, 2:325.

23. Cymbre Quincy Raub, "Robert Chambers and William Whewell: A Nineteenth-Century Debate over the Origin of Language," *Journal of the History of Ideas* 49, no. 2 (April–June 1988): 292.

24. Donald H. McNally, "Science and the Divine Order: Law, Idea, and Method in William Whewell's Philosophy of Science" (PhD diss., University of Toronto, 1982), 386.

25. Whewell, *Indications of the Creator*, xii–xiv, xviii. For more on Owen's negative reaction to the *Vestiges* and Whewell's interest in Owen's contributions to natural theology, see John Hedley Brooke, "Richard Owen, William Whewell, and the Vestiges," *British Journal for the History of Science* 10, no. 2 (July 1977).

26. Whewell, *Indications of the Creator*, 6–8.

27. Whewell, *Indications of the Creator*, 12–13, 16–17.

28. WW to JH, January 3, 1854, quoted in Todhunter, *Whewell*, 2:399.

29. Brooke, "Natural Theology and the Plurality of Worlds," 266.

30. Michael J. Crowe, *The Extraterrestrial Life Debate 1750–1900*, rev. ed. (Mineola, NY: Dover, 1999), 281. Crowe devotes an entire chapter of his book to "William Whewell: Pluralism Questioned." He offers one of the most detailed analyses of *Of the Plurality of Worlds* and the subsequent controversy.

31. Laura J. Snyder, "'Lord Only of the Ruffians and Fiends'? William Whewell and the Plurality of Worlds Debate," *Studies in History and Philosophy of Science* 38 (2007): 586.

32. See Brooke, "Richard Owen, William Whewell, and the Vestiges"; Snyder, "'Lord Only of the Ruffians and Fiends,'" 587; Yeo, "William Whewell, Natural Theology and the Philosophy of Science," 509.

33. Whewell, *Astronomy and General Physics*, 270.

34. [William Whewell], *Of the Plurality of Worlds: An Essay* (London: John W. Parker & Son, 1853), 38, 46, 78, 99, 102.

35. [Whewell], *Of the Plurality of Worlds*, 116, 137, 140, 146, 149, 151, 171, 173, 185–86, 188, 192, 203.

36. [Whewell], *Of the Plurality of Worlds*, 210–11.

37. [Whewell], *Of the Plurality of Worlds*, 211, 216–18, 237, 250.

38. Crowe, *Extraterrestrial Life Debate*, 298.

39. WW to Charles Darwin, January 2, 1860, "Letter no. 2634," Darwin Correspondence Project, accessed June 19, 2022, https://www.darwinproject.ac.uk/letter/?docId=letters/DCP-LETT-2634.xml.

40. WW to Charles Darwin, January 2, 1860, Darwin Correspondence Project, accessed June 19, 2022, https://www.darwinproject.ac.uk/letter/?docId=letters/DCP-LETT-2637.xml.

41. Janet Browne, *Charles Darwin: The Power of Place* (New York: Knopf, 2003), 107.

42. William Whewell, quoted in Charles Darwin, *On the Origin of Species by Means of Natural Selection* (London: John Murray, 1859), opposite title page.

43. Whewell's influence on Darwin in terms of his philosophy of science has been explored by Ruse and Honenberger; see Michael Ruse, "Darwin's Debt to Philosophy," *Studies in History and Philosophy of Science A* 6 (1975); Phillip Honenberger, "Darwin among the Philosophers: Hull and Ruse on Darwin, Herschel, and Whewell," *HOPOS* 8 (Fall 2018).

44. Darwin, *On the Origin of Species*, 488, 490.

45. McNally, "Science and the Divine Order," 395.

46. William Whewell, *On the Philosophy of Discovery: Chapters Historical and Critical. Including the Completion of the Third Edition of the Philosophy of the Inductive Sciences* (London: John W. Parker and Son, 1860), 354, 358–59, 363–64.

47. Brooke and Cantor, *Reconstructing Nature*, 142.

48. Michael Ruse, "William Whewell and the Argument from Design," *The Monist* 60 (1977): 245.

49. Roberts, "That Darwin Destroyed Natural Theology," 163.

9. Whewell, Bacon, and the History of Science

1. See, for instance, Steffen Ducheyne, "Whewell's Philosophy of Science," in *The Oxford Handbook of British Philosophy in the Nineteenth Century*, ed. W. J. Mander (Oxford: Oxford University Press, 2014). The accounts of Whewell's history of science almost always appear as part of studies of (aspects of) Whewell's philosophy of science.

2. Yehuda Elkana was one of the first to point to the relation between Whewell's *History* and *Philosophy*. See Yehuda Elkana, "William Whewell, Historian," *Revista di Storia della Scienza* 1 (1984): 149–97. Other valuable studies that engage Whewell as a historian of science are, in chronological order, Cantor,

"Between Rationalism and Romanticism"; H. Floris Cohen, "Beginning to Learn from the Past of Science: William Whewell," in *The Scientific Revolution: A Historiographical Inquiry* (Chicago: University of Chicago Press, 1994); and Yeo, "Using History," in *Defining Science*.

3. William Whewell, *The History of the Inductive Sciences, from the Earliest to the Present Time*, 3 vols. (London: John W. Parker, 1837), 1:ix.

4. Fisch, *William Whewell*, sec. 3.24; Cantor, "Between Rationalism and Romanticism," 70.

5. Steven L. Goldman, *Science Wars: The Battle over Knowledge and Reality* (Oxford: Oxford University Press, 2021), 117; Yeo, *Defining Science*, 152; Antonio Pérez-Ramos, *Francis Bacon's Idea of Science and the Maker's Knowledge Tradition* (Oxford: Clarendon Press, 1988), 24. See also Richard Yeo, "An Idol of the Market-Place: Baconianism in Nineteenth Century Britain," *History of Science* 23 (1995): esp. 252–53.

6. See WW to JH, April 9, 1836, quoted in Todhunter, 2:233–36; Whewell, *History*, 1:viii.

7. Whewell, *History*, 1:viii.

8. For an overview of this debate, see Snyder, "Whewell and the Reform of Inductive Philosophy," chap. 1 in *Reforming Philosophy*.

9. WW to RJ, July 27, 1834, quoted in Todhunter, *Whewell*, 1:90; emphases in original.

10. See the manuscripts in the folder "William Whewell: Notebooks and Drafts of a History of the Philosophy of Science, 1820–1860," R.18.17, TCL.

11. See William Whewell, "The Philosophy of the Progressive Sciences," fos. 1–13, 136, R./18.17/8, TCL; and WW to RJ, December 10, 1833, quoted in Todhunter, *Whewell*, 2:173–74.

12. See George Sarton, *A Guide to the History of Science* (Waltham, MA: Chronica Botanica, 1952), 49.

13. WW to RJ, [December 23, 1825], Add.Ms.c.51/25; WW to RJ, August 16, 1822, Add.Ms.c.51/14; WW to RJ, September 23, 1822, Add.Ms.c.51/15, all in TCL.

14. WW to RJ, October 17, 1825, Add.Ms.c.51/23; WW to RJ, August 16, 1822, Add.Ms.c.51/14; WW to RJ, September 23, 1822, Add.Ms.c.51/15, all in TCL.

15. WW to RJ, October 17, 1825, Add.Ms.c.51/23, TCL.

16. WW to JH, November 1, 1818, quoted in Todhunter, *Whewell*, 2: 28–29; emphasis added; Fisch, *William Whewell*, 41.

17. WW to RJ, April 24, 1831, Add.Ms.c.51/104, TCL.

18. The best account of what and who Whewell had in mind when he tried to show that science was not irreligious is still Pietro Corsi, *Science and Religion: Baden Powell and the Anglican Debate, 1800–1860* (Cambridge: Cambridge University Press, 1988), part 3, 143–224.

19. See, for instance, WW to RJ, April 24, 1831, Add.Ms.c.51/104, TCL.

20. This extravagant but apt phrase appears in David B. Wilson, "The Historiography of Science and Religion," in *Science and Religion: A Historical Introduction*, ed. Gary B. Ferngren (Baltimore: Johns Hopkins University Press, 2002), 15.

21. WW to RJ, February 26, 1827, quoted in Todhunter, *Whewell*, 2:82; WW to RJ, July 23, 1831, Add.Ms.c.51/110; WW to RJ, December 6, 1831, Add. Ms.c.51/122, both in TCL.

22. See William Whewell, "Religious Views," book 3 of *Astronomy and General Physics Considered with Reference to Natural Theology* (London: William Pickering, 1833). For a discussion, see Bernard Lightman's chapter in the present volume. See also Joan L. Richards, "The Probable and the Possible in Early Victorian England," in *Victorian Science in Context*, ed. Bernard Lightman (Chicago: University of Chicago Press, 1997); and Yeo, "Moral Scientists," chap. 5 in *Defining Science*.

23. In brief, for Whewell good science was inductive science that was grounded in careful reasoning, led to original discoveries and was formative of a pious character and indicative of divine design in nature as well as a supernatural reality beyond human knowledge.

24. WW to RJ, March 24, 1833, quoted in Todhunter, *Whewell*, 2:161.

25. For this point, see Yeo, *Defining Science*, 154.

26. Cantor, "Between Rationalism and Romanticism," 71.

27. William Whewell, *An Elementary Treatise on Mechanics*, vol. 1, *Containing Statics and Part of Dynamics* (Cambridge: J. Deighton & Sons; London: G. & W. B. Whittaker, 1819), vii.

28. WW to RJ, September 23, 1822, Add.Ms.c.51/15, TCL. Smith's *History of Astronomy*, despite its title, would today be understood as a work on the philosophy of science.

29. WW to RJ, August 16, 1822, Add.Ms.c.51/14, TCL. Whewell captured the contrast between the "metaphysics" and "logic" of science by writing that "the latter examines the accuracy of your mode of deducing conclusions from your principles and the former your way of getting your principles."

30. William Whewell, *The First Principles of Mechanics, with Historical and Practical Illustrations* (Cambridge: J. & J. J. Deighton; London: Whittaker, Treacher & Arnot, 1832), see, e.g., iv–v.

31. See William Whewell, "On the Logic of Induction," in *The Mechanical Euclid . . .* (Cambridge: J. and J. J. Deighton; London: John W. Parker, 1837).

32. The same arguably holds for Whewell's mature work on the history and philosophy of science. See Todhunter, *Whewell*, 1:107–8, 135.

33. WW to RJ, October 31, 1832, Add.Ms.c.51/143, TCL.

34. WW to HJR, November 19, 1826; WW to HJR, December 12, 1826, both quoted in Todhunter, *Whewell*, 2:75–76, 78; emphasis added.

35. Richard Yeo has made this point in the context of Whewell's philosophy of

science, but does not extend it to his analysis of Whewell's history of science; see Yeo, "William Whewell, Natural Theology and the Philosophy of Science"; Yeo, "Using History," chap. 6 in *Defining Science*.

36. Whewell, *Astronomy and General Physics*, vi; William Whewell, *Indications of the Creator. Extracts Bearing upon Theology, from the History and the Philosophy of the Inductive Sciences* (London: John W. Parker, 1845), viii.

37. WW to RJ, March 24, 1833, Add.Ms.c.51/154, TCL.

38. Whewell, *Astronomy and General Physics*, 304.

39. For instance, in his diary Whewell lists the "Vulgar errors of the nineteenth century," such as the "superiority of the age"; William Whewell, "Diary," September 26, 1828, R.18.9/14, TCL.

40. For this point, see Yeo, *Defining Science*, 149.

41. William Whewell, "Address. Delivered on June 25th, 1833," *Report of the Third Meeting of the British Association for the Advancement of Science, Held at Cambridge in 1833* (London: John W. Parker, 1834), xi–xii. Whewell himself would contribute one such "Report" in 1835, dealing with theories of electricity, magnetism and heat.

42. See WW to W. Vernon Harcourt, September 1, 1831, quoted in Todhunter, *Whewell*, 2:126–30; William Whewell, "Report on the Recent Progress and Present State of Mineralogy," in *Report of the First and Second Meetings of the British Association for the Advancement of Science; at York in 1831, and at Oxford in 1832* (London: John Murray 1833), 322–65.

43. Whewell, *History*, 1:xi, 5.

44. William Whewell, "Philosophy of the Progressive Sciences," f. 1–13, p. 136, R.18/17.8, TCL.

45. WW to RJ, November 13, 1833, Add.Ms.c.51/159, TCL. By this date Whewell had arranged for his load as head tutor to be substantially lightened.

46. See Whewell, "Philosophy of the Progressive Sciences," 136.

47. See, for instance, Fisch, "The Formative Years: From Didactics to Philosophy of Science (1812–1834)," chap. 2 in *William Whewell*; and Snyder, *Reforming Philosophy*, chap. 1. Whewell first expressed his mature position publicly in 1837, in "Remarks on the Logic of Induction," which was appended to his textbook *Mechanical Euclid*.

48. Whewell, *An Elementary Treatise on Mechanics*, iv, vi.

49. WW to RJ, March 5, 1829, Add.Ms.c.51/62, TCL.

50. WW to RJ, July 23, 1831, quoted in Todhunter, *Whewell*, 2:124. See also Whewell, *Astronomy and General Physics*, 12–17.

51. WW to RJ, April 3, 1832, Add.Ms.c/15/135, TCL.

52. WW to RJ, February [?], 1831, quoted in Todhunter, *Whewell*, 2:115–16.

53. William Whewell, "Modern Science—Inductive Philosophy," *Quarterly Review* 45 (1831): 377; emphasis added.

54. WW to RJ, February [?], 1831, quoted in Todhunter, *Whewell*, 2:116.

55. William Whewell, "Induction II. Astronomy, Botany, Chemistry (1830–[32?])," 28, R.18.17/11, TCL.

56. Whewell, *History*, 1:5.

57. WW to RJ, October 6, 1834, Add.Ms.c.51/176, TCL.

58. Whewell, *First Principles of Mechanics*, 1, v.

59. William Whewell, "Cambridge Transactions—Science of the English Universities," *British Critic and Theological Review* 9 (1831), 88.

60. Among the elements that Whewell himself listed as key to the historiography of the *History* were: "Facts and Ideas," "Successive Steps in Science," "Generalization," "Inductive Epochs; Preludes; Sequels," and "Stationary Periods"; Whewell, *History*, 1:6–16.

61. See Whewell, "Modern Science–Inductive Philosophy," 379–81, where he offered an account of induction that he claimed to be in agreement with that of his friend Herschel's. Perhaps somewhat surprisingly, the first time that Whewell mentioned the notion of "facts" and "ideas" as equally necessary to science was in WW to RJ, August 5, 1834, Add.Ms.c/51/174, TCL.

62. Whewell, "Modern Science–Inductive Philosophy," 391, 397.

63. WW to RJ, February 19, 1832, Add.Ms.c.51/129, TCL.

64. Interestingly, at exactly the same time, Whewell transported this notion of "idea" from architecture to physics, from pointed arch to "causation" and "reaction." See William Whewell, "Draft, 'On the Idea of Causation' and 'On the Idea of Reaction' [1830s?]," R.18.17/9, TCL.

65. [William Whewell], *Architectural Notes on German Churches, with Remarks on the Origin of Gothic Architecture* (Cambridge: J. & J. J. Deighton, 1830), iv.

66. See WW to RJ, August 21, 1834, Add.Ms.c.51/175, TCL.

67. Whewell's thinking on history and philosophy of science developed very much in tandem. During this same period Whewell was also in the process of formulating the key elements of his distinctive philosophical views on induction, later put forward in the *Philosophy*. See, in this regard, Claudia Cristalli's chapter in the present volume.

68. Whewell, *History of the Inductive Sciences*, 1:xii.

69. This two-volume work originally appeared as the first part of the third edition of *The Philosophy of the Inductive Sciences*. As Whewell explained in the preface, the book was "formerly published as a portion of the [*Philosophy*]; but [its] nature and subject . . . [is] more exactly described by the present title. . . . For this part of the work is mainly historical, and was, in fact, collected from the body of scientific literature, at the same time that the [*History*] was so collected."

70. WW to RJ, August 21, 1834, op. cit. note 66.

71. Whewell, "Philosophy of the Progressive Sciences," 1.

72. See Cantor, "Between Rationalism and Romanticism," 71–74.

73. William Whewell, *The History of Scientific Ideas. Being the First Part of the*

Philosophy of the Inductive Sciences. The Third Edition. In Two Volumes (London: John W. Parker and Son, 1858), 1:v.

74. WW to JH, April 9, 1836, quoted in Todhunter, *Whewell*, 2:235.

75. William Whewell, *Novum Organon Renovatum. Being the Second Part of the Philosophy of the Inductive Sciences. The Third Edition, with Large Additions* (London: John W. Parker, 1858), iii.

76. William Whewell, *The Philosophy of the Inductive Sciences, Founded upon Their History*, 2 vols. (London: John W. Parker, 1840), 1:ix; original emphasis.

77. Whewell, *History*, 1:viii.

78. Whewell, *Novum Organon Renovatum*, iii; Whewell, *History*, 1:ix.

10. Whewell's Kant and Beyond

This chapter owes much to productive discussions with Matteo Colombo, Alun David, Luca Timponelli, and Lukas Verburgt.

1. John Wettersten, *Whewell's Critics: Have They Prevented Him from Doing Good?* (Amsterdam: Rodopi, 2005), 31.

2. See Todhunter, *Whewell*; and Douglas, *Whewell*. See also Christopher Stray's chapter in the present volume.

3. Marion Stoll, *Whewell's Philosophy of Induction* (Lancaster, PA: Lancaster Press, 1929).

4. Robert Blanché, *Le rationalisme de Whewell* (Paris: Alcan, 1935).

5. Silvestro Marcucci, *L'"idealismo" scientifico di W. Whewell* (Pisa: Università degli Studi di Pisa, 1963).

6. Robert Butts, *William Whewell's Theory of Scientific Method* (Pittsburgh: University of Pittsburgh Press, 1968).

7. For classic accounts on Whewell's notion of consilience, see Mary B. Hesse, "Consilience of Inductions," in *The Problem of Inductive Logic*, ed. Imre Lakatos (Amsterdam: North Holland, 1968); Mary B. Hesse, "Whewell's Consilience of Inductions and Predictions [Reply to Laudan]," *The Monist* 55, no. 3 (1971); and Larry Laudan, "William Whewell on the Consilience of Inductions," *The Monist* 55, no. 3 (1971). For a discussion of Whewell's notion of induction in general philosophy of science, see Malcolm Forster, "The Debate between Whewell and Mill on the Nature of Scientific Induction," in *Handbook of the History of Logic*, vol. 10, *Inductive Logic*, ed. Dov M. Gabbay, Stephan Hartmann, and John Woods (Amsterdam: Elsevier, 2011).

8. Wettersten, *Whewell's Critics*, 21 and following.

9. See Fisch and Schaffer, *Whewell*.

10. Yeo, *Defining Science*; Snyder, *Reforming Philosophy*; Reidy, *Tides of History*.

11. Robert Butts, "Necessary Truth in Whewell's Theory of Science," *American Philosophical Quarterly* 2, no. 3 (July 1965): 161.

12. Snyder, "'The Whole Box of Tools'; Laura J. Snyder, "Renovating the

Novum Organum: Bacon, Whewell and Induction," *Studies in History and Philosophy of Science* 30, no. 4 (1999); Snyder, "Discoverers' Induction," *Philosophy of Science* 64, no. 4 (1997); Snyder, "'It's *All* Necessarily So': William Whewell on Scientific Truth," *Studies in History and Philosophy of Science* 25, no. 5 (1994). On Whewell's Baconianism, see also Verburgt's chapter in the present volume.

13. See Butts, *William Whewell's Theory*; Menachem Fisch, "Necessary and Contingent Truth in William Whewell's Antithetical Theory of Knowledge," *Studies in History and Philosophy of Science* 16, no. 4 (1985); Steffen Ducheyne, "Kant and Whewell on Bridging Principles between Metaphysics and Science," *Kant-Studien* 102, no. 1 (2011); Ducheyne, "Fundamental Questions"; Ducheyne, "Whewell, Necessity and the Inductive Sciences: A Philosophical-Systematic Survey," *South African Journal of Philosophy* 28, no. 4 (2009).

14. Ducheyne, "Kant and Whewell."

15. Margaret Morrison, "Whewell on the Ultimate Problem of Philosophy," *Studies in History and Philosophy of Science* 28, no. 3 (1997): 418.

16. Morrison, "Whewell on the Ultimate Problem," 417.

17. Henry M. Cowles, "The Age of Methods: William Whewell, Charles Peirce, and Scientific Kinds," *Isis* 107, no. 4 (2016).

18. Whewell, *The Philosophy of the Inductive Sciences, Founded upon Their History. A New Edition, with Corrections and Additions, and an Appendix, Containing Philosophical Essays Previously Published*, 2nd ed,, 2 vols. (London: John W. Parker, 1847), 1:iii; emphasis added; all subsequent citations to the *Philosophy* are to this edition, unless otherwise stated. On Whewell's history of science, see Verburgt's chapter in the present volume.

19. Whewell, *Philosophy*, 1:vii.

20. See, respectively, Verburgt's introduction and Maas's chapter in the present volume.

21. Walter F. Cannon, "John Herschel and the Idea of Science," *Journal of the History of Ideas* 22, no. 2 (1961): 220.

22. Charles Darwin, *The Life and Letters of Charles Darwin, including an Autobiographical Chapter*, 2 vols. (London: John Murray, 1887), 1:55: "During my last year at Cambridge, I read with care and profound interest Humboldt's 'Personal Narrative.' This work, and Sir J. Herschel's '*Introduction to the Study of Natural Philosophy*,' stirred up in me a burning zeal to add even the most humble contribution to the noble structure of Natural Science. No one or a dozen other books influenced me nearly so much as these two." See also n. 26.

23. See John Stuart Mill's letter to John Herschel: "You will find that the most important chapter of the book, that on the four Experimental Methods, is little more than an expansion & a more scientific statement of what you had previously stated in the more popular manner suited to the purpose of your 'Introduction' [to *A Preliminary Discourse on the Study of Natural Philosophy*]." Mill to JH, May 1, 1843, in Mill, *The Earlier Letters, 1812–1848*, vol. 13 of *The Collected Works of*

John Stuart Mill, ed. F. E. Mineka (Toronto: University of Toronto Press, 1963), 583, letter 397.

24. William Whewell, "Modern Science—Inductive Philosophy" [review of Herschel's *Preliminary Discourse*], *Quarterly Review* 45 (1831). On the complex relation between early Victorian philosophy of science and Baconianism, see Lukas M. Verburgt, "*The Works of Francis Bacon*: A Victorian Classic in the History of Science," *Isis* 112, no. 4 (2021).

25. Whewell, "Modern Science—Inductive Philosophy," 378–79: "Induction must be described to be 'the process of considering a class of phenomena, or two associated classes of phenomena, as represented by a general *law*, or single conception of the mind;' such a single law, or representation, is capable of including so many phenomena, in virtue of the natural connexions of our thoughts . . . [such a law] will derive its validity from its being a connected representation of the phenomena, and nothing more."

26. John Herschel, *A Preliminary Discourse on the Study of Natural Philosophy* (London: Longman, 1831): 102; emphasis added. Further on the same page, Herschel confirms: "This process is what we mean by induction." Darwin echoed Herschel's account of induction almost word by word while recounting the find of a tropical shell in a gravel pit by a local worker in Shrewsbury: "Nothing before had ever made me thoroughly realise, though I had read various scientific books, *that science consists in grouping facts so that general laws or conclusions may be drawn from them*"; Darwin, *Life and Letters*, 1:48; emphasis added.

27. Whewell, "Modern Science—Inductive Philosophy," 379.

28. Whewell, *Philosophy*, 1:ix.

29. William Whewell, *Astronomy and General Physics Considered with Reference to Natural Theology* (London: William Pickering, 1833), 300. See Todhunter, *Whewell*, 1:68: "Mr. Whewell holds the former [induction] to be of greater value in science, and to have a stronger tendency to religion than the latter [deduction]: this opinion he never relinquished." See also Giorgio Lanaro, *La teoria dell'induzione in William Whewell* (Milano: Franco Angeli, 1987), 86: "Within the study of nature, Whewell never tires of repeating, deduction remains subordinate to induction"; translation mine.

30. This point is duly recognized by Ducheyne, "Whewell, Necessity and the Inductive Sciences," 332–33.

31. On the integration of Whewell's essay on "Laws of Motion" in the *Philosophy*, see Ducheyne, "Whewell, Necessity and the Inductive Sciences," 344n29.

32. Yeo, *Defining Science*, 153–55.

33. Todhunter, *William Whewell*, 1:25.

34. Whewell, *The History of the Inductive Sciences, from the Earliest to the Present Time*, 3 vols. (London: John W. Parker, 1837), 1:x–xi.

35. For a list of scientific figures featuring in Whewell's correspondence, see Todhunter, *Whewell*, 1:xix–xxiv; and Diana Smith's chapter in the present volume.

Beside Herschel, one notes Sir George Biddell Airy, astronomer royal; the chemist Michael Faraday; the logician Augustus De Morgan; the geologist Charles Lyell; the statistician Adolphe Quetelet; and the explorer Alexander von Humboldt.

36. Todhunter, *Whewell*, 1:223–24, 239–30.

37. For a list, see Todhunter, *Whewell*, 1:230.

38. See William Whewell, "On the Fundamental Antithesis of Philosophy. Read Feb. 5, 1844," *Transactions of the Cambridge Philosophical Society* 8, no. 2 (1844).

39. I thank one of the anonymous reviewers for pushing me to emphasize this point.

40. Lanaro, *La teoria dell'induzione*, 45: "la riformulazione semplificata della fondamentale dottrina kantiana che considera la conoscenza come il risultato dell'unione di sensibilità e intelletto"; translation mine.

41. Whewell, *Philosophy*, 1:x.

42. Whewell explicitly referred to Goethe and Schelling (*Philosophy*, 1:29–33) but not Hegel, although his work and thought is very much present in the background of Whewell's characterization of German idealism.

43. Whewell, *Philosophy*, 1:17.

44. Whewell, *Philosophy*, 1:18.

45. Whewell, *Philosophy*, 1:24.

46. Whewell, *Philosophy*, 1:40; original emphasis.

47. Whewell, "On the Fundamental Antithesis," 178.

48. Whewell, *Philosophy*, 1:45.

49. Whewell, *Philosophy* (1840), 1:45.

50. Whewell, *Philosophy*, 1:45.

51. Whewell, *Philosophy*, 1:45.

52. Whewell, *Philosophy*, 1:iv. See also Snyder, *Reforming Philosophy*.

53. For an introduction to Locke's *Essay*, see for instance John W. Yolton, *Locke and the Compass of Human Understanding: A Selective Commentary on the Essay* (Cambridge: Cambridge University Press, 1970).

54. John Locke, *An Essay Concerning Human Understanding* (Oxford: Oxford University Press, 1975 [1689]), 2:xi, §17, 1–7.

55. Locke, *Essay*, 2:xii, §1, 29–33.

56. Whewell, *Philosophy*, 1:42.

57. Wettersten, *Whewell's Critics*, 39.

58. Immanuel Kant, *Critique of Pure Reason*, ed. and trans. Paul Guyer and Allen W. Wood (Cambridge: Cambridge University Press, 1998). In the following, I refer to the two editions of the *Critique* (1781, 1787) with *CpR* followed by the letters "A" and "B" respectively, followed by the page number.

59. Kant, *CpR*, A 24/B 39.

60. Michael Friedman, *Kant and the Exact Sciences* (Cambridge, MA: Harvard University Press, 1992), 74.

61. Whewell, *Philosophy*, 1:85.

62. William Whewell, "A Letter to the Author of the Prolegomena Logica, by the Author of the History and Philosophy of the Inductive Sciences," Add.266c.80.96[12], TCL. See also Whewell, *History*, 1:16, where this passage is reproduced.

63. Ducheyne, "Whewell, Necessity and the Inductive Sciences," 338.

64. Compare Snyder, "It's *All* Necessarily So," 794–881; and see Morrison, "Whewell on the Ultimate Problem," for a more psychological interpretation.

65. Whewell, *Philosophy*, 1:96.

66. Kant, *CpR*, A 70/B 95, A 80/B 106.

67. Whewell, *Philosophy*, 1:112; emphasis added.

68. See Whewell, *Philosophy*, 1:114.

69. Gary Hatfield, "Perception as Unconscious Inference," in *Perception and the Physical World: Psychological and Philosophical Issues in Perception*, ed. Dieter Heyer and Rainer Mausfeld (Chichester, UK: John Wiley & Sons, 2002).

70. Jutta Schickore, *The Microscope and the Eye: A History of Reflections, 1740–1870* (Chicago: University of Chicago Press, 2007), chap. 4.

71. Whewell, *Philosophy*, 1:116.

72. Wettersten, *Whewell's Critics*, 25. In fact, William James made even the perception of space a simple sensation, and the idea that psychological analysis had to focus on the "building blocks" of our mental life, i.e., individual sensations, became commonplace at the beginning of the twentieth century.

73. Whewell, *Philosophy*, 1:119.

74. Whewell, *Philosophy*, 1:119.

75. Whewell, *Philosophy*, 1:123.

76. Whewell acknowledged in a footnote his adoption of Bell's results in the treatise (*Philosophy*, 1:123).

77. For a recent defense of the possibility of a physiological grounding of our representations in a Kantian framework, see Paolo Pecere, "'Physiological Kantianism' and the 'Organization of the Mind': A Reconsideration," *Intellectual History Review* 31, no. 4 (2021).

78. Augustus De Morgan, "The Philosophy of the Inductive Sciences, Founded upon Their History," [review of Whewell's *Philosophy*], *The Athenaeum* 672 (September 12, 1840): 707.

79. William Whewell, "Remarks on the Review of the '*Philosophy of the Inductive Sciences*' in the *Athenaeum*, No. 672, Sep. 12, 1840," f. 7, Add.266.80.96[4], TCL.

80. See Verburgt's introduction to the present volume for an account of Whewell's reception.

81. Abbreviated references to Peirce's work will be integrated with the standard reference style adopted by Peirce scholars: "W" for *Writings* followed by the volume and page numbers.

82. Fisch, *William Whewell*, 109–10. I thank one of the anonymous reviewers for pointing out this passage to me.

83. Wettersten, *Whewell's Critics*, 100.

84. Tullio Viola, *Peirce on the Uses of History* (Berlin: De Gruyter, 2020).

85. Harro Maas and Mary S. Morgan, "Timing History: The Introduction of Graphical Analysis in 19th Century British Economics," *Revue d'Histoire des Sciences Humaines* 2 (2002). See also Thomas L. Hankins, "A 'Large and Graceful Sinuosity': John Herschel's Graphical Method," *Isis* 97, no. 4 (2006).

86. Chiara Ambrosio, "Diagrammatic Thinking, Diagrammatic Representations, and the Moral Economy of Nineteenth-Century Science," in *The Oxford Handbook of Charles Sanders Peirce*, ed. Cornelis de Waal (Oxford: Oxford University Press, 2024).

87. Charles S. Peirce, "On the Logic of Science," in *Writings of Charles S. Peirce: A Chronological Edition*, vol. 1, *1857–1866*, ed. by M. H. Fisch, C. J. W. Kloesel, et al. (Bloomington: Indiana University Press, 1982 [1865]), W1:211.

88. Peirce, "On the Logic of Science," W1:205; Snyder, "'Whole Box of Tools,'" 172n32.

89. Peirce, "On the Logic of Science," W1:205.

90. Peirce, "On the Logic of Science," W1:209.

91. Peirce, "On the Logic of Science," W1:208–9. Peirce quotes Whewell verbatim; see William Whewell, *Novum Organon Renovatum. Being the Second Part of the Philosophy of the Inductive Sciences. The Third Edition, with Large Additions* (London: John W. Parker and Son, 1858), 98.

92. Peirce, "On the Logic of Science," W1:210.

93. Charles S. Peirce, "On the British Logicians," in *Writings of Charles S. Peirce: A Chronological Edition*, vol. 2, *1867–1871*, ed. by M. H. Fisch, C. J. W. Kloesel, et al. (Bloomington: Indiana University Press, 1984 [1871]), W2:337.

94. Peirce, "On the British Logicians," W2:341.

95. Peirce, "On the British Logicians," W2:341.

96. Peirce, "On the British Logicians," W2:345.

97. Peirce integrated this aspect in his system by naming it "the outward clash" or the category of "secondness."

98. Charles S. Peirce, "Review of Fraser's *The Works of George Berkeley*," W2:484.

99. Peirce's most famous paper on this issue is his 1892 "The Law of Mind," which is part of the Monist Metaphysical Project, a series of papers appeared in *The Monist* in the years 1891–1892. See *Writings of Charles S. Peirce: A Chronological Edition*, vol. 8, *1890–1892*, ed. by M. H. Fisch, C. J. W. Kloesel, et al. (Bloomington: Indiana University Press, 2010), W8:135–57.

100. See for example Giovanni Maddalena, "Anti-Kantianism as a Necessary Characteristic of Pragmatism," in *Pragmatist Kant: Pragmatism, Kant, and Kantianism in the Twenty-First Century*, ed. Krzysztof Piotr Skowroński

and Sami Pihlström (Helsinki: Nordic Pragmatism Network, 2019); Gabriele Gava, *Peirce's Account of Purposefulness: A Kantian Perspective* (New York: Routledge, 2014).

11. Whewell, Gender, and Science

1. See, for example, Kathryn A. Neeley, *Mary Somerville: Science, Illumination, and the Female Mind* (Cambridge: Cambridge University Press, 2001), 13–15.

2. For Mary Somerville and her contribution to early nineteenth-century science, see Elizabeth Chambers Patterson, *Mary Somerville and the Cultivation of Science, 1815–1840* (The Hague: Nijhoff, 1983).

3. [William Whewell], "On the Connexion of the Physical Sciences. By Mrs. Somerville," *Quarterly Review* 51, no. 101 (March 1834): 55–56.

4. For the importance of a gentlemanly model of scientific masculinity at the BAAS in its early years, see Heather Ellis, *Masculinity and Science in Britain, 1831–1918* (London: Palgrave Macmillan, 2017).

5. See Jack Morrell and Arnold Thackray, *Gentlemen of Science: The Early Years of the British Association for the Advancement of Science* (Oxford: Clarendon Press, 1981).

6. William Whewell, *On the Free Motion of Points, and on Universal Gravitation, Including the Principal Propositions of Books I. and III. of the Principia; The First Part of* A Treatise on Dynamics. *Third Edition* (Cambridge: J. & J. J. Deighton; London: Whittaker & Arnot, 1836), v.

7. See Ellis, *Masculinity and Science in Britain*, esp. chap. 4.

8. On the "decline of science" debate, see, for instance, James A. Secord, *Visions of Science: Books and Readers at the Dawn of the Victorian Age* (Chicago: University of Chicago Press, 2014), 52–79. See also Max Dresow's chapter in the present volume.

9. Steven Shapin, "'A Scholar and a Gentleman': The Problematic Identity of the Scientific Practitioner in Early Modern England," *History of Science* 29, no. 3 (1991); Ellis, *Masculinity and Science in Britain*, 24–32.

10. Ellis, *Masculinity and Science in Britain*, esp. chap. 4.

11. Adam Sedgwick to Mrs Lyell, October 16, 1837, quoted in in J. W. Clark and T. M. Hughes, *The Life and Letters of the Reverend Adam Sedgwick*, 2 vols. (Cambridge: Deighton, Bell, 1890), 1:490.

12. ODNB, s.v. "Whewell, William," by Richard Yeo, accessed June 19, 2022, https:// doi.org/10.1093/ref:odnb/29200. For more on Whewell's social origins, see Christopher Stray's chapter in the present volume.

13. WW to his sisters, April 14, 1815, Add.Ms.a.301, TCL.

14. On the importance of Whewell's social origins in shaping his academic and scientific career, see Yeo, *Defining Science*, esp. 15–19.

15. See Shapin, "Scholar and a Gentleman."

16. Charles Lyell to WW, September 23, 1840, Add.Ms.a.208/130, TCL.

17. WW to R. I. Murchison, September 18, 1840, quoted in Todhunter, *Whewell*, 2:287.

18. Morrell and Thackray, *Gentlemen of Science*, 149.

19. Morrell and Thackray, *Gentlemen of Science*, 150.

20. Jan Golinski, "Humphry Davy's Sexual Chemistry," *Configurations* 7, no. 1 (1999): 20.

21. Adam Sedgwick, quoted in Morrell and Thackray, *Gentlemen of Science*, 151.

22. Charles W. J. Withers and Rebekah Higgitt, "Science and Sociability: Women as Audience at the British Association for the Advancement of Science, 1831–1901," *Isis* 99, no. 1 (2008): 14.

23. Withers and Higgitt, "Science and Sociability," 20.

24. Withers and Higgitt, "Science and Sociability," 21.

25. Harriet Martineau, *Harriet Martineau's Autobiography*, 2 vols. (London: Virago, 1983), 2:137.

26. *Times*, August 24, 1836, quoted in Withers and Higgitt, "Science and Sociability," 13.

27. Caroline Fox, quoted in Withers and Higgitt, "Science and Sociability," 22.

28. John Herschel, quoted in Clark and Hughes, *Adam Sedgwick*, 1:516.

29. Charles Babbage to Charles Daubeny, April 28, 1832, quoted in Morrell and Thackray, *Gentlemen of Science*, 137.

30. William Buckland, quoted in Morrell and Thackray, *Gentlemen of Science*, 151.

31. [Whewell], *"On the Connexion of the Physical Sciences,"* 55.

32. [Whewell], *"On the Connexion of the Physical Sciences,"* 55.

33. [Whewell], *"On the Connexion of the Physical Sciences,"* 55; original emphasis.

34. [Whewell], *"On the Connexion of the Physical Sciences,"* 56.

35. Todhunter, *Whewell*, 1:92.

36. [Whewell], *"On the Connexion of the Physical Sciences,"* 58.

37. [Whewell], *"On the Connexion of the Physical Sciences,"* 64.

38. [Whewell], *"On the Connexion of the Physical Sciences,"* 65.

39. [Whewell], *"On the Connexion of the Physical Sciences,"* 65.

40. Mary Orr, "Catalysts, Compilers and Expositors: Rethinking Women's Pivotal Contributions to Nineteenth-Century 'Physical Sciences,'" in *The Palgrave Handbook of Women and Science since 1660*, ed. C. G. Jones, A. E. Martin, and A. Wolf (Cham, Switzerland: Palgrave Macmillan, 2021), 507.

41. Todhunter, *Whewell*, 1:116.

42. William Whewell, quoted in Todhunter, *Whewell*, 1:116. Todhunter quotes from Whewell's reply to a reviewer of the *History*.

43. We know for certain that at least the first of the sonnets was composed

by Whewell as it reappears in his *Sunday Thoughts and Other Verses* (Cambridge: Cambridge University Press, 1847), 27.

44. [Whewell], *"On the Connexion of the Physical Sciences,"* 68.

45. [Whewell], *"On the Connexion of the Physical Sciences,"* 68.

46. [Whewell], *"On the Connexion of the Physical Sciences,"* 65.

47. [Whewell], *"On the Connexion of the Physical Sciences,"* 65.

48. [Whewell], *"On the Connexion of the Physical Sciences,"* 65.

49. WW to Mrs [Sarah] Austin, May 13, 1857, quoted in Janet Ross, *Three Generations of Englishwomen. Memoirs and Correspondence of Mrs. John Taylor, Mrs. Sarah Austin, and Lady Duff Gordon*, 2 vols. (London: John Murray, 1888), 2:50.

50. Ross, *Three Generations of Englishwomen*, 465.

51. WW to Mrs. Austin, quoted in Ross, *Three Generations of Englishwomen*, 492–93.

52. [Whewell], *"On the Connexion of the Physical Sciences,"* 66.

53. On the role of Mary Somerville's husband, William Somerville, in supporting and promoting her scientific work, see Brigitte Stenhouse, "Mister Mary Somerville: Husband and Secretary," *Mathematical Intelligencer* 43, no. 1 (March 2021).

54. Francis Galton, *Hereditary Genius: An Inquiry into Its Laws and Consequences* (London: Macmillan, 1869), 192.

55. Galton, *Hereditary Genius*, 193.

56. Frederic W. H. Myers, *Six Lectures on Great Men* (Keswick, UK: T. Bailey and Son, 1848).

57. WW to Frederick W. H. Meyers, June 18, 1848, quoted in Douglas, *Whewell*, 350.

58. WW to Frederick W. H. Meyers, June 18, 1848, 353.

59. WW to Frederick W. H. Meyers, June 18, 1848, 351–52.

60. WW to Frederick W. H. Meyers, June 18, 1848, 352. For more details on Whewell's views on the character of great scientists, and the broader context in which they were shaped, see, for instance, Richard Yeo, "Genius, Method, and Morality: Images of Newton in Britain, 1760–1860," *Science in Context* 2, no. 2 (1988).

61. William Whewell, *The Philosophy of the Inductive Sciences, Founded upon Their History*, 2 vols. (London: John W. Parker, 1840), 2:392.

62. Whewell, *Philosophy*, 2:389.

63. William Whewell, *The History of the Inductive Sciences, from the Earliest to the Present Time*, 3 vols. (London: John W. Parker, 1837), 1:xi–xii.

64. Whewell, *Philosophy*, 2:302–3.

65. Dena Goodman, *The Republic of Letters: A Cultural History of the French Enlightenment* (Ithaca, NY: Cornell University Press, 1994), 3.

12. William Whewell and the Palaetiological Sciences

1. Yeo, *Defining Science*, 53, 73.

2. William Whewell, *An Account of the Experiments Made at Dolcoath Mine, in Cornwall, in 1826 and 1828, for the Purpose of Determining the Density of the Earth* (Cambridge: J. Smith, 1828), 8. See also Yeo, *Defining Science*, 53; Snyder, *Reforming Philosophy*, 149.

3. For Whewell on mineralogical classification, see William Whewell, *An Essay on Mineralogical Classification and Nomenclature: With Tables of the Orders and Species of Minerals* (Cambridge: J. Smith, 1828); Aleta Quinn, "Whewell on Classification and Consilience," *Studies in History and Philosophy of Science Part C: Studies in History and Philosophy of Biological and Biomedical Sciences* 1 (2017). For Whewell's tidology, see Reidy, *Tides of History*; and Snyder, *Philosophical Breakfast Club*, 169–79. See also Aleta Quinn's and Michael S. Reidy's chapters in this volume.

4. Whewell terms *palaetiological* those sciences "in which the object is to ascend from the present state of things to a more ancient condition, from which the present is derived by intelligible causes." See Whewell, *The History of the Inductive Sciences, from the Earliest to the Present Time*, 3 vols. (London: John W. Parker, 1837), 3:481. So understood, the palaetiological science of geology excludes mineralogy and tidal theory.

5. Yeo, *Defining Science*, 73. In his response to Whewell's letter, Sedgwick cites many of the achievements just listed as qualifications for the position: "Have you not figured in the mines of Cornwall? . . . Have you not given the only philosophical view of . . . [mineralogy] that exists in our language? Have you not written the best review of Lyell's system that has appeared in our language? . . . You are just the man we want." Sedgwick also mentions that Whewell "pick[ed] geological rubbish" out of his eyes in 1820, although it is not clear what he is referring to. Adam Sedgwick to WW, October 12, 1836, quoted in in J. W. Clark and T. M. Hughes, *The Life and Letters of the Reverend Adam Sedgwick*, 2 vols. (Cambridge: Cambridge University Press, 1890), 1:464.

6. Martin Rudwick, *The Great Devonian Controversy: The Shaping of Scientific Knowledge among Gentlemanly Specialists* (Chicago: University of Chicago Press, 1985).

7. I write "man" here advisedly. During the 1830s, women were not admitted to the Geological Society of London, even as guests, although some women nonetheless managed to make significant contributions to geology. An example is the fossil hunter Mary Anning. See Hugh Torrens, "Mary Anning of Lyme: The Greatest Fossilist the World Ever Knew," *British Journal for the History of Science* 28 (1995). See in this regard the chapter by Heather Ellis in the present volume.

8. Yeo, *Defining Science*, 8.

9. For more on Whewell as a priest, see Michael Ledger-Lomas's chapter in this volume.

10. For more on Whewell and Cambridge University, see Percy Williams, "Passing on the Torch: Whewell's Philosophy and the Principles of English University Education," in Fisch and Schaffer, *Whewell*; and Sheldon Rothblatt's chapter in the present volume.

11. Whewell, *History*, 3:481.

12. Rudwick, *Great Devonian Controversy*, 3.

13. M. J. S. Hodge, "The History of the Earth, Life, and Man: Whewell and Palætiological Science," in Fisch and Schaffer, *Whewell*, 260. Notably, Whewell, in his early writings, seems to have had some reservations about whether geology belonged among the inductive sciences given its current state of development; see William Whewell, "Modern Science—Inductive Philosophy," *Quarterly Review* 45 (1831): 390.

14. Fundamental Ideas are the foundations of scientific knowledge in an area. They are what allows us to have real knowledge of the world, by furnishing the relations that structure the facts of our experience. For more detailed explications see Snyder, *Reforming Philosophy*, 40–42; and Claudia Cristalli's chapter in the present volume.

15. Whewell, *The Philosophy of the Inductive Sciences, Founded upon Their History*, 2 vols. (London: John W. Parker, 1840), 2:112; original emphases.

16. Hodge, "History of Earth, Life, and Man," 266.

17. Hodge, "History of Earth, Life, and Man," 266.

18. Aleta Quinn, "William Whewell's Philosophy of Architecture and the Historicization of Biology," *Studies in History and Philosophy of Biological and Biomedical Sciences* 59 (2016).

19. Whewell, in his 1838 address to the Geological Society, indicates that geology is presently in a condition not so different from "astronomy at the time of Kepler, when the accumulated observations of twenty centuries resisted all [attempts] . . . to construct a science of physical astronomy." See William Whewell, "Address to the Geological Society, delivered on the Anniversary, on the 16th of February, 1838, by the Reverend William Whewell, M.A., F.R.S., President of the Society," *Proceedings of the Geological Society of London* 2 (1838): 632. He goes on to observe that after Kepler, the science of dynamics came together relatively quickly; and perhaps something similar can be hoped for in the case of geology. But see Whewell, *History*, 3:554, for a splash of cold water.

20. Whewell, *Philosophy*, 2:104.

21. Whewell, *Philosophy*, 2:102.

22. Whewell, *Philosophy*, 2:122.

23. Hodge, "History of Earth, Life, and Man," 264.

24. Hodge, "History of Earth, Life, and Man," 264.

25. Whewell, *History*, 3:581.

26. Whewell, *History*, 3:582.

27. Whewell, *History*, 3:581.

28. Whewell, *Philosophy*, 2:145.

29. Whewell, *Philosophy*, 2:145.

30. To better understand Whewell's position, it is useful to keep in mind that early nineteenth-century geologists often described the task of geology as recovering a past prehuman world, or later, a succession of worlds. Such language implied a succession of causal orders separated by discontinuities, not the more dynamic continuum of later conceptions of deep history. For more on these early conceptions of geohistory, see Martin Rudwick, *Bursting the Limits of Time: The Reconstruction of Geohistory in the Age of Revolution* (Chicago: University of Chicago Press, 2005); Ralph O'Connor, *The Earth on Show: Fossils and the Poetics of Popular Science (1802–1856)* (Chicago: University of Chicago Press, 2008).

31. This is a reconstruction of Whewell's position. Nowhere does he explicitly state that originations involve the advent of new causal laws, yet this seems to follow from his identification of "the present order of things" with a particular set of laws. See, for example Whewell, "Address to the Geological Society." As with much in this section, I follow Hodge in this interpretation. See Hodge, "History of Earth, Life, and Man."

32. Whewell, *Philosophy*, 2:145.

33. Whewell, *Philosophy*, 2:145.

34. Whewell, *Philosophy*, 2:147.

35. Whewell, *Philosophy*, 2:148.

36. Whewell, *Philosophy*, 2:148; original emphasis.

37. Whewell, *Philosophy*, 2:148.

38. William Whewell, "Lyell's *Principles of Geology*," *British Critic and Quarterly Theological Review* 9 (1831): 194. For more on Whewell's religious outlook, see Bernard Lightman's and Michael Ledger-Lomas's chapters in this volume.

39. Whewell, *Philosophy*, 2:116.

40. Whewell, *History*, 3:588.

41. Whewell, *History*, 3:581.

42. For Whewell's relationship with Babbage, see Snyder, *Philosophical Breakfast Club*, 26, 36, 203–10.

43. David Brewster, "On the Decline of Science in England," *Quarterly Review* 43 (1830): 326–27.

44. See Jack Morrell and Arnold Thackray, *Gentlemen of Science: Early Years of the British Association for the Advancement of Science* (Oxford: The Clarendon Press, 1981), 47–52.

45. William Whewell, "Cambridge Transactions—Science of the English Universities," *British Critic and Quarterly Theological Review* 9 (1831): 90.

46. Whewell, "Science of the English Universities," 73.

47. Whewell, "Science of the English Universities," 73.

48. Whewell, "Science of the English Universities," 73.

49. Humphrey Davy, *Six Discourses Delivered before the Royal Society* (London:

John Murray, 1827), 51. The fossils of the Kirkland Cave were a jumble of bones belonging to hyenas, elephants, hippopotamuses, and some smaller animals. Buckland's accomplishment was to show that these animals lived in Britain before the supposed Deluge, and to restore for the first time an antediluvian habitat, an accomplishment that captured the public imagination and vaulted Buckland to scientific fame. For more on this episode, see Rudwick, *Bursting the Limits of Time*, 622–37.

50. Whewell, "Science of the English Universities," 74.

51. Whewell, "Science of the English Universities," 74.

52. Whewell, "Science of the English Universities," 74.

53. Whewell, "Lyell's *Principles of Geology*," 180–81.

54. See William Whewell, "Lyell's *Principles of Geology, Volume 2*," *Quarterly Review* 93 (1832).

55. Michael Bartholomew, "The Singularity of Lyell," *Journal of the History of Science* 17 (1979): 288–89.

56. Whewell, "Lyell's *Principles of Geology*," 199.

57. See Whewell, *History*, 3:548. Whewell goes on to state that Lyell's *Principles* "may . . . be looked upon as the beginning of Geological Dynamics, at least among us [Britons]. Such generalisations and applications as it contains give the most lively interest to a thousand observations respecting rivers and floods, mountains and morasses, which otherwise appear without aim or meaning" (3:552).

58. Whewell, "Lyell's *Principles of Geology*," 195.

59. Whewell, *History*, 3:552.

60. Whewell, *History*, 3:554.

61. Whewell, "Lyell's *Principles of Geology*," 199.

62. Whewell, *Philosophy*, 2:126. The embedded quotation belongs to Lyell. See Charles Lyell, *Principles of Geology, Being an Attempt to Explain the Former Changes of the Earth's Surface, by Reference to Causes Now in Operation* (London: John Murray, 1830), 328.

63. Whewell, *History*, 3:615.

64. Charles Lyell, *Life, Letters and Journals of Sir Charles Lyell*, 2 vols. (London: John Murray, 1881), 1:234.

65. Whewell, *Philosophy*, 2:126.

66. Whewell, *Philosophy*, 2:127.

67. Whewell, *History*, 3:616.

68. Isaac Newton, The Principia, Mathematical Principles of Natural Philosophy: A New Translation, trans. I. B. Cohen & A. Whitman (Berkeley: University of California Press, 1999), 794.

69. Thomas Reid, *Essays on the Intellectual Powers of Man*, critical ed., ed. D. Brookes (Edinburgh: Edinburg University Press, 2002), 80. Reid's explication of Newton's rules for reasoning was first published in 1785.

70. Whewell, *Philosophy*, 2:441; original emphasis.

71. Whewell, *Philosophy*, 2:441–42.

72. Whewell, *Philosophy*, 2:442.

73. Whewell, *Philosophy*, 2:446.

74. For Whewell on consilience, see Snyder, *Reforming Philosophy*, 175–77, 183–85; Quinn, "Whewell on Classification and Consilience"; and Quinn's and Cristalli's chapters in this volume.

75. Snyder suggests that Darwin's argument was not necessarily consilient in Whewell's sense; Snyder, *Reforming Philosophy*, 198–99. Perhaps this is because, as Hodge argues, Darwin did not have the criterion of consilience in mind when composing the *Origin*. See M. J. S. Hodge, *Before and after Darwin: Origins, Species, Cosmogonies, and Ontologies* (Aldershot, UK: Ashgate, 2008), 163–82.

76. Yeo, *Defining Science*, 73.

77. For some suggestive remarks from Whewell's correspondence with James D. Forbes, see Hodge, "History of Earth, Life, and Man," 287.

78. Bert Hansen, "The Early History of Glacial Theory in British Geology," *Journal of Glaciology* 9 (1970): 139.

13. Whewell and Scientific Classification

1. Ursula Klein, "Shifting Ontologies, Changing Classification: Plant Materials from 1700 to 1830," *Studies in History and Philosophy of Science* 36, no. 2 (2005); Aaron Novick, "On the Origins of the Quinarian System of Classification," *Journal of the History of Biology* 49, no. 1 (2015); Aleta Quinn, "Charles Girard: Relationships and Representation in Nineteenth Century Systematics," *Journal of the History of Biology* 50, no. 3 (2017); Oliver Rieppel, *Phylogenetic Systematics* (Boca Raton, FL: CRC Press, 2016); Raphaël Sandoz, "Whewell on the Classification of the Sciences"; Mary P. Winsor, "Non-essentialist Methods in Pre-Darwinian Taxonomy," *Biology and Philosophy* 18, no. 3 (2003).

2. The difficulties apply to the various permutations of "natural/real" and "system/classification/arrangement/relationship/affinity," etc.

3. Willi Hennig, "Cladistic Analysis or Cladistic Classification?," *Systematic Zoology* 24, no. 2 (1975): 246.

4. P. D. Magnus, "No Grist for Mill on Natural Kinds," *Journal for the History of Analytic Philosophy* 2, no. 4 (2014); Magnus, "John Stuart Mill on Taxonomy and Natural Kinds," *HOPOS: The Journal of the International Society for the History of Philosophy of Science* 5, no. 2 (2015); Gordon McOuat, "The Origins of 'Natural Kinds': Keeping 'Essentialism' at Bay in the Age of Reform," *Intellectual History Review* 19, no. 2 (2009).

5. J. S. L. Gilmour, "Taxonomy and Philosophy," in *The New Systematics*, ed. J. B. S. Haldane (Oxford: Clarendon Press, 1940).

6. Compare Winsor, "Non-essentialist Methods"; Snyder, *Reforming Philosophy*.

7. Whewell, *The Philosophy of the Inductive Sciences, Founded upon Their Histo-*

ry (London: John W. Parker, 1840), 1:385, 452, 463; all citations to the *Philosophy* in this chapter will be to the first edition; John Stuart Mill, *A System of Logic: Ratiocinative and Inductive* (London: Harrison, 1843), 2:344–46.

8. Douglas, *Whewell*, 98, 122.

9. WW to HJR, July 11, 1825, R./2.99/24, TCL.

10. WW to HJR, August 15, 1825, R.2.99/25, TCL.

11. George Rapp, "William Whewell: Professor of Mineralogy [and Crystallography], Cambridge University 1828–1834," *Earth Sciences History* 33, no. 1 (2014).

12. Whewell, *Philosophy*, 1:385, 387, 443. Note that *element* refers to a basic constituent that combines with other constituents in a regular way. It did not mean a chemical element defined via the number of protons, as in twentieth-century physical theory. See Hasok Chang, *Is Water H₂O?* (Dordrecht: Springer, 2012).

13. William Whewell, *An Essay on Mineralogical Classification and Nomenclature: With Tables of the Orders and Species of Minerals* (Cambridge: J. Smith, 1828), xxi. See Snyder, *Reforming Philosophy*; and Sandoz, "William Whewell on the Classification of the Sciences."

14. Herbert D. Deas, "Crystallography and Crystallographers," 135. See William Whewell, "A General Method of Calculating the Angles Made by Any Planes of Crystals, and the Laws According to Which They Are Formed. Read Nov. 25, 1824," *Philosophical Transactions of the Royal Society of London* 115, no. 1 (1825).

15. Rapp, "William Whewell," 7.

16. Whewell, *Philosophy*, 1:463–69.

17. Joan Richards, "Observing Science in Early Victorian England: Recent Scholarship on William Whewell," *Perspectives on Science* 4, no. 2 (1996): 235; and Yeo, *Defining Science*, 52–56.

18. See Reidy, *Tides of History*; Ducheyne, "Fundamental Questions." See also Reidy's chapter in the present volume.

19. Carol Cleland, "Methodological and Epistemic Differences between Historical Science and Experimental Science," *Philosophy of Science* 69, no. 3 (2002); Derek Turner, *Making History* (Cambridge: Cambridge University Press, 2007). See Dresow's chapter in this volume for a more complete account of Whewell on the historical sciences.

20. Whewell's historical approach to architecture is evident, for instance, in a letter to Hugh James Rose: "I look at the features of the architecture and pronounce very authoritatively what the date and history of the building ought to be"; WW to HJR, February 20, 1835, R./2.99/41, TCL. See Gillin's chapter in the present volume for a more complete account of Whewell on architecture. See Quinn, "William Whewell's Philosophy of Architecture," for an analysis of Whewell's reasoning about how structural elements necessarily follow from a single key characteristic of Gothic architecture, the pointed arch. See also William J. Ashworth, "William Whewell, Fundamental Ideas and Gothic Architecture,"

Studies in Victorian Architecture and Design 6 (2019); and Yanni, "On Nature and Nomenclature," for analysis of how Whewell's methodology and theory differed from other views about historical architecture.

21. Whewell, *Philosophy*, 1:463–64.

22. Whewell, *Philosophy*, 1:468.

23. Sandoz, "William Whewell on the Classification of the Sciences," provided an extremely cogent analysis of Whewell's classification of the sciences, arguing that Whewell's method of classifying on the basis of scientific *methodology* constitutes a unique innovation distinct from Aristotle's classification on the basis of the scientific *object* and Bacon's classification on the basis of *faculty of the mind.* The latter discussion is particularly helpful in clarifying the relationships among Whewell's "Fundamental Ideas," concepts, and Kantian conditions of experience.

24. William Whewell, *Architectural Notes on German Churches. A New Edition. To Which Is Now Added, Notes Written during an Architectural Tour in Picardy and Normandy* (Cambridge: J. & J. J. Deighton, 1835), xi.

25. Whewell, *Philosophy*, 2:103.

26. Whewell, *Philosophy*, 2:279.

27. William Whewell, *The History of the Inductive Sciences, from the Earliest to the Present Time*, 3 vols. (London: John W. Parker, 1837), 3:487.

28. See the chapter by Max Dresow in the present volume.

29. Whewell, *Philosophy*, 2:103. See also Whewell, *History*, 3:492–538; and William Whewell, "Lyell's *Principles of Geology*," *British Critic and Quarterly Theological Review* 9 (1831); Whewell, "Lyell's *Principles of Geology, Volume Two*," *Quarterly Review* 93 (1832).

30. Winsor, "Non-essentialist Methods"; and Mary P. Winsor, "Considering Affinity: An Ethereal Conversation (Part One of Three)," *Endeavour* 39, no. 1 (2005).

31. Gordon McOuat, "Species, Rules and Meaning: The Politics of Language and the Ends of Definitions in 19th Century Natural History," *Studies in History and Philosophy of Science* 27, no. 4 (1996): 508.

32. Charles Darwin et al., "Report of a Committee Appointed 'to Consider the Rules by Which the Nomenclature of Zoology May Be Established on a Uniform and Permanent Basis,'" in *Report of the Twelfth Meeting of the British Association for the Advancement of Science* (London: John Murray, 1843).

33. As well as the International Code of Zoological Nomenclature (ICZN) and the International Code of Nomenclature for algae, fungi, and plants (ICN; prior to 2011, the International Code of Botanical Nomenclature), the International Code of Phylogenetic Nomenclature (PhyloCode) retains use of the type method for fixing the reference of species names.

34. Darwin et al., "Report," 110–11.

35. Darwin et al., "Report," 121.

36. Matt Haber argued that the type method constitutes an independent dis-

covery, together with empirical application, of Kripke's theory of reference. See Matt Haber, "How to Misidentify a Type Specimen," *Biology and Philosophy* 27, no. 6 (2012); Saul Kripke, "Naming and Necessity," *Semantics of Natural Language*, ed. Donald Davidson and Gilbert Harman (Dordrecht: Reidel, 1972).

37. Whewell, *Philosophy*, 1:476.

38. Darwin et al., "Report," 121.

39. See Cristalli's chapter in this volume for the relation of concepts to "Fundamental Ideas" in Whewell's philosophy of science.

40. The following account is developed more fully and somewhat differently in Aleta Quinn, "Whewell on Classification and Consilience," *Studies in History and Philosophy of Science Part C: Studies in History and Philosophy of Biological and Biomedical Sciences* 64 (2017). Sandoz, "William Whewell on the Classification of the Sciences," provides an analysis I take to be compatible, with greater attention to broader philosophical influences and impacts, particularly with respect to classifying the sciences themselves.

41. Whewell analyzed botany in chapters 1, 2, and 4 of book 8 of the *Philosophy*. Chapter 3 covers mineralogy, building on Book 7's analysis of morphology, including crystallography. Book 9 covers biology.

42. See also William Whewell, "*On the Connexion of the Physical Sciences*. By Mrs. Somerville," *Quarterly Review* 51 (1834).

43. Whewell (*Philosophy*, 2:4) objected to the term *physiology* because the etymology of the root word suggests that physiologists are concerned with "nature" as a whole.

44. Whewell, *Philosophy*, 2:4–5. See Philip Sloane, "Whewell's Philosophy of Discovery and the Archetype of the Vertebrate Skeleton: The Role of German Philosophy of Science in Richard Owen's Biology," *Annals of Science* 60, no. 1 (2003), for analysis of Whewell's interactions with Richard Owen and the development of physiology and comparative anatomy in nineteenth-century Britain. Drawing on correspondence in the archive at Trinity College Library, Sloane showed that Whewell relied on Owen in revising his treatment of biology in the *Philosophy* when preparing the second (1847) edition. Whewell had been criticized for relying too heavily on Cuvier; see Richard Owen to WW, February 11, 1839, February 19, 1839, both in Add. MS a. 9.210.55–57, TCL. In turn Owen thanked Whewell for clarifying "the scientific character of teleological reasoning" (Richard Owen to WW, March 26, 1840, f. 1, Add. MS a. 210.61, TCL) and solicited comments on Owen's anatomical work; Owen to WW, March 14, 1847, March 19, 1847, both in Add. MS a. 210.72–73, TCL.

45. Whewell, *History*, 3:470; Whewell, *Philosophy*, 2:4–5. Whewell was quoting Immanuel Kant, *Kritik der Urtheilskraft* (Berlin: Lagarde und Friederich, 1790), 292. The translation appears to be Whewell's own. See Andrew Cooper, "Reading Kant's Kritik der Urteilskraft in England, 1796–1840," *British Journal for the History of Philosophy* 29, no. 3 (2021).

46. *Change* here refers to transformation in a conceptual sense, in considering the possible space of morphologies, in light of (whatever are) the causal laws by which individual forms are governed. *Change* may be interpreted temporally in the ontogenic sense: a change in the history of crystal growth of a particular mineral, a change in the ontogeny of a particular organism. Whewell did not intend *change* in what we would call an evolutionary (e.g., Darwinian) sense.

47. Whewell, *Philosophy*, 1:518.

48. William Whewell, "Chap 5. Attempts at More Extensive Applications of the Idea of Affinity," ca. 1835, R.18.10/20, TCL.

49. On Whewell's natural theology, see Bernard Lightman's chapter in the present volume.

50. This point is sometimes overlooked; see Francis Arthur Bather, "Biological Classification: Past and Future," *Quarterly Journal of the Geological Society of London* 83, no. 2 (1927). That Whewell viewed natural affinity—the key concept proper to the classificatory sciences—as a *method* fits nicely with Sandoz's arguments that Whewell viewed classification of the sciences in terms of scientific methodologies.

51. See Quinn, "Charles Girard," for analysis of Charles Girard's exposition of this type of argument.

52. Augustin Pyramus de Candolle, *Théorie élémentaire de la botanique; ou, Exposition des principes de la classification naturelle et de l'art de d'écrire et d'étudier les végétaux* (Paris: Déterville, 1813), 207, 220.

53. De Candolle, *Théorie*, 208–9.

54. Whewell, *Philosophy*, 1:471.

55. Whewell, *Philosophy*, 1:518.

56. Just as Whewell separated physiological biology from botany and zoology, BAAS distinguished Section 4, natural history, from Section 5, physiology, anatomy, medicine. "Whewell, William, Letters and Printed Material Received, 1819–1833," R./1.76/73.1, TCL.

57. *Report of the Third Meeting of the British Association for the Advancement of Science* (London: John Murray, 1834).

58. "Whewell, William, Letters and Printed Material Received, 1819–1833," R.1.76/51.1, R./1.76/20.1, R./1.76/28.1, R./1.76/52.1, TCL.

59. Whewell, *History*, 3:227.

60. Snyder, *Reforming Philosophy*.

61. Rieppel, *Phylogenetic Systematics*; John S. Wilkins, *Species: The Evolution of an Idea*, 2nd ed. (Abingdon, UK: Routledge, 2018); and Wilkins, Frank Zachos, and Igor Ya. Pavlinov, eds., *Species Problems and Beyond: Contemporary Issues in Philosophy and Practice* (Boca Raton, FL: CRC Press, 2022).

62. See, for instance, Zina B. Ward, "William Whewell, Cluster Theorist of Kinds," *HOPOS: The Journal of the International Society for the History of Philosophy of Science* 13, no. 2 (2023).

14. Whewell and Moral Philosophy

1. John Rawls, preface to Henry Sidgwick, *The Methods of Ethics*, 7th ed. (Indianapolis: Hackett, 1981 [1874]), v.

2. Rawls, preface, v.

3. See Jeremy Bentham, *An Introduction to the Principles of Morals and Legislation* (London: T. Payne and Son, 1789); John Stuart Mill, *Utilitarianism*, in *The Classical Utilitarians*, ed. John Troyer (Indianapolis: Hackett, 2003); R. M. Hare, *Freedom and Reason* (New York: Oxford University Press, 1963).

4. Mill, *Utilitarianism*, chap. 4.

5. James and Stuart Rachels, *The Elements of Moral Philosophy*, 7th ed. (New York: McGraw Hill, 2012), 112.

6. William Paley, *The Principles of Moral and Political Philosophy*, 2 vols. (London: R. Faulder, 1785).

7. John Stuart Mill, "Dr. Whewell on Moral Philosophy," in vol. 2 of *Dissertations and Discussions: Political, Philosophical, and Historical*, 3 vols. (New York: Henry Holt, 1874).

8. See William Whewell, *The Elements of Morality, Including Polity. In Two Volumes. Third Edition, with a Supplement* (London: John W. Parker and Son, 1854); all references to the *Elements of Morality* in this chapter are to this edition. Unless otherwise specified, references are to volume 1.

9. Paley, *Principles of Moral and Political Philosophy*, 1:vi.

10. Joseph Butler, *Three Sermons on Human Nature and Dissertation on Virtue. Edited by W. Whewell, D.D. with a Preface and a Syllabus of the Work* (Cambridge: Deightons; London: John W. Parker, 1848).

11. William Whewell, "Editor's Preface," in Butler, *Three Sermons on Human Nature*, x–xi.

12. Whewell, "Editor's Preface," xxxvi.

13. Terence Irwin, *The Development of Ethics*, 3 vols. (Oxford: Oxford University Press, 2009), 3:387.

14. Whewell, *Elements of Morality*; William Whewell, *Lectures on the History of Moral Philosophy in England* (Cambridge: John Deighton; London: John W. Parker and Son, 1852).

15. Whewell, *Elements of Morality*, 2.

16. Whewell, *Elements of Morality*, 1.

17. Whewell, *Elements of Morality*, 8.

18. Whewell, *Elements of Morality*, 8–9.

19. Whewell, *Lectures on the History of Moral Philosophy*, xii; original emphasis.

20. Whewell, *Lectures on the History of Moral Philosophy*, xiii; original emphasis.

21. Whewell, *Lectures on the History of Moral Philosophy*, xiv.

22. For these claims, see in particular Whewell, *Elements of Morality*, 14–15.

23. Whewell, *Elements of Morality*, 16–17; emphases in the original.

24. Mill, "Dr Whewell on Moral Philosophy," 135.

25. J. B. Schneewind, *Sidgwick's Ethics and Victorian Moral Philosophy* (Oxford: Oxford University Press, 1977), 130 (quote), 131.

26. Whewell, *Lectures on the History of Moral Philosophy*, 202.

27. Whewell, *Lectures on the History of Moral Philosophy*, 203–5.

28. John Stuart Mill, "Remarks on Bentham's Philosophy," in *The Classical Utilitarians*, ed. John Troyer (Indianapolis: Hackett, 2003), 257.

29. Snyder, *Reforming Philosophy*, chap. 4.

30. Mill, "Dr. Whewell on Moral Philosophy," 141; original emphasis.

31. For a discussion of this aspect of Butler, see Robert Louden, "Butler's Divine Utilitarianism," *History of Philosophy Quarterly* 12, no. 3 (July 1995).

32. Whewell, *Lectures on the History of Moral Philosophy*, 216.

33. Whewell, *Lectures on the History of Moral Philosophy*, 210.

34. Whewell, *Lectures on the History of Moral Philosophy*, 215.

35. Bentham, *Introduction to the Principles of Morals and Legislation*, chap. 19, sec. 4, 311.

36. Whewell, *Lectures on the History of Moral Philosophy*, 223–25; original emphasis.

37. Mill, "Dr. Whewell on Moral Philosophy," 167.

38. Mill, *Utilitarianism*, 97.

39. Mill, "Dr. Whewell on Moral Philosophy," 172–73.

40. Mill, "Dr. Whewell on Moral Philosophy," 177.

41. Sidgwick, *Methods of Ethics*, xvii.

42. Sidgwick, *Methods of Ethics*, 97–98.

43. The key chapter in which Sidgwick articulates his own position (and the most important chapter in the *Methods*) is book 3, chap. 13.

44. Sidgwick, *Methods of Ethics*, book I, chap. 3, "Ethical Judgments."

45. Sidgwick, *Methods of Ethics*, 3, 337, 338 (quote).

46. Sidgwick, *Methods of Ethics*, 338.

47. T. H. Green, *Prolegomena to Ethics*, ed. David Brink (Oxford: Clarendon Press, 2003); F. H., Bradley, *Ethical Studies* (London: Henry King, 1876).

48. Other notable but to my mind less philosophically illuminating discussions are: Becher, "William Whewell's Odyssey"; Sergio Cremasci, "'As Boys Pursue the Rainbow': Whewell's Independent Morality vs. Sidgwick's Dogmatic Intuitionism," in *Proceedings of the Second World Congress on Henry Sidgwick*, ed. Placido Bucolo, Roger Crisp, and Bart Schultz (Catania, Italy: Universita degli Studi Catania, 2011); Ashworth, *Trinity Circle*, chap. 5.

49. J. B. Schneewind, "Whewell's Ethics," in *Studies in Moral Philosophy*, American Philosophical Quarterly Monograph Series, no. 1, series ed. Nicholas Rescher (Oxford: Basil Blackwell, 1968); Schneewind, *Sidgwick's Ethics and Victorian Moral Philosophy* (Oxford: Oxford University Press, 1977).

50. Alan Donagan, "Sidgwick and Whewellian Intuitionism: Some Enigmas," *Canadian Journal of Philosophy* 7, no. 3 (September 1977).

51. Snyder, *Reforming Philosophy*, chap. 4.

52. Terence Irwin, *The Development of Ethics*, 3 vols. (Oxford: Oxford University Press, 2009), vol. 3, secs. 1119–25.

15. Whewell and Liberal Education

1. See Lynn Pasquerella, *What We Value: Public Health, Social Justice, and Educating for Democracy* (Charlottesville: University of Virginia Press, 2022).

2. See Stuart Russell, "Living with Artificial Intelligence," BBC Reith Lectures, 2021, https://www.bbc.co.uk/programmes/m001216k.

3. George Malcolm Young, *Portrait of an Age: Victorian England* (London: Oxford University Press, 1976), 27.

4. Some of this interest becomes apparent from Whewell's correspondence with James D. Forbes, professor of natural philosophy at the University of Edinburgh from 1833 to 1860.

5. For context, see Peter Searby, *A History of the University of Cambridge*, vol. 3, *1750–1870* (Cambridge: Cambridge University Press, 1997), chaps. 2 and 12–14.

6. William Whewell, *Of a Liberal Education in General, and with Particular Reference to the Leading Studies of the University of Cambridge* (London: John W. Parker, 1845), 227.

7. On Whewell's mathematical career, see Tony Crilly's chapter in the present volume.

8. For the philosophical underpinnings of this distinction, see Claudia Cristalli's chapter in this volume.

9. William Whewell, *On the Principles of English University Education* (London: John W. Parker, 1837), 65.

10. Yeo, *Defining Science*, 224.

11. On the Whewell-Paley connection see the chapters by Bernard Lightman and David Phillips in this volume.

12. M. V. Wilkes, "Herschel, Peacock, Babbage and the Development of the Cambridge Curriculum," *Notes and Records of the Royal Society* 44, no. 2 (1990): 212.

13. Noel Annan, *Leslie Stephen, the Godless Victorian* (New York: Random House, 1984), 178.

14. Todhunter, *Whewell*, 2:348, 359.

15. Adam Sedgwick, *A Discourse on the Studies of the University of Cambridge* (Leicester, UK: Leicester University Press, 1969 [1833]), 41.

16. John Stuart Mill, *Autobiography* (Oxford: Oxford University Press, 1924 [1873]), 189–90.

17. Whewell, *On the Principles of English University Education*, 169–71.

18. Whewell, *On the Principles of English University Education*, 34, 37.

19. Sheldon Rothblatt, *Tradition and Change in English Liberal Education: An Essay in History and Culture* (London: Faber and Faber, 1976), 126–31.

20. E. L. Youmans, *The Culture Demanded by Modern Life* (New York: D. Appleton, 1867), 13–14.

21. Youmans, *Culture Demanded by Modern Life*, x, 3–7.

22. William Hamilton, "Thoughts on the Study of Mathematics as a Part of Liberal Education," *Edinburgh Review* 62 (1836). The essay was written in response to Whewell's *Thoughts on the Study of Mathematics as Part of a Liberal Education* (1835).

23. Geoffrey N. Cantor, "Between Rationalism and Romanticism: Whewell's Historiography of the Inductive Sciences," in Fisch and Schaffer, *Whewell*, 77.

24. John Stuart Mill, *Utilitarianism*, 2nd ed. (London: Longman, 1864), 20.

25. Whewell, *On the Principles of English University Education*, 148, 144–45, 17, 12, 41, 34, 145; Whewell, *Of a Liberal Education*, 107; WW to JCH, March 13, 1842, quoted in Douglas, *Whewell*, 264.

26. WW to JCH, March 13, 1842, quoted in Douglas, *Whewell*, 264.

27. See Sheldon Rothblatt, *The Revolution of the Dons: Cambridge and Society in Victorian England* (Cambridge: Cambridge University Press, 1981 [1968]), 198–210, 228, 231.

28. "Mr. [Richard] Whitcombe" to WW, April 29, 1817, quoted in Todhunter, *Whewell*, 1:9.

29. Whewell, *Of a Liberal Education*, 218.

30. See Rothblatt, *Tradition and Change*, chap. 2.

31. Whewell, *On the Principles of English University Education*, 19.

32. George W. Stocking Jr., *Victorian Anthropology* (New York: Free Press, 1987), 32.

33. Thomas Babington Macaulay, *The History of England from the Accession of James II* (New York: Publishers Plate Renting, 1885 [1848]), 1:385–86.

34. William Whewell, *Of a Liberal Education in General; and with Special Reference to the University of Cambridge. Second Edition* (London: John W. Parker, 1850), 107.

35. Snyder, *Reforming Philosophy*, 249–50.

36. Whewell, *On the Principles of English University Education*, 30.

37. Whewell, *On the Principles of English University Education*, 24, 21–24.

38. Whewell, *On the Principles of English University Education*, 18.

39. Whewell, *On the Principles of English University Education*, 26.

40. Cantor, "Between Rationalism and Romanticism," 78.

41. Stocking, *Victorian Anthropology*, 37.

42. Snyder, *Reforming Philosophy*, 257, 265.

43. See Matthew Arnold, "Culture and Anarchy: An Essay in Political and Social Criticism," first published in *Cornhill Magazine*, 1867–68.

44. Snyder, *Reforming Philosophy*, 233.

45. Frederic Myers to WW, [n.d.], quoted in Douglas, *Whewell*, 335.

46. See WW to JCH, March 13, 1846, quoted in Douglas, *Whewell*, 263–64.

47. See *Manliness and Morality: Middle-Class Masculinity in Britain and America*, ed. J. A. Mangan and James Walvin (London: Palgrave Macmillan, 1987).

48. See Paul Schrimpton, *The "Making of Men": The Idea and Reality of Newman's University in Oxford and Dublin* (Leominster, UK: Gracewing, 2014).

49. Whewell, *Of a Liberal Education* (1845), 107. Compare Simon Schaffer, "Scientific Discoveries and the End of Natural Philosophy," *Social Studies of Science* 16, no. 3 (1986).

50. Whewell, *On the Principles of English University Education*, 91–92.

51. WW to Frederic Myers, June 18, 1848, quoted in Douglas, *Whewell*, 350–53.

52. See Harvey W. Becher, "William Whewell and the Preservation of a Liberal Education in an Age of Challenge," *Rocky Mountain Social Science Journal* 12 (1975).

53. "But where are the snows of yesteryear?" from *Le Grand Testament*, "Ballade des Dames du Temps Jadis" (1461); translation by Dante Gabriel Rossetti.

16. "In the Chapel of Trinity College"

1. Joseph Barber Lightfoot, *In Memory of William Whewell, DD: A Sermon* (London: Macmillan, 1866), 9–11, 14, 18.

2. Ashworth, *Trinity Circle*, 101.

3. See Bernard Lightman's chapter in this volume.

4. John Hedley Brooke, "Indications of a Creator," 171.

5. Ashworth, *Trinity Circle*, 5 and elsewhere.

6. William Gibson, "The British Sermon 1689–1901: Quantities, Performance, and Culture," in *The Oxford Handbook of the British Sermon, 1689–1901*, ed. Keith A. Francis and Gibson (Oxford: Oxford University Press, 2014), 3–4.

7. William Whewell, Sermon, St. Paul's Cathedral, January 29, 1860, R.6.17/29, TCL.

8. Whewell, Sermon, St. Paul's Cathedral, January 29, 1860.

9. William Whewell, *Sermons, Preached in the Chapel of Trinity College, Cambridge* (London: John W. Parker, 1847), v.

10. JCH to WW, June 1, 1834, Add.MS.a.206/162, TCL; Thomas Arnold, *Sermons Preached in the Chapel of Rugby School* (London: B. Fellowes, 1833).

11. M. E. Bury and Joseph Pickles, eds., *Romilly's Cambridge Diary 1848–1864* (Cambridge: Cambridgeshire Records Society, 2000), 233.

12. Gibson, "British Sermon," 3.

13. Keith Francis, "Paley to Darwin: Natural Theology versus Science in Victorian Sermons," in Francis and Gibson, *Oxford Handbook of the British Sermon*, 444–61; see also Lightman's chapter in the present volume.

14. J. P. T. Bury, ed., *Romilly's Cambridge Diary, 1832–42* (London: Cambridge University Press, 1967), 56.

15. Bury, *Romilly's Cambridge Diary, 1832–42*, 57.

16. Bury, *Romilly's Cambridge Diary, 1832–42*, 38.

17. See Ashworth, *Trinity Circle*, 69.

18. William Whewell, *On the Foundations of Morals. Four Sermons Preached before the University of Cambridge, November 1837. The Second Edition* (Cambridge: J. and J. J. Deighton, 1839), 37. See Lightman's chapter for their critique of Paley.

19. William Whewell, Sermon, May 27, 1861, R.6.17/34, TCL. For similar advocacy, see Lewis Hensley, *A Sermon Preached in the Chapel of Trinity College, Cambridge, December 15, 1856 on the Occasion of the Commemoration of Benefactors* (Cambridge: Macmillan, 1857), 15.

20. JCH to WW, July 1, 1834, Add.Ms.a.206/162, TCL.

21. Bury, *Romilly's Cambridge Diary, 1832–42*, 74; M. E. Bury and J. D. Pickles, eds., *Romilly's Cambridge Diary, 1842–47* (Cambridge: Cambridgeshire Records Society, 1994), 52.

22. Bury, *Romilly's Cambridge Diary, 1832–42*, 43, 22; Bury and Pickles, *Romilly's Cambridge Diary, 1848–1864*, 77.

23. Jonathan Smith and Christopher Stray, eds., *Cambridge in the 1830s: The Letters of Alexander Chisholm Gooden, 1831–1841* (Woodbridge, UK: Boydell, 2003), 67, 162.

24. W. Rouse Ball, *Notes on the History of Trinity College* (Cambridge, UK: Macmillan, 1899), 161; V. H. H. Green, *Religion at Oxford and Cambridge* (London: SCM Press, 1964), 233; Ashworth, *Trinity Circle*, 87; Bury and Pickles, *Romilly's Cambridge Diary, 1842–47*, 116.

25. Smith and Stray, *Cambridge in the 1830s*, 102; original emphasis.

26. Ashworth, *Trinity Circle*, 86–90.

27. William Whewell, *Remarks on Some Parts of Mr Thirlwall's Letter on the Admission of Dissenters to Academical Degrees* (Cambridge: J. and J. J. Deighton, 1834), 9. See also Whewell, *Additional Remarks upon Some Parts of Mr Thirlwall's Two Letters on the Admission of Dissenters to Academical Degrees* (Cambridge: J. & J. J. Deighton, 1834), 6–7, 11–12.

28. Rouse Ball, *Notes on the History of Trinity*, 162.

29. Bury, *Romilly's Cambridge Diary, 1832–42*, 142.

30. Smith and Stray, *Cambridge in the 1830s*, 110.

31. [William Makepeace Thackeray], *A Few More Words to Freshmen, by the Rev. T.T.* (Cambridge, 1841), 4, 7.

32. Charles Astor Bristed, *Five Years in an English University* (New York: G. P. Putnam, 1853), 23, 53; original emphasis.

33. William Whewell, *On the Principles of English University Education. The Second Edition, Including Additional Thoughts on the Study of Mathematics* (London: John W. Parker, 1838), 111.

34. Whewell, *On the Foundations of Morals*, 59; original emphasis.

35. Whewell, *On the Foundations of Morals*, 56–57.

36. Whewell, *On the Foundations of Morals*, 169.

37. William Whewell, Sermon, October 19, 1862, R.6.17/40, TCL.

38. J. Willis Clark, *Old Friends at Cambridge* (London: Macmillan, 1909), 48, 63.

39. John Kerr, *Memories Grave and Gay* (London: Thomas Nelson, 1911), 363.

40. See Bury, *Romilly's Cambridge Diary, 1832–42*, 224–25.

41. Harriet Scholefield, *Memoir of James Scholefield* (London: Seeley, Jackson, 1855), 27, 36.

42. Bury, *Romilly's Cambridge Diary, 1832–42*, 222, 233.

43. See Robert O. Preyer, "The Romantic Tide Reaches Trinity: Notes on the Transmission of New Approaches to Traditional Studies at Cambridge, 1820–1840," in *Victorian Science and Victorian Values*, ed. James Paradis and Thomas Postlewait (New Brunswick, NJ: Rutgers University Press, 1985); Yeo, *Defining Science*, 65–71.

44. Whewell, *Sermons*, iii.

45. Douglas, *Whewell*, 292–93.

46. William Whewell, Sermon, May 27, 1861, R.6.17/34, TCL.

47. Bury and Pickles, *Romilly's Cambridge Diary, 1842–47*, 4; Bury and Pickles, *Romilly's Cambridge Diary, 1848–1864*, 86, 200.

48. Rouse Ball, *Notes on the History of Trinity*, 166.

49. Bury, *Romilly's Cambridge Diary, 1832–42*, 228, 231; original emphasis.

50. Bury and Pickles, *Romilly's Cambridge Diary, 1842–47*, 79, 47.

51. Bury and Pickles, *Romilly's Cambridge Diary, 1848–1864*, 427.

52. Whewell, *Sermons*, 23–24.

53. Whewell, *Sermons*, 13, 17.

54. Whewell, *Sermons*, 37.

55. William Whewell, Sermon, 1864, R.6.17/44, TCL.

56. Francis Cowley Burnand, *Records and Reminiscences: Personal and General*, 4th ed. (London: Methuen, 1905), 181.

57. Whewell, Sermon, April 25, 1863, R.6.17/41, TCL.

58. John Grote, *The Commemoration Sermon, Preached in Trinity College, Cambridge, December 15, 1848* (Cambridge: John Deighton, 1848), 7.

59. W. H. Thompson, *Old Things and New* (Cambridge: J. Deighton, 1853), 16–17. See Ashworth, *Trinity Circle*, 38–40, for a good summary of Hare's and Thirlwall's contribution.

60. Bury and Pickles, *Romilly's Cambridge Diary, 1842–1847*, 6.

61. Whewell, *Sermons*, 44–45, 51; original emphasis.

62. Ashworth, *Trinity Circle*, 69–70.

63. Whewell, *Sermons*, 74.

64. William Whewell, Sermon, [1863–64], R.6.17/42, TCL.

65. Whewell, *Sermons*, 100.

66. William Whewell, Sermon, November 8, 1861, R.6.17/36; Sermon, October 20, 1861, R6.17/35, both in TCL; original emphasis.

67. Whewell, *Sermons*, 109.

68. Ashworth, *Trinity Circle*, 108.

69. Ashworth, *Trinity Circle*, 102.

70. Joseph Williams Blakesley, *Catholicity and Protestantism: The Commemoration Sermon* (Cambridge: J. and J. J. Deighton, 1842), 8, 21–23.

71. Whewell, *Sermons*, 142, 327.

72. Bury and Pickles, *Romilly's Cambridge Diary, 1842–47*, 48, 72–75.

73. William Whewell, *Strength in Trouble: A Sermon Preached in the Chapel of Trinity College, Cambridge on February 23, 1851* (London: John W. Parker, 1851), 14–15.

74. See, e.g., Thompson, *Old Things and New*, which contains a lengthy diatribe against Newman's theology.

75. Whewell, *Sermons*, 157; Whewell, *Strength in Trouble*, 11.

76. Whewell, Sermon, March 10, 1861, R.6.17/33, TCL.

77. Hensley, *Sermon*, 14.

78. Whewell, *Sermons*, 188, 5.

79. Whewell, Sermon, May 27, 1861, R6.17/34, TCL.

80. William Whewell, *A Sermon Preached on Trinity Monday, June 15, 1835, before the Corporation of the Trinity House, in the Parish Church of Saint Nicholas, Deptford, and Printed at Their Request* (London: Charles Whittingham, 1835), 16.

81. William Whewell, *The Christian's Duty towards Transgressors: A Sermon Preached in the Chapel of the Philanthropic Society St George's Fields, on Sunday, 16 May 1847, Being the Annual Commemoration of the Society's Establishment* (London: J. Samuel, 1847), 18–19.

82. Whewell, *Christian's Duty*, 20.

83. Whewell, *Sermons*, 160.

84. Ashworth, *Trinity Circle*, 103.

85. Whewell, *Sermons*, 171–73.

86. Whewell, *Sermons*, 190, 268.

87. Whewell, *Sermons*, 197, 201.

88. Whewell, Sermon, January 29, 1860, R.6.17/29, TCL.

89. William Whewell, Sermon, February 16, 1862, R.6.17/37, TCL.

90. See, e.g., Lightman's chapter in this volume for his riposte to *Vestiges*.

91. William Whewell, Sermon, April 29, 1860, R.6.17/31, TCL

92. William Whewell, Sermon, May 18, 1862, MS R.6.17/38, TCL.

93. See Simon Goldhill, "Genealogy, Translation and Resistance: Between the Bible and the Greeks," in *Victorian Responses to the Bible and Antiquity: The Shock of the Old*, ed. Goldhill and Ruth Jackson Ravenscroft (Cambridge: Cambridge University Press, 2023).

94. J. R. M. Butler, *Henry Montagu Butler: Master of Trinity College, Cambridge, 1886–1918* (London: Longmans, 1925), 10, 19–20.

95. Butler, *Henry Montagu Butler*, 60–62.

96. Butler, *Henry Montagu Butler*, 43.

17. Whewell on Church and State

1. For background, see Peter Searby, *A History of the University of Cambridge*, vol. 2, *1750–1870* (Cambridge: Cambridge University Press, 1997), chaps. 12–14.

2. Robert Peel, "Tamworth Manifesto," in *Peel*, ed. J. R. Thursfield (London: Macmillan, 1891).

3. Christopher Wordsworth to WW, October 12, 1841, quoted in Douglas, *Whewell*, 226.

4. Jack Morrell and Arnold Thackray, *Gentlemen of Science: Early Years of the British Association for the Advancement of Science* (Oxford: Clarendon Press, 1981), 114.

5. Whewell to his sister Ann, [?] 1822, quoted in Douglas, *Whewell*, 74.

6. D. R. Fisher, *The History of Parliament: The House of Commons 1820–32*, 7 vols. (Cambridge: Cambridge University Press, 2009), 2:95.

7. WW to Rev. Wilkinson, May 21, 1822, quoted in Douglas, *Whewell*, 76.

8. Fisher, *History of Parliament*, 2:103, 5:341.

9. J. W. Clark, *Old Friends at Cambridge and Elsewhere* (London, 1900), 10.

10. Clark, *Old Friends*, 17.

11. WW to Adam Sedgwick, May 27, 1834, quoted in Douglas, *Whewell*, 163.

12. WW to Connop Thirlwall, 23 September [1834], quoted in Douglas, *Whewell*, 166.

13. William Whewell, *Remarks on Some Parts of Mr Thirlwall's Letter on the Admission of Dissenters to Academical Degrees* (Cambridge: J. & J. J. Deighton, 1834), 5–6, 8–9.

14. William Whewell, *Additional Remarks upon Some Parts of Mr Thirlwall's Two Letters on the Admission of Dissenters to Academical Degrees* (Cambridge: J. & J. J. Deighton, 1834), 22.

15. WW to Adam Sedgwick, May 27, 1834, quoted in Douglas, *Whewell*, 163.

16. William Whewell, *Two Introductory Lectures to Two Courses of Lectures on Moral Philosophy, Delivered in 1839 and 1841* (Cambridge: John W. Parker, 1841), iv.

17. Thomas Turton, *Thoughts on the Admission of Persons without Regard to Their Religious Options to Certain Degrees in the Universities of England* (Cambridge: John Smith; London: John W. Parker, 1834), 25.

18. Martha Garland, *Cambridge before Darwin: The Ideal of a Liberal Education, 1800–1860* (Cambridge: Cambridge University Press, 1980), 78.

19. WW to sister, May 8, 1835, quoted in Douglas, *Whewell*, 173.

20. WW to Lord Lytteton, March 10, 1840, quoted in Douglas, *Whewell*, 201.

21. D. A. Winstanley, *Early Victorian Cambridge* (Cambridge: Cambridge University Press, 1947), 105.

22. WW to James Marshall, December 27, 1842, quoted in Douglas, *Whewell*, 282; original emphasis.

23. William Whewell's journal, n.d., quoted in Winstanley, *Early Victorian Cambridge*, 112, 120.

24. WW to HJR, November 19, 1826, R.2.99/26, TCL, quoted in Snyder, *Reforming Philosophy*, 23.

25. William Whewell, *Newton and Flamsteed: Remarks on an Article in Number CIX of the Quarterly Review* (Cambridge: J. & J. J. Deighton; London: John W. Parker, 1836), 4.

26. S. P. Rigaud to WW, June 25, 1836, Add. MS a. 2011/79, TCL.

27. Morrell and Thackray, *Gentlemen of Science*, 25.

28. Morrell and Thackray, *Gentlemen of Science*, 143.

29. WW to Vernon Harcourt, September 22, 1831, quoted in Morrell and Thackray, *Gentlemen of Science*, 75.

30. See Whewell, *The History of the Inductive Sciences, from the Earliest to the Present Time*, 3 vols. (London: John W. Parker, 1837), 2:280–82.

31. [David Brewster], "Review of Whewell's *History of the Inductive Sciences*," *Edinburgh Review* 66 (1837): 150–51.

32. Morrell and Thackray, *Gentlemen of Science*, 295.

33. JH to WW, September 20, 1831, *Electronic Enlightenment. Letters and Lives*, https://www.e-enlightenment.com/.

34. WW to JDF, February 14, 1835, quoted in Todhunter, *Whewell*, 2:294.

35. See Todhunter, *Whewell*, 1:95.

36. See Todhunter, *Whewell*, 1:150–51.

37. WW to Adolphe Quetelet, February 3, 1835, quoted in Todhunter, *Whewell*, 2:201.

38. William Whewell, *The Elements of Morality, Including Polity*, 2 vols. (London: John W. Parker, 1845), 2:475.

39. Wilfred Airy, ed., *Autobiography of Sir George Biddell Airy* (Cambridge: Cambridge University Press, 1896), 106.

40. John Campbell Shairp, Peter Guthrie Tait, and Anthony Adams-Reilly, eds., *Life and Letters of James David Forbes* (London: Macmillan, 1873), 175.

41. WW to R. I. Murchison, November 22, 1835, *Electronic Enlightenment*.

42. Morrell and Thackray, *Gentlemen of Science*, 355, 361, 368.

43. Reidy, *Tides of History*, 284. See also Reidy's chapter in the present volume.

44. R. I. Murchison to WW, March 1, 1844, *Electronic Enlightenment*; original emphasis.

45. See Michael Reidy, *Tides of History*, 71–284, and his chapter in this volume.

46. See, in this regard, the chapters by Verburgt and Cristalli in this volume.

47. See Todhunter, *Whewell*, 1:79.

48. William Whewell, "Researches on the Tides—6th Series. On the Results of an Extensive System of Tide Observations Made on the Coasts of Europe and America in June 1835," *Philosophical Transactions of the Royal Society of London* 126, no. 2 (1836): 289, 291.

49. Whewell, "Researches on the Tides—6th Series," 304.

50. WW to his sister, December 19, 1834; WW to his sister, March 1, 1835, both quoted in Douglas, *Whewell*, 162, 172.

51. Reidy, *Tides of History*, 133–36.

52. Reidy, *Tides of History*, 146.

53. Morrell and Thackray, *Gentlemen of Science*, 378.

54. Todhunter, *Whewell*, 1:339.

55. J. W. Clark, *Old Friends*, 11.

56. Todhunter, *William Whewell*, 1:342.

57. Whewell, *Elements of Morality*, 401.

58. WW to Rev. H. Murchison, 2 October, 1840, quoted in Todhunter, *Whewell*, 2:290–92.

59. Morrell and Thackray, *Gentlemen of Science*, 425.

60. See Bernard Lightman's chapter in the present volume.

61. WW to JDF, January 4, 1864, quoted in Todhunter, *Whewell*, 2:435.

62. William Whewell, "Editor's Preface," in *Butler's Six Sermons on Moral Subjects. A Sequel to the Three Sermons on Human Nature. Edited by W. Whewell, D.D. with a Preface and a Syllabus of the Work* (Cambridge: John Deighton; London: John W. Parker, 1849), ix.

63. Brooke, "Indications of a Creator: Whewell as Apologist and Priest," 155, 157.

64. William Whewell, quoted in Becher, "William Whewell's Odyssey," 17.

65. Todhunter, *Whewell*, 1:153.

66. Whewell, *Two Introductory Lectures*, 37, 49. On Whewell's moral philosophy, see David Phillips's chapter in the present volume.

67. Whewell, *Elements of Morality*, 208, 406, 525, 533, 251, 257.

68. Whewell, *Elements of Morality*, 208, 406, 525, 533, 251, 257.

69. "Dr. Whewell's Moral Works," *British Quarterly Review* 38 (1863): 398.

70. WW to Frederic Myers, September 6, 1845, quoted in Douglas, *Whewell*, 325–26.

71. Whewell, *Sermons Preached in the Chapel of Trinity College, Cambridge* (1847), 279.

72. Whewell, *Elements of Morality*, 391, 395; original emphasis.

73. See William Whewell, "Mathematical Exposition of Some Doctrines of Political Economy. Read March 2 and 14, 1829," *Transactions of the Cambridge Philosophical Society* 3 (1830). On Whewell's views on political economy, see Harro Maas's chapter in the present volume.

74. Yeo, *Defining Science*, 106. For a similar view, see Paul Oslington, "Natural Theology, Theodicy, and Political Economy in Nineteenth-Century Britain: William Whewell's Struggle," *History of Political Economy* 49, no. 4 (2017): 596.

75. William Whewell, *Of a Liberal Education in General; and with Particular Reference to the Leading Studies of the University of Cambridge* (London: John W. Parker, 1845), 127.

76. William Whewell, quoted in Searby, *History of the University of Cambridge*, 450.

77. Whewell, *Liberal Education*, 118–19.

78. William Whewell, *Suggestions Respectfully Offered to the Cambridge University Commissioners*, January 18, 1858, Letters and Papers Concerning College Statutes, Trinity College Documents, 9, quoted in Winstanley, *Early Victorian Cambridge*, 352n1.

79. Winstanley, *Early Victorian Cambridge*, 174.

80. Winstanley, *Early Victorian Cambridge*, 345.

81. Clark, *Old Friends*, 63.

82. Becher, "Whewell's Odyssey," 27.

83. See Douglas, *Whewell*, 400.

84. WW to JDF, January 28, 1863, quoted in Todhunter, *Whewell*, 2:429.

85. A. J. Engel, *From Clergyman to Don: The Rise of the Academic Profession in Nineteenth-Century Oxford* (Oxford: Oxford University Press, 1983). On changes in nineteenth-century Cambridge, see also Sheldon Rothblatt, *The Revolution of the Dons: Cambridge and Society in Victorian England* (New York: Faber and Faber, 1968); and his chapter in this volume.

18. The Whewell Papers at Trinity College Library

1. Rev. William Whewell D.D., Master of Trinity College, Cambridge. Copy will, 17 Dec. 1865, 30 Trinity College 12, Trinity College Archive.

2. Whewell, Copy will. Ann Whewell appears to have married William Newton in the mid-1860s, as she was mentioned as Ann Whewell in the will, but signs her name Ann Newton by December 1866.

3. ODNB, s.v. "Wright, William Aldis," by D. N. Smith, rev. David McKitterick, accessed May 15, 2022, https://doi.org/10.1093/ref:odnb/18370.

4. For Wright's papers by and relating to Edward FitzGerald, see Add. MS a. 5–20 and Add. MS a. 282–284, TCL. For Wright's papers, see Add. MS a. 39–43, 88 (various papers); Add. MS b. 63–66 (Bible revision); Add. MS c. 68–71 (correspondence); and WAW (papers).

5. W. H. Thompson to J. L. Hammond, November 5, 1872, Add. MS a. 77/258, TCL.

6. Montague Rhodes James, *The Western Manuscripts in the Library of Trinity College, Cambridge* (Cambridge: Cambridge University Press, 1902), 3:xii.

7. Ann Newton to J. L. Hammond, August 16, 1869, Add. MS a. 77/164, TCL.

8. Todhunter, *Whewell*, 1:v.

9. Isaac Todhunter entered St. John's College, Cambridge in 1844 and was senior wrangler in 1848. He won the Smith's Prize and the Burney Prize and was elected to the fellowship in 1849. A respected mathematician, he was awarded the Adams Prize in 1871 and elected to the council of the Royal Society that year. He was interested in the history and philosophy of science and served as examiner for the Moral Sciences Tripos three times. It is not clear why Todhunter was chosen for this work, other than that he was immensely respected and shared some of Whewell's interests. See Stray's chapter in the present volume.

10. Todhunter, *Whewell*, 1:vi.

11. Todhunter, *Whewell*, 1:vii.

12. Todhunter, *Whewell*, 1:vii–viii.

13. Samuel Muller, Johan Feith, and Robert Fruin, *Handleiding voor het ordenen en beschrijven van archieven* (Groningen, the Netherlands: Erven B. van der Kamp, 1898).

14. Isaac Todhunter, "Catalogue of Dr Whewell's Papers," [1872–1874], Add. MS a. 70, TCL.

15. Douglas, *Whewell*, v. Todhunter's view of the division of the biography was later represented by his son Lt. Col. H. W. Todhunter in a letter to Trinity College librarian A. F. Scholfield in November 1919 when donating a volume of copies of Whewell's letters: "When my father accepted that task he imagined that the whole work would be left to him—but subsequently he was asked to confine himself to the scientific side to leave room for Mrs. Stair Douglas' Life of Whewell. The result was that my father did a great amount of work in arranging and examining Whewell's papers which was wasted and I belief [*sic*] he felt deeply that he was hereby compelled to bring out under his name a strikingly inadequate memoir." H. W. Todhunter to the Trinity College Librarian [A. F. Scholfield], November 18, 1919, Add MS c. 82/89, TCL.

16. Douglas, *Whewell*, vi.

17. Douglas, *Whewell*, v.

18. Douglas, *Whewell*, vi.

19. Contents list of Whewell's papers sent by W. A. Wright to [Janet Mary Douglas], [18—], Add. MS a. 58/59, TCL.

20. Ann Newton to J. L. Hammond, November 6, 1872, Add. MS a. 77/146, TCL.

21. Information from a copy letter to T. H. Marshall, unsigned, March 18, 1970 with Ann Newton's collection of William Whewell letters to John Whewell, Alice Lyon, and Elizabeth, Martha, and Ann Whewell, 1811–1840, Add. MS c. 191/1–66, TCL. The letters are cataloged as Add. MS c. 191–193, TCL.

22. Now cataloged as William Whewell letters to his family, 1815–1823, Add. MS a. 27, TCL.

23. Now cataloged as Ann Newton's collection of William Whewell letters to his sister Ann Whewell, 1845–1850, Add. MS c. 190, TCL.

24. W. H. Thompson to J. L. Hammond, August 27, 1870. Add. MS a. 77/260, TCL.

25. The Additional Manuscripts series was originally ordered by size, Add. MS a to Add. MS d, with "a" as the largest. This division of material has since been abandoned and all new acquisitions are now added to "a."

26. Todhunter, "Catalogue."

27. "Index to the Whewell Letters in O & Autographs & Letters Filed in Boxes & Placed in Adversaria Class in Lower Library," [c 1930?], O.15.51, TCL.

28. Walter F. Cannon, Typescript outline of the papers of William Whewell, July 1961, O.15.50/2, TCL.

29. See Papers of William Whewell, Trinity College, Cambridge, Archives, https://archives.trin.cam.ac.uk/index.php/papers-of-william-whewell (or tinyurl.com/whewell).

30. The Royal Commission on Historical Manuscripts, *Report on the Papers of William Whewell, D. D., F.R.S. (1794–1866), Mathematician and Master of Trinity College, Cambridge* (London: Royal Commission on Historical Manuscripts, 1973).

31. Newton, William Whewell letters, 1845–1850; Ann Newton's collection of William Whewell letters to John Whewell, Alice Lyon, and Elizabeth, Martha, and Ann Whewell, 1811–1840, Add. MS c. 191, TCL; Ann Newton's collection of Ann Whewell letters received from William Whewell, Cordelia Whewell, and Frances Affleck, 1840–1861, Add. MS c. 192, TCL; Ann Newton's collection of Ann Whewell letters received from William Whewell and Frances Affleck, with miscellaneous letters, 1809–1869, Add. MS c. 193, TCL; Whewell letters to his family, 1815–1823.

32. WW to JCH, September 26, 1819, Add. MS a. 215/3, TCL.

33. John Ruskin to WW, May 2, [18—], Add. MS c. 90/84, TCL.

34. Michael Faraday to WW, May 20, 1840, O.15.49/13, TCL.

35. "Swing" to WW, [ca. 1830?], R.1.76/45, TCL.

36. George Biddell Airy to WW, February 14, 1863, Add. MS a. 200/48, TCL.

37. For Whewell's letters to Richard Jones, 1817–1854, see Add. MS c. 51, TCL.

38. WW's letters to JCH, 1818–1854, Add. MS a. 215, TCL.

39. Correspondence between WW and Michael Faraday, 1831–1860, O.15.49, TCL.

40. Gift of D. J. Blaikley in 1914 and 1916.

41. Michael Faraday to WW, May 15, 1834. O.15.49/6, TCL.

42. Draft of a letter to be sent by the Librarian to J. O. Halliwell, 3 Feb. [1845], O.15.45/1/141, TCL.

43. William Whewell tidology papers, 1772–1856, [18—], R.6.20, TCL.

44. William Cecil, 1st Baron Burghley to Robert Petre, [sixteenth century], Add. MS c. 65/18, TCL.

45. *A Prodigy of Universal Genius: Robert Leslie Ellis, 1817–1859*, ed. Lukas M. Verburgt (Cham, Switzerland: Springer, 2022).

46. The letters of other members of the Ellis family may be found as part of multiple shelf marks: Add. MS a. 68, 81, 218, 220, 223; Add. MS c. 65, 66, 67, all in TCL.

47. See note 31.

48. Ann Newton's collection of Ann Whewell letters received, 1840–1861, [18—]; Ann Newton's collection of Ann Whewell letters received, 1809–1869.

49. Ann Newton to J. L. Hammond, February 5, 1873, Add. MS a. 77/171, TCL.

50. Letters from John W. Clark to J. L. Hammond, March [1866], Add. MS a. 78/38–40, TCL.

51. Letters from various persons to J. L. Hammond, 1874–1879, Add. MS a. 77/207–272, TCL.

52. J. L. Hammond papers relating to Isaac Todhunter's *Whewell*, 1872–1876, Add. MS a. 36, TCL.

53. William Whewell, Draft of *The elements of morality, including polity*, vols. I–II, 1844–1845, R.6.1–2; Whewell, Drafts of parts of *The elements of morality, including polity* and other miscellaneous papers, [c 1848], R.6.10; Whewell, Letters and notes relating to *The elements of morality, including polity*, 1844–1849, [18—], R.6.13; Whewell, Notes relating to *The elements of morality*, 1842–1844, R.6.15/16–21; Whewell, Proof sheets, *The elements of morality*, 1845, R.6.15/22, all in TCL.

54. *The History of the Inductive Sciences*: Drafts and notes, R.18.10, Draft, R.18.11–13, both in TCL.

55. *The Philosophy of the Inductive Sciences*: Drafts and notes, R.18.10; Drafts, R.6.18, both in TCL.

56. Sermons: drafts and notes, R.6.17; Drafts, R.6.18, both in TCL.

57. Sermons preached in the Chapel of Trinity College, Cambridge: draft, R.6.19/8; Page proofs, R.6.19/7, both in TCL.

58. Partial draft of *The Platonic Dialogues for English Readers*, 1845–1854, R.6.19/9, TCL.

59. Todhunter, *Whewell*, 1:376.

60. William Whewell's papers relating to English hexameters, 1839–1874, R.18.14, TCL.

61. "The Worship of this Sabbath Morn," by Dorothy Wordsworth, 11 Aug. 1843, R.18.15/13/32, TCL.

62. "Account of a Charade on the Name of the Learned Professor Whewell as Represented at Castle Ashby, December 29 1837, 1858," R.18.15/13/2–3, TCL.

63. See Trinity Library Catalogue, https://lib-cat.trin.cam.ac.uk/.

64. William Whewell, Etymological notes, lists of words beginning with "A" and "H" [19th cent.], R.6.4; Letters, notes, and printed material relating to etymology [19th cent.], R.6.5; Etymological notes, lists of words beginning with "E" [19th cent.], R.6.6; Etymological notes, lists of words beginning with "D" [19th cent.], R.6.7; Page proofs of *Lectures on systematic morality*, draft prefaces to *Butler's sermons*, and etymological notes [19th cent.], R.6.8, all in TCL.

65. William Whewell tidology papers, 1772–1856, [18—], R.6.20, TCL.

66. "V.C. 1842–3" notebook, 1842–1843, Add. MS a. 66/1; and Vice-Chancellor's cash book, 1842–1843, Add. MS a. 80/9, both in TCL.

67. William Whewell diaries, 1820–1840. Add. MS a. 80/2–6; and William Whewell diary, July–September 1829, Add. MS a. 80/8, both in TCL.

68. Journal, 1841–1853, O.15.45, TCL.

69. William Whewell notes on books read, 1817–1840, R.18.9; and Notes on books read, Oct.-Dec. 1817, R.18.6/1, both in TCL.

70. William Whewell sketchbooks, 1821–1853, [18—], R.6.12; and Whewell, Sketchbook "N," 1828, Add.MS.a.83/1, both in TCL.

71. William Whewell, "W" account book, Add. MS a. 219/4; and "P" account book, Add. MS a. 219/5; Whewell memorandum books and miscellaneous writings, R.18.17a; Whewell memorandum books, Add. MS a. 83/2–4; Whewell memorandum books, Add. MS a. 83/2–4; Whewell memorandum books, Add. MS a. 83/2–4, all in TCL.

Selected Bibliography

The following list has been compiled to provide a starting point for further research on Whewell's life and work, including aspects not covered in the present book. It contains sources used in the chapters as well as items not cited or discussed in the book that are relevant for Whewell scholarship.

For an overview of Whewell's oeuvre, see the online supplement, "List of Whewell's Published Work."

Ashworth, William J. *The Trinity Circle: Anxiety, Intelligence, and Knowledge Creation in Nineteenth-Century England*. Pittsburgh: University of Pittsburgh Press, 2021.

Becher, Harvey W. "William Whewell and Cambridge Mathematics." *Historical Studies in the Physical Sciences* 11, no. 1 (1980): 1–48.

Becher, Harvey W. "William Whewell's Odyssey: From Mathematics to Moral Philosophy." In Fisch and Schaffer, *Whewell*, 1–29.

Brooke, John Hedley. "Indications of a Creator: Whewell as Apologist and Priest." In Fisch and Schaffer, *Whewell*, 149–73.

Brooke, John Hedley. "Natural Theology and the Plurality of Worlds: Observations on the Brewster-Whewell Debate." *Annals of Science* 34, no. 3 (1977): 221–86.

Buchwald, Jed Z. "The Emerging Dominance of the Wave Theory." In *The Rise of the Wave Theory of Light: Optical Theory and Experiment in the Early Nineteenth Century*, 291–310. Chicago: University of Chicago Press, 1989.

Butts, Robert E. *William Whewell's Theory of Scientific Method*. Pittsburgh: University of Pittsburgh Press, 1968.

Cannon, Susan Faye. "The Whewell-Darwin Controversy." *Journal of the Geological Society of London* 132 (1976): 377–84.

Cantor, Geoffrey N. "Between Rationalism and Romanticism: Whewell's Historiography of the Inductive Sciences." In Fisch and Schaffer, *Whewell*, 67–87.

Cowles, Henry M. "The Age of Methods: William Whewell, Charles Peirce, and Scientific Kinds." *Isis* 107, no. 4 (2016): 722–37.

Deas, Herbert D. "Crystallography and Crystallographers in England in the Early Nineteenth Century: A Preliminary Survey." *Centaurus* 6, no. 2 (1959): 129–48.

Douglas, Janet Mary Stair. *The Life and Selections from the Correspondence of William Whewell, D.D., Late Master of Trinity College, Cambridge.* London: C. Kegan Paul, 1881.

Ducheyne, Steffen. "Fundamental Questions and Some New Answers on Philosophical, Contextual and Scientific Whewell: Some Reflections on Recent Whewell Scholarship and the Progress Made Therein." *Perspectives on Science* 18, no. 2 (2010): 242–72.

Ducheyne, Steffen. "Whewell's Philosophy of Science." In *The Oxford Handbook of British Philosophy in the Nineteenth Century*, edited by W. J. Mander, 71–88. Oxford: Oxford University Press, 2014.

Elkana, Yehuda, ed. *William Whewell: Selected Writings on the History of Science.* Chicago: University of Chicago Press, 1984.

Fisch, Menachem. "'The Emergency Which Has Arrived': The Problematic History of Nineteenth-Century British Algebra—A Programmatic Outline." *British Journal for the History of Science* 27 (1994): 247–76.

Fisch, Menachem. *William Whewell, Philosopher of Science.* Oxford: Clarendon Press, 1991.

Fisch, Menachem, and Simon Schaffer, eds. *William Whewell: A Composite Portrait.* Oxford: Clarendon Press, 1991.

Henderson, James P. *Early Mathematical Economics: William Whewell and the British Case.* Lanham, MD: Rowman and Littlefield, 1986.

Jenkins, Alice. "Space and the Languages of Science." In *Space and the "March of Mind": Literature and the Physical Sciences in Britain 1815–1850*, 113–38. Oxford: Oxford University Press, 2007.

Maas, Harro. "'A Hard Battle to Fight': Natural Theology and the Dismal Science, 1820–50." *History of Political Economy* 40, no. 5 (2008): 143–67.

Marsden, Ben. "'The Progeny of These Two Fellows': Robert Willis, William Whewell and the Sciences of Mechanism, Mechanics and Machinery in Early Victorian Britain." *British Journal for the History of Science* 37, no. 4 (2004): 401–34.

Morrell, Jack, and Anthony Thackray. *Gentlemen of Science: Early Years of the British Association for the Advancement of Science.* Oxford: Oxford University Press, 1981.

Oslington, Paul. "Natural Theology, Theodicy, and Political Economy in Nineteenth-Century Britain: William Whewell's Struggle." *History of Political Economy* 49, no. 4 (2017): 575–606.

Quinn, Aleta. "William Whewell's Philosophy of Architecture and the Historicization of Biology." *Studies in History and Philosophy of Science Part C* 59 (2016): 11–19.

Reidy, Michael S. *Tides of History: Ocean Science and Her Majesty's Navy.* Chicago: University of Chicago Press, 2008.

Ruse, Michael. "Darwin's Debt to Philosophy: An Examination of the Influence of the Philosophical Ideas of John F.W. Herschel and William Whewell on the Development of Charles Darwin's Theory of Evolution." *Studies in History and Philosophy of Science Part A* 6, no. 2 (1975): 159–81.

Sandoz, Raphaël. "Whewell on the Classification of the Sciences." *Studies in History and Philosophy of Science* 60 (2016): 48–54.

Schaffer, Simon. "The History and Geography of the Intellectual World: Whewell's Politics of Language." In Fisch and Schaffer, *Whewell*, 201–31.

Schaffer, Simon. "Scientific Discoveries and the End of Natural Philosophy." *Social Studies of Science* 16, no. 3 (1986): 387–420.

Siegel, Daniel M. "The Background to Maxwell's Electromagnetic Theory." In *Innovation in Maxwell's Electromagnetic Theory*, 5–28. Cambridge: Cambridge University Press, 1991.

Snyder, Laura J. "Discoverers' Induction." *Philosophy of Science* 64, no. 4 (1997): 580–604.

Snyder, Laura J. "'It's All Necessarily So': William Whewell on Scientific Truth." *Studies in History and Philosophy of Science Part A* 25, no. 5 (1994): 785–807.

Snyder, Laura J. *Reforming Philosophy: A Victorian Debate on Science and Society.* Chicago: University of Chicago Press, 2006.

Snyder, Laura J. "'The Whole Box of Tools': William Whewell and the Logic of Induction." In *Handbook of the History of Logic.* Vol. 4, *British Logic in the Nineteenth Century*, edited by Dov. M. Gabbay and John Woods, 163–228. Amsterdam: Elsevier, 2008.

Todhunter, Isaac, ed. *William Whewell, D.D., Master of Trinity College, Cambridge: An Account of His Writings with Selections from His Literary and Scientific Correspondence.* 2 vols. London: Macmillan, 1876.

Wettersten, John R. *Whewell's Critics: Have They Prevented Him from Doing Good?* Edited by James A. Bell. Amsterdam: Rodopi, 2005.

Yanni, Carla. "On Nature and Nomenclature: William Whewell and the Production of Architectural Knowledge in Early Victorian Britain." *Architectural History* 40 (1997): 204–21.

Yeo, Richard, ed. *The Collected Works of William Whewell.* 16 vols. Bristol, UK: Thoemmes Press, 2001.

Yeo, Richard, *Defining Science: William Whewell, Natural Knowledge and Public*

Debate in Early Victorian Britain. Cambridge: Cambridge University Press, 1993.

Yeo, Richard. "William Whewell: A Cambridge Historian and Philosopher of Science." In *Cambridge Scientific Minds*, edited by Peter Harman and Simon Mitton, 51–63. Cambridge: Cambridge University Press, 2002.

Yeo, Richard. "William Whewell, Natural Theology and the Philosophy of Science in Mid-Nineteenth Century Britain." *Annals of Science* 36, no. 5 (1979): 493–516.

List of Contributors

James Clackson is professor of comparative philology at the Faculty of Classics, University of Cambridge, and a fellow and director of studies at Jesus College, Cambridge.

Tony Crilly is emeritus reader in mathematical sciences at Middlesex University.

Claudia Cristalli is currently a postdoctoral fellow at Tilburg University.

Max Dresow is currently a visiting assistant professor at Macalester College, and a fellow at the Minnesota Center for Philosophy of Science, University of Minnesota.

Heather Ellis is Vice-Chancellor's Fellow at the School of Education, University of Sheffield.

John Gascoigne is emeritus professor of history at the University of New South Wales.

Edward J. Gillin is lecturer in the History of Building Sciences and Technology at the Bartlett School for Sustainable Construction, University College London.

Michael Ledger-Lomas is a visiting research fellow in the Department of Theology and Religious Studies at King's College London.

Bernard Lightman is professor of humanities at York University.

Harro Maas is professor in history of economics at the Walras-Pareto Center for the History of Economic and Political Thought at the University of Lausanne.

Ben Marsden is senior lecturer at the School of Divinity, History, Philosophy and Art History of the University of Aberdeen.

David Phillips is professor of philosophy and chair of the Philosophy Department at the University of Houston.

Aleta Quinn is assistant professor of philosophy at the Department of Politics and Philosophy, University of Idaho.

Michael S. Reidy is professor of history at Montana State University.

Sheldon Rothblatt is emeritus professor of history and former director of the Center for Studies in Higher Education at the University of California, Berkeley.

Diana Smith is assistant archivist at Trinity College Library, Cambridge.

Christopher Stray is an honorary research fellow, Swansea University.

Lukas M. Verburgt is currently an independent scholar, based in the Netherlands.

Index